T0305581

Embodied Engineering

NEW AFRICAN HISTORIES

SERIES EDITORS: JEAN ALLMAN, ALLEN ISAACMAN, AND DEREK R. PETERSON

David William Cohen and E. S. Atieno Odhiambo, *The Risks of Knowledge*

Belinda Bozzoli, *Theatres of Struggle and the End of Apartheid*

Gary Kynoch, *We Are Fighting the World*

Stephanie Newell, *The Forger's Tale*

Jacob A. Tropp, *Natures of Colonial Change*

Jan Bender Shetler, *Imagining Serengeti*

Cheikh Anta Babou, *Fighting the Greater Jihad*

Marc Epprecht, *Heterosexual Africa?*

Marissa J. Moorman, *Intonations*

Karen E. Flint, *Healing Traditions*

Derek R. Peterson and Giacomo Macola, editors, *Recasting the Past*

Moses E. Ochonu, *Colonial Meltdown*

Emily S. Burrill, Richard L. Roberts, and Elizabeth Thornberry, editors, *Domestic Violence and the Law in Colonial and Postcolonial Africa*

Daniel R. Magaziner, *The Law and the Prophets*

Emily Lynn Osborn, *Our New Husbands Are Here*

Robert Trent Vinson, *The Americans Are Coming!*

James R. Brennan, *Taifa*

Benjamin N. Lawrance and Richard L. Roberts, editors, *Trafficking in Slavery's Wake*

David M. Gordon, *Invisible Agents*

Allen F. Isaacman and Barbara S. Isaacman, *Dams, Displacement, and the Delusion of Development*

Stephanie Newell, *The Power to Name*

Gibril R. Cole, *The Krio of West Africa*

Matthew M. Heaton, *Black Skin, White Coats*

Meredith Terretta, *Nation of Outlaws, State of Violence*

Paolo Israel, *In Step with the Times*

Michelle R. Moyd, *Violent Intermediaries*

Abosede A. George, *Making Modern Girls*

Alicia C. Decker, *In Idi Amin's Shadow*

Rachel Jean-Baptiste, *Conjugal Rights*

Shobana Shankar, *Who Shall Enter Paradise?*

Emily S. Burrill, *States of Marriage*

Todd Cleveland, *Diamonds in the Rough*

Carina E. Ray, *Crossing the Color Line*

Sarah Van Beurden, *Authentically African*

Giacomo Macola, *The Gun in Central Africa*

Lynn Schler, *Nation on Board*

Julie MacArthur, *Cartography and the Political Imagination*

Abou B. Bamba, *African Miracle, African Mirage*

Daniel Magaziner, *The Art of Life in South Africa*

Paul Ocobock, *An Uncertain Age*

Keren Weitzberg, *We Do Not Have Borders*

Nuno Domingos, *Football and Colonialism*

Jeffrey S. Ahlman, *Living with Nkrumahism*

Bianca Murillo, *Market Encounters*

Laura Fair, *Reel Pleasures*

Thomas F. McDow, *Buying Time*

Jon Soske, *Internal Frontiers*

Elizabeth W. Giorgis, *Modernist Art in Ethiopia*

Matthew V. Bender, *Water Brings No Harm*

David Morton, *Age of Concrete*

Marissa J. Moorman, *Powerful Frequencies*

Ndubueze L. Mbah, *Emergent Masculinities*

Judith A. Byfield, *The Great Upheaval*

Patricia Hayes and Gary Minkley, editors, *Ambivalent*

Mari K. Webel, *The Politics of Disease Control*

Kara Moskowitz, *Seeing Like a Citizen*

Jacob Dlamini, *Safari Nation*

Alice Wiemers, *Village Work*

Cheikh Anta Babou, *The Muridiyya on the Move*

Laura Ann Twagira, *Embodied Engineering*

Marissa Mika, *Africanizing Oncology*

Holly Hanson, *To Speak and Be Heard*

Embodied Engineering

Gendered Labor, Food Security, and Taste in Twentieth-Century Mali

∽

Laura Ann Twagira

OHIO UNIVERSITY PRESS ∽ ATHENS, OHIO

Ohio University Press, Athens, Ohio 45701
ohioswallow.com
© 2021 by Ohio University Press
All rights reserved

Printed in the United States of America Ohio University Press books are printed on
acid-free paper ∞ ™

31 30 29 28 27 26 25 24 23 22 21 5 4 3 2 1

Library of Congress Cataloging-in-Publication Data
Names: Twagira, Laura Ann, author.
Title: Embodied engineering : gendered labor, food security, and taste in twentieth-
century Mali / Laura Ann Twagira.
Other titles: New African histories series.
Description: Athens, Ohio : Ohio University Press, 2021. | Series: New African histo-
ries | Includes bibliographical references and index.
Identifiers: LCCN 2020050210 (print) | LCCN 2020050211 (ebook) | ISBN
9780821424414 (hardcover) | ISBN 9780821447338 (pdf)
Subjects: LCSH: Mali. Office du Niger. | Women in agriculture—Mali. | Ag-
ricultural development projects—Mali. | Agricultural processing—Mali. |
Women—Mali—Economic conditions.
Classification: LCC HD6077.M42 T83 2021 (print) | LCC HD6077.M42 (ebook) |
DDC 333.76082096623—dc23
LC record available at https://lccn.loc.gov/2020050210
LC ebook record available at https://lccn.loc.gov/2020050211

In loving memory of Sekou Diarra,
Adam Bah Dagno,
and Fatouma Coulibaly

Contents

Illustrations

Acknowledgments

I am enormously grateful to the late Sekou Diarra, without whom this book would never have been have written. He was a renowned hunter whose knowledge about the world and curiosity to always learn more continues to inspire me. I am further indebted to his wife and my host mother, Hawa Fomba. She taught me about women's agricultural expertise and cooking but, more importantly, about *mogoya* and generosity. It was in working and living with the Diarra family that as a scholar I first became interested in learning from Malian women.

This project took shape during a research seminar taught by Temma Kaplan at Rutgers University, and she has since remained a steadfast supporter. I further pursued the project under the guidance of Barbara Cooper, whose encouragement pushed me to pursue the story of the pots and whose insights helped shape this book. I am also grateful for the support and mentorship of Julie Livingston, Bonnie Smith, Carolyn Brown, and Al Howard. When it became clear to me that this was a history of technology, Michael Adas connected me with Rosalind Williams and Clapperton Chakanetsa Mavhunga at MIT, who were both very generous to a dissertation writer delving into an entirely new literature. I also thank Mike Siegel from the Rutgers Cartography Lab.

In Mali, my research would not have been possible without the help of numerous friends and colleagues. I thank Oumou Sidibe, Gregory Mann, Modibo Sidibe, Isaïe Dougnon, and Emily Burrill for helping me to get my bearings in Bamako. In Bamako, I also thank colleagues at the Institut d'Economie Rurale, the Université de Bamako, and Point-Sud. Brandon County introduced me to Souleyman "Bonheur" Doumbia and Labassy "Labass" Gnono. Bonheur's help was essential in setting up my first interviews in Markala. He and his wife, Mariam Kelepily, also hosted me when I returned to Bamako for additional archival work. Labass translated and transcribed a number of interviews and connected me with Almamy

Thiènta, whose connections in Sokolo and Kokry were invaluable. I am also grateful for the help of numerous staff members from the National Archives of Mali, but I especially thank Timothée Saye, Souleymane Koné, and Youssouf Dagno. Outside of the archives, Youssouf Dagno connected me with family members living in in Markala and Niono. I also thank Youssouf's sister Mariam "Nènè" Dagno. In Ségou, I am grateful to Mme. Fatmata Maïga Sidibe, who made the archives for the Office du Niger open to me beyond regular hours. My interviews were greatly aided by the assistance of several Office du Niger staff members: Aïssata Coulibaly, Bintou Diarra, Assanatou Dieunta, Bintu Dieunta, Fatoumata Guindo Tamboura, Bintu Kané Dagnoko, Assane Keita Diallo, and N'Faly Samake. In Niono, I thank Abdoulaye Bah and his family for their hospitality. I also thank the Kassambara family in Markala and the Sammessekou family in Kokry. I will forever be grateful to Aïssata "Maba" Kassonke for her help with my interviews but also for offering her friendship over the several long months that we both spent away from family. Above all, I am thankful to the numerous women and men who generously sat down with me to speak about their lives at the Office du Niger.

I completed additional archival work for this book in France and Italy. At the UN Food and Agricultural Archives, I thank Fabio Ciccarello. Also in Rome, at the archives of the White Fathers, I am grateful for the assistance of Father François Richard, Father Dominique Arnaud, Father Odon Kipili, and Father Fritz Stenger. I am additionally grateful for the assistance of several staff members of the Archives Nationales d'Outre Mer in Aix-en-Provence, France.

While a scholar-in-residence at the Schomburg Center for Research in Black Culture, my thinking was greatly aided by colleagues, including Brent Hayes Edwards, Shannen Dee Williams, Candacy Taylor, Tiffany Gill, Kim Hall, Philip Misevich, and Anthony Di Lorenzo. I am also grateful for the research assistance provided by the library staff, but I am especially grateful for the help of Auburn Nelson and Melay Araya, as well as for logistical support from Sister Aisha al-Adawiya. This manuscript was further improved during my time at Wesleyan University and especially while a fellow at the Center for the Humanities. I thank Megan Glick, Natasha Korda, Heather Vermeulen, Catherine Damman, the late Christina Crosby, and the students in my seminar on Body Histories in Africa for their engagement with my work. I also thank my Wesleyan colleagues Jennifer Tucker, Paul Erickson, Erik Grimmer-Solem, Victoria Smolkin, Courtney Fullilove, Rick Elphick, Mike Nelson, and my two student research assistants, Aimée Wilkerson and Orelia Jonathan.

Many other colleagues and friends graciously read different chapters or versions of the manuscript or helped me to think through my research and encouraged me to keep writing. I am especially grateful to Judi Byfield, who read an earlier version of the entire manuscript and gave me invaluable feedback. I also thank Rochisha Narayan, Robin Chapdelaine, Mahriana Rofheart, Ousseina Alidou, Dorothy Hodgson, Indrani Chatterjee, Renée Larrier, Renée DeLancey, Dora Vargha, Bridget Gurtler, Tal Zalmanovich, Lindsay Braun, Molly Giblin, Mario da Penha, Trina Hogg, Rhiannon Stephens, Abosede George, Arianna Huhn, James McCann, Diana Wylie, Casey Golomski, Shelby Carpenter, Rachel Maines, Stephan Miescher, Drew Thompson, Sarah Hardin, Josh Grace, Robyn d'Avignon, Dave Newman Glovsky, Julie Landweber, Jenna Nigro, Jacqueline-Bethel Tchouta Mougoué, Chikwenye Ogunyemi, Priscilla Murolo, Mary Dillard, and Amrys Williams. Thanks especially to Batamaka Some for his language instruction during SCALI and his colleagueship, *i ni ce, i ni baara ji.*

I am grateful to have had the research support of the Fulbright-Hays Dissertation Research Fellowship, the Kranzberg Fellowship from the Society for the History of Technology, the Bernadotte E. Schmitt Grant from the American Historical Association, and a residency fellowship at the Schomburg Center for Research in Black Culture supported by the Ford Foundation. I received additional support from the History Department of Rutgers University, which included a Mellon Research and Training Grant to attend the Oral History Summer Institute at the University of California-Berkeley. A Foreign Language and Areas Studies Summer Fellowship also allowed me to attend the Summer Cooperative African Language Institute at the University of Illinois at Urbana-Champaign. In addition, I received support from the Center for African Studies at Rutgers University, the Wesleyan History Department, and Wesleyan University.

Finally, thank you to Benjamin Twagira for your emotional support, always on-point analytical insights, and unwavering encouragement as this project slowly progressed from an idea to a book. *Ndagukunda cyane koko.*

A Note on Language

I conducted my interviews in Bamana (Bambara)—one of the primary languages spoken in Mali and at the Office du Niger—and, to a lesser extent, in French. However, Bamana was often a second (or third) language for my interviewees. In some cases, interviews were conducted in a mix of Bamana and French, which typifies the linguistic flexibility of daily life in Mali. As a result, the spelling of names for people, places, plants, and food varies in everyday use. For example, Bamana and French orthographies are often employed interchangeably, such as in the city name Segu/Ségou or the name Kulibali/Coulibaly.

I have chosen to embrace this linguistic flexibility in the text. For place names, related to an interviewee's recollection, I have employed locally accepted spellings. In other cases, I have employed the French names for Office du Niger towns as recorded in the archives. For my interviewees' names, I use the orthography that they have given me. For the names of plants, tools, and other words related to food production, I have employed spellings given to me by an interviewee, or those taken from a Bamana language dictionary. In some cases, when a term was not included in a formal Bamana language reference, I have used the spelling of the word as found in the French archival record.

The translation of my interviews was a collective effort. In Markala, I was aided on occasion by Souleymane "Bonheur" Doumbia. More often, Aïssata "Maba" Kassonke assisted me during the course of an interview, when my Bamana was not understood or I missed a point from an interviewee. And sometimes we learned an older word together. A small number of these interviews was translated into French and transcribed in both the Bamana and French by Labassy "Labass" Gnono.

All translations from the French are mine, unless otherwise noted.

Abbreviations

AOF   Afrique-Occidentale française (French West Africa)

FR   French West African franc (unit of currency under colonial rule)

CFA   Communauté Financière d'Afrique (unit of currency that replaced the FR in the mid-1940s)

CRM   Centre du Riz Mécanisé (Mechanized Rice Center)

OPAM   Office des produits agricoles au Mali (Malian Office for Agricultural Products)

STIN   Service temporaire des irrigations du Niger (Temporary Irrigation Services for the Niger River)

WFP   World Food Program

Introduction

IN THE first half of the twentieth century, young women and girls in Mali sang about the quality of their food, specifically their sauce: "Oh! it's so good, oh! it's so good / The fish sauce / Served nicely over fonio! . . . The mix of okra and rice is a tasty meal."[1] As they sang in call and response, each girl danced and proclaimed her own sauce-making prowess. As the line "Oh! it's so good" hints, sexuality and sensuality mingled with the sauce. The song animated the nighttime, but it also playfully communicated a significant message about the value of women's labor and the centrality of food to rural life. Women and girls made delicious food, and it was a feat to be lauded. Indeed, food production and preparation were not easy tasks. It was taxing manual labor involving wood collection, growing and gathering ingredients for the sauce, and pounding grains before the fire was even lit under the cooking pot.

The song also suggests that women's food production and preparation was the subject of popular interest and discourse. A folktale from the opening of the century further addresses how women accomplished the difficult task of preparing an appetizing meal. It is a story about a pot and speaks to the ways that women created and managed a technological infrastructure for food production. In the tale, a woman asks her female neighbor to borrow a cooking pot. The neighbor obliges the request, and after a few days the first woman returns the original pot along with a second smaller one. In giving her neighbor the two pots, the first woman calls the smaller one the "daughter" of the big pot. Not long after this episode the same woman returns again to borrow the big cooking pot, but this time fails to return it. When the neighbor inquires after her pot, she is told the pot is dead. The audience for the story is prompted, along with the neighbor, to ask, "How is it that pots die?" The first woman replies: "They have daughters."[2] The big pot's transformation into a mother serves as a comic (albeit dark) explanation for the first woman's failure to return the pot.

What does this tale have to do with women's expertise and cooking in twentieth-century Mali? First, for all women a cooking pot was an ordinary yet essential object. The story signifies the importance of a *women's* technological infrastructure but also its potentially shifting nature. Both women characters in the tale negotiate their access to a changing number of cooking pots just as women throughout the twentieth century have innovated and managed transformations in the range of tools and technologies available to them for food production and preparation. If women wanted to make good food, they needed pots and a host of other women's technologies.

Importantly, this story about pots connects women and their means of cooking with physical labor, specifically the experience of childbirth. Food preparation is a similarly distinct woman's task associated with sexuality, social reproduction, environmental fertility, and women's embodied labors. Moreover, the "daughter pot" in the story alludes not only to women's difficult and potentially dangerous labor (both in childbirth and in food preparation) but also to the complex relations between women who often must work together. Indeed, at first glance, the story seems to be one of failed female cooperation. Yet the humor of the story suggests a more positive interpretation: that is, if women heed its lesson. Collective female labor, especially during food shortages, was a critical element of women's work. It was also essential to the assurance of food security. Finally, calling one of the pots a "daughter" suggests the transmission of feminine knowledge and social continuity, specifically for the generations of women over the course of the century who prepared the quotidian meal.

Food was the talk of rural life, and it was at the center of gender politics throughout the century. In moments of leisure, young women bragged to one another about their sexual allure and food preparation skills. For them, cooking was a point of individual pride. Other songs and stories from the first decades of the century feature heroic women saving towns from famine, hospitable wives who are generous with meals, and magical cooking pots. But they also highlight young girls who refuse to cook and female lovers who deceive young men with gifts of food.[3] Women's food labors, so central to daily life, were sometimes a source of social contention. Fittingly, the supernatural aspects of many of these tales anticipated the ways women's cooking and use of their pots could appear fantastical later in the century. In those years, women faced severe ecological crises and the oftentimes counterproductive intrusions of colonial and postcolonial states in agriculture and the food supply. It would not have been amiss to ask: How could they possibly make a meal? Women's mundane yet *extraordinary* daily food production and preparation labors over the course of the twentieth century are the subject of this book.

The women who did this work were creative technological actors. As I argue here, women in rural Mali engineered a complex and highly adaptive food production system that depended on female labor power and made use of modest technologies. Theirs is a history that showcases rural domestic space as an arena of technological innovation. Women and what they did with their pots mattered a great deal. Certainly, Malian women incorporated labor-saving techniques and technologies into their work routines, but they were equally concerned with the production and collection of nutritious ingredients, the availability of resources such as wood fuel, and the pleasure of eating. Their technological work was a complex interplay of skill, knowledge, social meaning, leisure, and survival. Importantly, women's embodied techniques were central to their ability to ensure food production. Beyond subsistence, women's embodied expertise made the preparation of tasty and culturally meaningful meals possible, helped manage their labor time, and significantly displayed the value of their labor—like the young women who sang about their fish sauce. Together, this women's work is embodied engineering.

Cooking in Mali was very much a "technique of the body," to quote Marcel Mauss.[4] Women bent their backs to tend gardens and gather wild foods. They also walked into the bush to collect and carry wood;[5] they wielded the pestle and stirred the pot. Importantly, embodied labor, skill, and knowledge were idealized in rural Malian society. At the same time, these labors were compounded by recurring ecological crises such as drought or deforestation. For example, the regular preparation of food might have seemed like a remarkable feat in early twentieth-century rural communities with memories of a severe famine prompted by a drought from 1913 to 1914. Dogon children born during those years were sometimes named *Ogulum* or "I survived the drought."[6] Undoubtedly, women's food production tasks during periods of shortage were physically taxing but all the more materially and symbolically significant. As Naminata Diabate has shown in her work on women's agency and the female body, when women collectively signal distress—as Malian women displayed in the last quarter of the century by hiding rice under their clothes to prevent state seizure during a major drought—their embodied actions, while betraying great vulnerability, are paradoxically powerful.[7] It is in such "exceptional and biopolitical conditions" that "their bodies seem to be all that they have left."[8] Diabate's examination of biopolitics in Africa is instructive here in its insistence on agency and the power of the female body, even in moments of significant political or environmental crisis.

Malian women's relationships with the natural world have not been ideal. They have faced unpredictable flooding, drought, and locust swarms, all of which threatened harvests and potentially produced fatigue and hunger. Yet, women also have been agents in shaping their natural environment and their relationships with that world. They sought nourishing, flavorful foods in the wilds, collected edible plants from rivers, and planted food-bearing trees to make sure there would be something to eat. In times of abundant grain harvests, women celebrated the collective physical labor that produced the surplus and prepared millet beer to heighten the pleasure of eating. Feast or famine, women's material experiences and those of their communities have been intimately connected to changes in the environment.[9] Moreover, eating was always a sensory experience.[10]

The creation of this embodied rural world was a technological affair. Specifically, women's embodied interactions—or, to borrow Donna Haraway's formulation, "conversations"—with their natural world incorporated the technological.[11] Just as daily food labor was a task of the body, the use of technological objects such as a grinding stone, a garden hoe, a fishing basket, or a cooking pot was a sensorial and embodied experience. Women's quotidian actions and gestures were purposeful and drew together their environmental and technological expertise. Writing about the ancient world, Marcia-Anne Dobres articulates a broader claim about women as technological agents: "Ancient technicians were sensual and experiential beings who made sense of the world—and made sense of themselves—as they made and used material culture during the mundane routines of everyday practice. This body was mindful, sensual and a gendered *conduit* through which technicians materialised, negotiated and transformed their world—and through such means *made* things meaningful."[12] Similarly, women in Mali made filling and flavorful food by drawing upon their material and embodied expertise with the natural and technological world. In their labors they also produced the social meanings of women's work, but in so doing, their actions reinforced gendered social expectations for women.

Women were expected to cook.[13] Yet, through this quotidian production of food, the vast majority of women in Mali took a formative role in shaping the region's twentieth-century history from colonial-era French West Africa to its postcolonial transformations in 1960 and the succeeding decades. Malian women lived in a region that was predominantly rural. As a result, government administrations in need of finance sought to direct the agricultural economy, and early French administrations aimed to turn the region into a colonial breadbasket. An ecological crisis prompting the 1913–14 famine complicated these ambitions. Yet in the midst of hunger, and facing the threat of famine, women's labors produced significant food security. Succeeding government interventions and agricultural development programs were often heavy handed, and over several decades of forced labor regimes, industrial development, decolonization, rural socialism, military dictatorship, and severe drought, the region saw dramatic (and in some cases rapid) transformation. The gendered dimensions of such wide, sweeping change were significant. Specifically, food supply was politicized in these years, making women's daily food labors extremely important.

The state presence in daily life is important to this history, but it did not determine women's daily activities. The gendered division of labor, women's food production and preparation technologies, the environment, rural animation, and local taste preferences were all essential to both the provision of and the meanings associated with food. At different points in time over the century, one or more of these factors became more pronounced than the others in ensuring rural well-being. Yet, each was an important element of the overall foodscape or enviro-technological habitus: by which I mean the context in which women gained access to food resources and prepared daily meals. Certainly, the power of the state was heightened at distinct moments in women's lives; nevertheless, the gradual social, environmental, and technological changes that affected women the most unfolded across the major political breaks and eras.

Women experienced political and economic shifts most immediately through changes in the environment and access to food resources but more specifically through changes in women's technologies. For example, in the late 1940s, women in rural Mali began cooking in metal pots rather than clay ones. It was a seemingly modest transformation in the daily labor of food preparation but one with wide-reaching import. Following several decades of intensified colonial demands for wood, deforestation was a real concern. This was particularly problematic for women because wood fuel for cooking was harder to come by. The adoption of a metal pot—which cooked faster and used less wood fuel—addressed this ecological concern and saved women much time devoted to wood collection. Indeed, this midcentury moment speaks to the ways that rural women in Mali ensured food security, now a ubiquitous term in development circles. Women's ability to reengineer their own labor was essential to making sure the daily meal was in the bowl.

The history presented here rejects a persistent image of women in Africa as subjects without access to or knowledge of technology. As I demonstrate, women in Mali were, in fact, rural food engineers. By contrast, prior narratives about women and development have centered the "status of African women," a generic catchall concept, which emerged as a policy concern first among colonial observers and administrators and later among development experts, scholars, and postcolonial government officials. These narratives broadly presented "third-world" women as a development "problem."[14] With particular regard to women and technology in Africa, one well-known development scholar from the 1970s, Ester Boserup, observed that women's increasing labor burdens were an obstacle to improving their status and argued that the solution was the introduction of new technologies.[15] It was not a perspective that framed women as active agents. Rather, when women entered the discussion of technology, the way to address women's concerns was understood simply to be a matter of transfer.[16] Unfortunately, the representation of African women as not having access to technology was coupled with an image of African women as overburdened by labor: these images have continued to frame outside perceptions of their history.

Women's need for wood fuel is an illustrative example. During the 1970s, the wider Sahel region experienced widespread drought, and pictures of rural Sahelian women suffering from the crisis exacerbated stereotypes about overburdened women on the continent. In particular, popular depictions of a lone woman walking a long distance to collect and carry a heavy load of wood fuel for cooking dramatized deforestation and marked women as sympathetic victims for international aid. Previous scholars of gender and development have already critiqued this widely circulated type of image, pointing to several problematic assumptions: that women are closer to nature because of their gendered labor, ideal environmental managers due to their assumed sacrifice, and also victims whose plight dramatizes environmental decline.[17] It is worth noting that these images also suggest Africa was a region lacking in modern infrastructure and domestic technologies that would otherwise alleviate women's labor. As Jane Guyer has already argued, such widespread representations imply that African women's work has remained static over the twentieth century.[18] These are stubborn stereotypes. Less widely circulated images from this period feature metal house goods, harkening back to the major midcentury technological shift previously mentioned and raising questions about women's technological work during the Great Sahel Drought (see figure I.1). Despite such evidence, colonial and postcolonial observers, administrators, and experts have refused to see African women as environmental and technological agents in their own right.

FIGURE I.1. A woman carrying a metal basin during the Great Sahel Drought, 1973. Courtesy of the Library of Congress. Prints and Photographs Division, lot no. 11515 (17). Photo collection credited to the Food and Agricultural Organization and the World Food Program.

By specifically looking at the women who were a part of the major agricultural program called the Office du Niger (hereafter referred to as "Office"), the history of rural Malian women as embodied engineers comes into focus. The Office was arguably the most important development intervention by the French in West Africa and later occupied a significant role in the economy and politics of independent Mali. The scheme was established under colonial rule in the 1930s—in what was then the French Soudan—to produce primarily cotton and secondarily rice for export. This twin vision of industrial cotton cultivation and regional food production fueled grand economic promises, and as Chéiban Coulibaly has observed, inspired a specific development mythology in Mali.[19]

Irrigation and so-called modern agricultural technologies were central to the design of the Office. It drew upon the Niger River to feed a canal network that radically altered the surrounding agricultural landscape, and its founding disrupted rural communities across much of the colony and neighboring French Upper Volta (contemporary Burkina Faso). The construction of a large dam on the Niger River was accompanied by the digging of canals and heavy machinery. It was a costly project. The French colonial government invested impressive sums into building the Office during a global depression. Indeed, it had more financial resources to build an irrigation system, roads, and towns than most officials in the empire had for colony-wide public works. Yet, French colonial critics argued that its founder, Emile Bélime, and other planners paid too much attention to the technical infrastructure to the detriment of people's material conditions. Indeed, for decades the Office failed to produce a profit for either its farmers or the colonial government.[20]

Nevertheless, administration officials believed modernization at the Office (read: "improved" farming) would bring about the intensification of production. This shift in the organization of agriculture would be accomplished through the importation of Western technological know-how and materials. Imported steel and cement from France went into building the Markala dam that fed the irrigation system. Thousands of men were conscripted by the colonial government to build the dam and carve wide canals into the countryside. As was the case in other areas of Africa, labor-intensive technologies were a hallmark of the colonization process rather than labor-saving technologies.[21] Male laborers and a small number of industrial machines dug canals and cleared vast tracts of land for expansive fields. In so doing, they also cleared bushes and trees that women relied upon for sauce ingredients. It was a new material world.

By the end of the century the population of the project was made up of an initial generation of settlers who had arrived by force under the colonial government's coercive recruitment scheme, voluntary male workers and their families who arrived following the Second World War and were attracted by the economic promises of industrial agriculture, the similarly hopeful nationalist settlers of the 1960s, and desperate migrants seeking refuge from the environmental catastrophes of the 1970s and 1980s. The Office was diverse, and as it expanded over several decades, settlers came from rural farming, fishing, and even herding communities from across the region. During all this time, the Office formally ignored women as possible target populations for agricultural development yet relied significantly on their labor and expertise. In rereading the history of the Office in light of women's own technological experiences, I draw attention to their sense of what constituted development rather than the goals of technocrats and other experts. Indeed, women's assessments of project agriculture and technology considered daily meals produced, their nutritional value, and importantly *taste*, rather than production for a colonial or postcolonial export economy. From this perspective, it was the texture of daily life and the quality of food that served as the measure of development and food security.

In the midst of this rapidly shifting environment, women's ability to adapt their labor and technical infrastructure to new conditions was critical to food security in project towns and villages. Unlike many development experts, they did not see their labor as a problem, rather it was an element of their engineering response. The technologies most directly associated with women—or more appropriately called women's "things"—included modest domestic tools such as the mortar and pestle, cooking pots, and various new metal household goods. A range of industrial agricultural machines also entered the changing rural landscape at the Office. For example, women adapted the first threshers into their food labors, bridging the domestic and industrial nature of twentieth-century agriculture. In short, women who came to the Office integrated the material realities of the colonial and postcolonial irrigation project into an existing system of food production and reengineered it in the process. Much about women's lives at the Office is distinct from that of other women in the region. Yet these distinctions are further instructive when considering the diversity of rural women's experiences.

AFRICAN WOMEN'S HISTORY AND THE HISTORY OF TECHNOLOGY

This book brings African gender history together with the field of science and technology studies (STS). In so doing, I draw on the insights of African gender scholars who have highlighted the central and shifting role of the household and domestic space for political and economic life.[22] In Mali, the household has long been a political space, and it has been recognized as such by both colonial administrators and postcolonial legislators who have intervened in cases of inheritance, runaway wives, and divorce. The same concerns were entangled with local and regional economies, as control over the labor of young wives, enslaved women, and other female servants provoked tensions and negotiations.[23] Food production and preparation similarly situated women's domestic lives and labors at the center of local and regional economies and politics throughout the twentieth century.

This study's analysis is further informed by the focus that scholars of STS bring to the dynamic and textured interactions between the social world, technological artifacts, and the built environment.[24] While the study of technology in twentieth-century Africa has only recently gained sustained interest from historians, scholars in the field have highlighted new and shifting African technological networks, forms of labor, and cultures across its decades.[25] At the Office, multiple technological cultures emerged, and they were distinctly gendered. Women certainly created and maintained a distinctive technological space at the project. Yet the Office was widely associated with men.[26] For example, male farmers' earliest interactions with project technologies were generally marked by avoidance and disdain. Birama Diakon, who has examined the introduction of the plow in Mali and at the Office specifically, has further shown how male farmer's perceptions of the agricultural tool shifted dramatically over several decades, from a belief that the plow harmed fields to its adoption as a *Malian* technology. As this shift unfolded, they remade the plow into a new symbol of a masculine farmer ethic.[27]

Diakon's study of technology at the Office focuses primarily on men, but he is concerned with women and technology. In addition to the social life of the plow, he also follows the introduction and adaption of modified threshers at the project. As Diakon notes, in the late 1980s and 1990s male farmers encouraged blacksmiths and other new iron workers to produce threshers that also winnowed, a task customarily performed by women. Winnowing ordinarily entitled women to a portion of the harvest. Since the new machines would often perform this task, women lost those harvest rights at the Office. In Diakon's analysis, the new threshers excluded women, and mechanization resulted in a material loss for them.[28] He rightly points out women's grievances in relation to these machines. However, women's technological work was not confined to their engagement with one machine. In focusing on the plow and the modified thresher, Diakon challenges scholars to appreciate the innovation of African settlers, specifically male farmers and blacksmiths, at the Office. However, both technologies were largely controlled by men, even if they also entered the women's technological infrastructure. Neither object allows us to fully see women's technological work and creativity.

What, for example, would a study of the life of the cooking pot reveal? Scholars of technology must attend to the modest and domestic technologies associated with African women. The potential analytical insights to be gleaned from an analysis of mundane technologies are highlighted by Suzanne Moon: "The very ubiquity of [every day or uncontroversial] technologies make them the invisible background of social life, not noticed or written about in any depth, and rarely a subject of interest or passion for contemporary informants."[29] Yet, the pots, buckets, and other ordinary household items employed by women at the Office enabled them to make the scheme's otherwise unwieldy irrigation and industrial apparatus actually function.

Women's experiences also tell us that the histories of industrial and domestic technologies at the Office overlapped. For example, women turned canals into domestic water resources, even though the irrigation system was meant to water the project's fields. Previous scholars of technological systems have already shown that large-scale technological infrastructures have the potential to alter domestic life.[30] Users of those systems, in turn, alter the technology itself—an example of the coproduction of technology, gender, and society.[31] Understanding the role of users, in particular, broadens the analytical lens beyond the category of engineer or designer—a framing that too often excluded most Africans from twentieth-century histories of technology.[32] Moreover, as Nelly Oudshoorn and Trevor Pinch articulate, "Granting agency to users, particularly women, can thus be considered central to the feminist approach to user-technology relations."[33] Seeing women as users of pots, buckets, canals, threshing machines, and a host of other technologies allows us to expand the technology story of the Office into the domestic arena. As women at the Office negotiated technological changes related to food production, they integrated elements of the scheme into their labor routines, highlighting their role as users of diverse technologies, but also as engineers of a wider food production system. In the process, aspects of women's work and identity were transformed. Women also found new ways to showcase the centrality of their labor to rural life.

It must be noted that this book does not aim to romanticize women's technological work nor to minimize the challenges they faced. Technologies like the thresher certainly had a negative impact on women's livelihoods and even their rural social status. However, as Judy Wajcman reminds us, feminist analysis must move beyond the debate about women's relationship to technology as either dystopian or utopian.[34] Women did exert agency at the Office, but the shifting contexts for women's actions and their multiple meanings matter. Agency is a seductive concept for historians interested in women (and men) who are not often framed as historical protagonists, and I carefully emphasize the import of women's actions. Indeed, as Lynn Thomas elaborates, asserting agency for our subjects is not the end of the historical argument but an opening for more refined analysis.[35]

Because women's innovation is most evident in their adoption of technologies for domestic use, their role as technical actors has not always been visible to historians. There are a few notable exceptions for African women artisans. For example, Sarah Brett-Smith's work on Malian women producers of mud cloth explores their creativity with a focus on the designs of produced cloth as a means for women's specific expression, often relating to bodily concerns such as circumcision and childbirth.[36] Additionally, the art historian Barbara Frank has studied how familial relationships between female potters in Mali aided in the transfer of specialized craft knowledge.[37] Engaging more directly with the field of STS, the archaeologist Shadreck Chirikure has framed African female potters (and male blacksmiths) as creators of laboratories. In his analysis, their spaces of scientific and technological work extended from the sites of extraction to their workshops. For women potters, the workshop was in the household.[38] Eugenia Herbert has examined women's roles in both metallurgy and pottery production in comparative African societies with a focus on the deeply gendered metaphors associated with what she terms "rituals of transformation."[39] What her work demonstrates is that African languages of technology are deeply gendered. For example, furnaces for iron production often mimicked the shape of the female body, and rituals surrounding smelting referenced fertility and birthing. Metallurgy, while most often practiced by men, was reliant on metaphorical female body power.

Yet, technology has generally been coded as masculine, and women's interventions have been relegated to realms outside of technology.[40] When Ralph Austen and Daniel Headrick wrote about the so-called technology gap in Africa, their overall argument rightly questioned classical tropes in the history of technology. Specifically, they suggested that technologies were not always portable. In one example, they point out that an innovation such as the plow would not have been particularly useful in precolonial Africa, thus its nonadoption was not a sign of technological "backwardness." In the 1980s it was a point well made. Austen and Headrick nevertheless suggested that African women reared children in a way that reinforced "technological conservatism," which they believed to be a broad social pattern in Africa: "By carrying and holding their infants off the ground more than other people, African mothers limit their babies' contacts with the world of objects. . . . In other words, child-rearing is human-energy-intensive and anti-materialistic. For the growing child, the results are a higher degree of inter-personal relations but less experience in manipulating the material world."[41] The association of African childrearing practices with technological conservatism is unwarranted. Rather, the embodied experience of childrearing produced a particularly embodied technology. Carrying children in a sling, or otherwise secured to the back, would have enabled women to continue to gather foodstuffs or wood fuel, for example, even while caretaking.[42] It was, in fact, a materially responsive technology. Nevertheless, the idea that African women lack technical capacities or access to "technology" is pervasive even today. It is an idea rooted in a gendered ideology that depicts women, and especially African women, as untechnical, which has, in turn, influenced the study of technology in Africa.[43]

In the context of debates about dependency theory—or more recently in the context of development—focusing on a gap or absence (as Austen and Headrick did) means that the technologies most often employed by women do not even show up in the frame of analysis. Yet, women in Africa have readily adopted new domestic technologies, from metal pots to diesel-operated grain grinders.[44] In using these technologies, women claim a specifically gendered technological space. If the starting point is that African women are caught in a technological gap because of their gender and location, it is not easy to discern their actual engagements with technology on a range of scales. Moreover, it discounts women's continued use of other modest technologies that are often considered unsophisticated: from slings to pots to the iconic mortar and pestle. In concentrating this research on women's technologies and infrastructure—as well as on their labor, techniques, and knowledge—the analysis in this book reveals a dynamic history of gender and improvisation in rural life.

REEVALUATING THE OFFICE DU NIGER

While technology was central to the Office's design, embodied labor was essential to the scheme's construction and operation. Amidu Magasa highlighted this exploitative history in his foundational study of the colonial Office. The title of this history, *Papa-commandant a jeté un grand filet devant nous*, refers to the callous but mundane violence of an administrator brandishing a whip.[45] It is an apt metaphor. The construction of the irrigation infrastructure, Office towns, and roads was primarily accomplished using forced labor. Indeed, the building of the Office displayed the worst of colonial rule in French West Africa. Beginning in 1935, the Office—still under construction—was settled by conscripted individual men and families who were subsequently moved by force to new and unfamiliar project towns. Many women and children migrated, but men predominated. Office administrators tightly controlled the daily labor of all project residents, as well as the sales of all cotton and rice. In addition, guards restricted the movement of settlers wishing to travel outside project territory. These first years brought abrupt and intrusive change for most households prompting many women to flee. This early absence of women has been remapped onto the historiography. Strikingly, most studies of the Office have ignored the initial demographic reality and its implications for labor history at the project.

During the Second World War labor coercion and violence only intensified across the colony under the Vichy government in France. More families were forcibly settled at the project, and its farmers—as elsewhere in French West Africa—were pressured to intensify agricultural production for export in the interests of the French empire. One result was chronic local food shortages. Officially, forced recruitment and labor ended in 1946, but these and other coercive practices continued.[46] Life at the Office during these years continued to be physically taxing and without much celebration for the irrigated harvests. Still, as several key studies have shown, Office settlers survived these difficult years and worked to make the project their own.[47]

Certainly, they worked as part of a project that was an important visible marker of colonial state power, or what Brian Larkin terms the "colonial sublime."[48] The Office also served as a monument to the evolutionary trappings inherent in ideas of modernization and eventually to the obvious "failures" of that same project. Indeed, for most of its history the Office has been criticized for failing to accomplish its goals for development, whether economic, environmental, or social.[49] The Office was an imposing institution, and the power dynamics rooted in its political structure and material form had obvious implications for the daily lives of its residents, extending the reach of state authority through technology.[50] In time, the project canals, roads, and irrigated fields came to be familiar elements of people's surroundings, but they did not go unnoticed. They were new and highly visible.[51] For example, running water could be found everywhere, and then would be abruptly cut off at the end of the agricultural campaign, leaving empty ditches that ran alongside towns. Much about the colonial Office was unquestioningly intrusive: where to live, how to work, even what to eat. Yet officials could hardly demonstrate mastery over every aspect of daily life. As Clapperton Chakanetsa Mavhunga and others have shown, when technologies from the West are transported to Africa, they are often reappropriated by the colonized (or the formerly colonized) for their own purposes.[52] For example, at the Office, men, women, and children bathed, fished, and swam in the irrigation canals. They also learned to redirect the water to nonproject fields, farm with the plow, and make use of large-scale threshing machines to harvest grain for domestic consumption.

The post–Second World War years marked several important shifts. First, the institution benefited from direct funding from France (and the United States through Marshall Plan funds). In previous decades the French government assumed that colonial revenues would finance economic and welfare betterment projects.[53] Continued poor production results at the Office encouraged officials to invest further in its technological apparatus. From the 1940s to the 1950s increased mechanization and motorized cultivation—with the addition of wage labor—altered the nature of project farming.[54] It was an increasingly industrialized rural world. During the same period the Office continued to expand, bringing more families to engage in semimechanized farming. In these and older villages, more women remained at the project. Many women still had little say in the decision to migrate, even if their husbands chose to go to the Office. However, the growing presence of women ushered in a period of greater food security.

The Malian state became independent in 1960, and incoming president, Modibo Keita, shared the colonial belief in the power of technology to bring about transformation in rural Malian society.[55] Keita strongly supported continued investment in the Office scheme. However, he also promoted the reorganization of all agricultural labor along socialist lines with material implications for the organization of women's labor. In 1968 Keita was overthrown and succeeded by the military leader Moussa Traoré. Like Keita, Traoré placed great emphasis on the Office as a national economic development project. Over three decades, from the 1960s to the 1980s, women faced the challenges of collectivization, militarized authority, severe rationing of the harvest, and, with the decline of national revenues, a deteriorating project infrastructure. To make matters worse, the Traoré years were marked by severe and recurrent drought and hunger during the Great Sahel Drought.

In these years, the Office actively engaged in what Gabrielle Hecht termed the "technopolitics" of the emerging Cold War.[56] The international political struggles between Mali, its former colonizer France, and the new global powers (the Soviet Union, China, and the United States) were played out through technology. The Office acquired tractors and vehicles from the Soviet Union and technocratic expertise from the Chinese. Office administrators also sought financial and other aid from international groups such as the World Bank.[57] As Hecht points out, both Soviet- and Western-inspired models for development emphasized large-scale industrialization.[58] Major dam projects such as at the Office fit this pattern of development.[59] Not surprisingly, uneven power relationships continued to permeate the Office at all levels: between the Office and its farmers and between the Office and its postcolonial partners.

Given the prominence of the Office in colonial and postcolonial development planning, this scheme has been subject to several scholarly studies assessing its failures, successes, and possibilities.[60] As other scholars have established, the definition of development is fragmented and historically contingent, but the emphasis on improving Africa with respect to a supposedly more advanced France (or later the Soviet Union) has been relatively consistent.[61] At the Office the targets of development have shifted from African (read: male) farmers, to the environment, the market economy, and ultimately to the technological infrastructure of the project itself. Indeed, the lack of clarity about what or who was actually being marked for improvement underscores the ambiguous nature of development itself.

Yet, Monica van Beusekom has demonstrated that agricultural methods at the Office were not simply imposed by outside experts: project agriculture was a negotiated practice. Staff members consulted with farmers who advocated for policies to better suit their needs, resulting in the practical use of farmers' knowledge about soil type, seed selection, and so on. One example of this flexibility is that rice eventually took precedence over cotton because of farmer preferences.[62] Van Beusekom is right to emphasize the exchanges between men at the Office. However, in her study, agricultural science is presented as a distinctly male endeavor. Men farmed, maintained the canals, and operated the machines. They also engaged the European experts on matters of agricultural science. At the same time, women's concerns about food resources and their labor time were critical in shaping men's farming practices. When women are at the center of the analysis, it becomes clear that women made as much use of the industrial landscape of the Office as men did. Food production had always been highly technical, and women brought that sensibility to their work when they readapted new infrastructures and technologies to their shifting foodscape and daily routines. Even in moments of financial, ecological, and technical decline, women reengineered the Office—an agricultural program designed for men's cash-crop farming—into a common food resource.

REENGINEERING FOOD SECURITY AND TASTE

The policies and programs of the Office consistently privileged agricultural production over consumption. Yet, elements of the project's design were meant to ensure regional food security. In the 1920s Emile Bélime integrated rice cultivation into his planning for the scheme, which he had originally intended solely for cotton production. Bélime did so only in the face of political pressure to address the recurrent problems of food shortages across French West Africa.[63] Yet the resulting rice harvests were destined primarily for export to Senegal's growing urban markets. One outcome of the channeling of food resources to territories well beyond the project area was that the consumption needs and preferences of its residents—the very people who had produced the project's rice harvest—slipped from the view of planners.

Before Bélime advanced his plans for large-scale irrigation, French West Africa had already gained a reputation among colonial scientists as a region in environmental crisis. They feared that the broader Sahel region was undergoing desertification and, worse, they thought it resulted from local farmers' cultivation and land management practices.[64] The French in particular believed that the reputedly barren environment was related to low population density. In fact, they raised alarms that interrelated environmental and population infertility was the cause of recurrent famine.[65]

When Bélime redesigned the Office to include rice he claimed that irrigation would rejuvenate the environment, produce food, and repopulate the region. In this example of colonial thinking, assuring agricultural abundance and prosperity was a matter of Western technology and undifferentiated but laboring bodies. In practice, the early operation of the Office provoked ecological crises and food shortages. Farmers faced uncontrolled flooding in their towns and fields, which harmed the harvest and contributed to outbreaks of dysentery and malaria.[66] Closer to the end of the twentieth century, government officials and Office administrators once again claimed that irrigation would solve the resurgent problems of drought and famine. In the intervening years, women who stayed at the Office learned how to produce and prepare food in the very particular and changing agricultural environment of the scheme. As it happened, irrigation was not a quick and easy solution to the chronic food problems experienced by Office residents. It was women's labor that ensured food security and social reproduction.

Nevertheless, the contours of the scheme inevitably shaped the nature of women's responses to food crises. Elements of the agro-industrial environment of the Office are perhaps more aptly described as "environmental infrastructure." The term suggested by Emmanuel Kreike takes into account both human and environmental agency in the process of ecological transformation, whether read as an improvement or degradation of the environment.[67] The resulting Office landscape was a product of this human-ecological interaction and created as much by the colonial technologies of irrigation as by the forces of the Niger River. Of similar hydraulic landscapes Kreike writes, "The creation of irrigation works has never been a one-time investment that produced a permanent and enduring outcome in which humans subjugated Nature forever. The need for repairs, maintenance, and upgrades was incessant; reservoirs, canals, and ditches need to be kept free from silt accumulations, and bunds, embankments, and dikes need to be kept in repair; failure to do so leads to collapse."[68] This description certainly applies to the challenges faced by farmers and project staff in maintaining the Office. Kreike's broader point about the environment as a form of infrastructure further resonates with the ways in which Office women engineered food production. They intentionally sought to make use of the project's particular agro-industrial resources and even to alter them. Project land was formally owned by the colonial state and later by successive postcolonial regimes.[69] Even male farmers had few formal rights to the plots they farmed. However, men's symbolic claims to the economic fruits of the Office were hardly disputed by the state. By contrast, women used marginal spaces for their cultivation but in so doing materially claimed those spaces for women.[70]

Women's central role in food production, rather than cash-crop farming, places them at the center of debates over the role of the Office du Niger in regional food security. As an object of study, the Office has been both a proposed solution to famine in Africa and a development project with a local history of hunger among its residents. This study concludes in the mid-1980s, not long after Amartya Sen refocused scholarly inquiry on famine toward the social, rather than environmental causes asserting that the problem was not the lack of food but social and political access to resources.[71] Around a decade later, Parker Shipton surveyed the state of research on famine in Africa and stressed that the rural producers of food remained the most at risk, especially women.[72] Historical memory in Mali suggests that while women have been critical to rural subsistence, they have also long been among the most vulnerable to hunger. For example, a folk story published in 1905 by Charles Monteil, a colonial official, narrates the fate of a regional ruler who casts out his favorite wife. She is left to glean for millet in the fields, a survival practice remembered by women in later decades.[73] One interpretation of the tale is that all women are potentially vulnerable to hunger. Yet, they are not simple victims. In the story, this mistreatment is ultimately punished. Other records of famine and food shortage for the early colonial period offer a complex understanding of women, vulnerability, and hunger. They produce food security for themselves and others out of necessity through their seasonal labors and the cultural emphasis on food sharing. Indeed, food production is a field in which women assert authority. A significant historical shift occurred with the establishment of the Office, which left women with reduced access to wild foods during moments of hunger. In response they turned the technological apparatus of the colonial Office into the new wilds. During the Great Sahel Drought, international food aid would similarly replace wild food, but it was marked locally as the most detested famine food rather than aid.

It is worth noting that the hunger of the outcast wife highlighted in the aforementioned story is unrelated to famine. Rather it represents an ongoing concern with gender, access to land, and the distribution of resources. As Diana Wylie has argued, an overemphasis on famine has had the effect of masking the related problems of chronic hunger and malnutrition.[74] Indeed, the Great Sahel Drought and related famine provoked countless studies in the midst of the crisis and immediate aftermath. Several of these works rightly countered long-standing colonial arguments about underproductive environments and African mismanagement of the land. They also pointed to the rise of cash cropping (a defining feature of the Office) under colonial rule and the commodification of the rural economy to the detriment of food production.[75] At the same time, the volume of studies framed the moment as an exceptional crisis.[76] More recently, Vincent Bonnecase has critiqued the intellectual and statistical production of the Sahel as a space of poverty and hunger during this period.[77] In the absence of strictly defined famine in the succeeding years, the provision of food aid slowed, and the resulting development policies emphasized environmental rehabilitation. Nevertheless, food remained a daily concern for many in the region.

What constituted food security in rural Mali? Throughout the twentieth century, "food security" has been a contested term and a subject for political debate.[78] At the Office food security was produced in large measure through women's labors. In the early years of the scheme, the predominantly male households did not produce enough food to eat. Chronic hunger plagued residents for many decades and recurred in the postcolonial years. Yet, it was the quality and not simply the quantity of available food that concerned women. Quality for the politician and the development worker related to the nutritional content of food. But for local women and men, quality was also measured by the taste for specific textures, smells, flavors, and the sensations of fullness. The taste sensorium is multiple and engages sight as well as smell.[79] All these bodily senses play a role in determining what is "tasty" (or even what counts as "food"). What people wanted to eat at the Office also related to cultural identity, even as their identities shifted at the Office. For example, women prepared familiar meals (originally made from millet) using the staple crop of their new environment: rice. In time, as rice became a more desirable food across the region, the Office gained a reputation for its crop, and Office women's rice repertoires expanded, thus reshaping what constituted food at the Office.

Over several decades at the Office, women often struggled to produce meals that conformed to local standards of high-quality food. The question of producing satisfying and flavorful food was also an issue that reflected women's felt sense of hard work and status. Indeed, Office women worked against assumptions that a proper village life did not exist at the project. Their work was important not only for ensuring everyday survival but also for creating respectable rural status at the Office. For many years, the difficulties of producing palatable meals made the maintenance of rural hospitality challenging. At the same time, women's changing food production practices altered the flavors of standard meals.[80] For women, maintaining control over food and daily life also meant attending to taste.[81] In short, food was a measure of social well-being much broader than the term "food security" generally implies in policy circles.[82]

The well-being of Office residents was profoundly challenged during the Great Sahel Drought. In the midst of extreme hunger women and men living at the Office struggled to feed themselves but also to provide aid to their regional neighbors. Strangers suffering from famine flocked to the Office for its rice. While a proliferation of international institutions and NGOs responded to the crisis, they failed to offer food security in local terms.[83] As Benedetta Rossi has shown, for Niger local responses to the drought drew on deep historical roots. While historical hierarchies determined access to resources in Niger and played a role in the production of hunger, they also produced social networks of support. For the most part, these ties of mutual aid were largely unseen by government and international aid workers responding to the regional crisis.[84] In Mali, long-standing cultural ideologies of food sharing similarly shaped the responses of Office women and men who hosted strangers (most of whom were women) in need of rice to take home. Office residents were obligated to share even when the numbers arriving overwhelmed them. They were further compelled to aid the strangers because the rice Office farmers produced in their irrigated fields was good "food" when compared to the substandard and barely edible international aid. Even in the midst of a famine, quality and taste were essential elements of food security. As an enjoyable embodied experience, eating always mattered.

METHODS AND OUTLINE

Taking the embodied techniques of women at the Office du Niger as an archive, this book traces a history of women's intellectual production that overlaps with patterns of consumption, taste, and the aesthetics of women's technological production. And it is also rooted in the value of female labor. This is an embodied form of knowledge that was accentuated in moments of extreme crisis and redefined the very meanings of development and food security. The writing of this history was inspired by a rich field of scholarship that centers the body as a site for the production of gendered knowledge.[85] I draw from a diverse array of primary sources to tease out a history with little direct documentation. Significantly, I rely on oral testimony and observations of women's work and physical movements, paying close attention to women's experiences *as women* at the Office.[86] Like Susan Geiger's study, the interviews cited here speak to collective experiences. This is due in no small measure to the fact that women living in distant villages across the Office du Niger offered similar testimony about labor routines, their use of specific tools and machines, and strategies for survival during the Great Sahel Drought. As Geiger finds for women political activists in Tanzania, the women's testimony from the Office demonstrates their *creative* collective action.[87]

As interviews were conducted with women (and men) at the present-day Office, many of the insights contained in this study were gained through observations of courtyard cooking spaces, village canals, and agricultural machines in use, and by picking vegetables in women's gardens, eating food in their homes, and exchanging gifts of soap, tea, chickens, and cola nuts. This social and material interaction with women and men at the Office often prompted specific memories of harvest celebrations, a remarkable market trip, or the arrival of the first industrial machines. Objects like metal pots and serving bowls, an old plow, or noisy nearby threshers also prompted fruitful conversations about technology. Focusing discussion on these tools, or talking while walking along a familiar path or working in a garden, allowed my interviewees to express their memories and historical interpretations in material and spatial form. It is an expanded conception of oral history beyond the interview that is drawn from the work of environmental historian Tamara Giles-Vernick.[88] As Barbara Cooper has also argued, oral history practice is often performance, and many women and men interviewed here gestured to give embodied expression to specific memories.[89]

This study also draws from regional folk stories collected in the first decades of colonial rule. The suggestion here in reading these sources is that a popular conversation about food, labor, technology, and gender emerges. Luise White writes about the significance of rumor in East Africa: "[People] construct and repeat stories that carry the values and meanings that most forcibly get their points across."[90] She elaborates that it is the circulation of the stories, in particular, that allows for their wider historical significance.[91] The stories examined here offer formulaic elements about food, hunger, women's cooking, and pots that convey specific local understandings of early twentieth-century rural life in Mali. Similarly, new stories about hosting guests and women hiding food began to circulate during the Great Sahel Drought, suggesting a renewed rural conversation about women's labor, hunger, and generosity. My analysis is complemented by reading oral traditions about the origins of agriculture and ecological change, which like the popular folk stories continue to circulate and resonate with historical transformations at the colonial and postcolonial Office.[92]

Also consulted for this study were European travel narratives and ethnographic writings, missionary records, botanical research reports, and official records from the Office du Niger, both colonial and postcolonial government archives, and records from the Food and Agricultural Organization (FAO) and World Food Program (WFP). While many of the formal state and institutional archives contain little documentation on women's lives, recorded food shortages and demographic crises highlight particular moments of embodied experience. The official records for the Office du Niger and agricultural development prior to its creation also document the diversity of regional food resources and cultivation practices, estimated regional food production figures, Office census figures, as well as the technical workings of the project's infrastructure, the recruitment and management of labor, and complaints of male farmers. In analyzing these sources alongside the memories and practices of women and men at the Office, this book documents the history of Mali's female food engineers.

Chapter 1 examines the gendered agricultural landscape of rural Mali in the early decades of the twentieth century and provides the theoretical framework for embodied engineering. At the outset of the twentieth century, rural life centered on food. It was a specifically embodied experience. Annual agricultural festivals, in particular, encouraged the excessive consumption of food and beer, all prepared by women. During these festivals, the sense of bodily fullness was made meaningful in material contrast to social memories of hunger and famine. Even in moments of food shortages, however, taste remained a central element to the food that was produced. Over the course of these first decades of the century, French policies intruded on rural life and the specific gendered labor routines of women whose work supported the year-round preparation of daily meals.

The remaining chapters follow the experiences of both women and men at the Office from the first years of settlement in the early 1930s up to the Great Sahel Drought (1969–73) and succeeding years of international food aid and development interventions. Chapter 2 focuses on recruitment for the Office scheme and the food shortages characteristic of the first decade of settlement. The colonial body politics of labor recruitment and the problems of food production at the Office were intimately connected. Until the mid-1940s families at the Office did not produce enough food to eat. Significantly, the colonial emphasis on production for export had the unforeseen consequence of suppressing the qualitative aspects of food cultivation that were relevant to local taste. Women were conspicuously missing from Office towns during these years. The absence of women's bodies and their labor was understood locally as a demographic crisis. Living at the Office required too much of women's bodies, and in the absence of their labor, the production of food and its association with animated rural life suffered.

After several decades of colonial rule, women radically reengineered the way they produced food. Chapter 3 showcases how women who stayed at the Office reorganized their food production to better manage a new agro-industrial environment and make it livable. They began to carve out space for food crop production along the edges of the industrial fields and plant valuable food trees. They also established social and market networks with women living in communities at the edges of the Office. With portions of the money earned from cash-crop production, women purchased the foodstuffs that they could not produce from the denuded Office landscape. Over time, living at the Office meant "farming for money." Women also worked to re-create a social agricultural world. In these years, Office women reshaped the natural world of the Office and their shifting socioeconomic realities.

Chapter 4 looks at the same period with special attention to the interplay between small-scale domestic technologies and large-scale industrial ones in the daily preparation of food. It was in the late 1940s that women began using new metal pots, buckets, and other modest household technologies to ease their daily labor. This transformation of household infrastructure enabled women to mitigate the negative effects of a depleted-resource landscape. Many women also sought informal employment alongside industrial threshers, which afforded them access to machine-processed rice. While European staff assumed men and not women were targets of technological development, women actively engaged with new technologies (modest and industrial) as it suited their needs. They also shaped how these new technologies took on gendered meaning in daily life and imbued them with social value.

In 1960 an independent Malian government took over the Office. Chapter 5 looks at this shift and the subsequent crisis of the Great Sahel Drought. When the region was hit by ecological crisis, successive governments looked to the Office to produce food for the nation. Unsurprisingly, state policies favoring urban markets and militarized rationing failed to provide meaningful food security. Yet, Office residents turned the project into a refuge for hunger migrants. Food aid was still critical to survival, but its poor quality required women to once again reengineer their cooking just to make edible meals. Guards strictly monitored the harvests, and women in need of something to cook would smuggle rice, hiding it in cloth, from state-run fields. They accomplished this task by feigning pregnancy with their bundled rice babies. It was another modest but effective technological solution that reveals the depth of women's embodied techniques and knowledge.

1 ᔒ Making the Generous Cooking Pot, ca. 1890–1920

AT THE beginning of the twentieth century, food production and preparation ordered much of daily life in the French Soudan. Women's labors in this realm assured the quotidian meal and the social practice of food sharing. A folktale from this period featuring a magical cooking pot reveals the considerable value of this labor and the importance of generosity as a cultural ideology in the region. The story opens as a hyena, a character that often served to comically highlight human missteps, sets out to find fortune.[1] While traveling, the hyena searches for something to eat and comes across a monkey and a tortoise. The hyena hopes in vain that the two animals will help gather fruits from a baobab tree and edible water lilies, but the monkey stays in the branches of a tree and the tortoise disappears into the water of a marsh. Their refusal to help the hyena find food signals a social ill for the listeners of the tale. On the same journey, the hyena finds a lump of butter, the mythical food first offered to humankind in Bamana traditions and a symbolic substance associated with women and fertility.[2] Unfortunately, the hyena loses the butter in an attempt to cook it directly over a fire for want of a pot. Not long into the hyena's search for fortune, how to find food and prepare it has become a central element of the tale. Following these failed attempts to procure sustenance, the hyena discovers a small cooking pot in the marsh where the tortoise disappeared.[3] The hyena asks the pot for its name and receives the reply "the Generous Cooking Pot." When asked to offer generosity, the pot miraculously produces cooked rice, couscous, and meat.

Then, as was custom in the region, the hyena—like a human traveler—seeks a local host. In this story it is an elderly hostess. Shortly after finding the magical pot, the hyena demonstrates its powers to the old woman, who is similarly astounded. Rather than employ the pot to cook for her guest, however, she decides to present the powerful object to the king. The hyena, endeavoring to retrieve the pot from the king, returns to the marsh and finds a sword that attacks anyone asking for its name. When the hyena offers the sword to the king, its attack distracts him long enough for the hyena to recover the pot and flee. It is a tale that reveals much about daily life in the region. Indeed, for the hyena and the tale's audience, fortune was found in the generosity of the cooking pot.[4]

Strangely, it is a woman who gives up the pot once it is presented to her, and perhaps it is this irony the audience is meant to take away from the story. Managing the pot and generously sharing food was the realm of women, not political leaders, whose raids and wars in the late nineteenth century provoked food shortages. Yet the relationship between women, abundance, and the state is complicated in the long history of the region. Interestingly, in the story the woman is elderly and no longer able to produce children, suggesting a connection between female fertility and food production.[5] Nevertheless, the old woman's character does testify to the great power of a technological object strongly associated with women and the provision of food. In the story, the pot repeatedly produces a rich meal from what appears to be nothing. The magical cooking pot is the very essence of hospitality. It also symbolizes the essential role of food production in rural life and how state politics and violence complicate food production. But even more, the pot represents the desire to be fully satiated. This cultural ideal was manifest during yearly agricultural festivals, when women prepared seemingly endless amounts of food and beer. Yet it was not an easy feat for women to emulate the pot's generosity in times of scarcity or political upheaval.

The French Soudan was predominantly rural, but the conditions for agricultural production were changing during the first decades of the century. Indeed, women's gathering, cultivation, and culinary skills were vital as farming households faced environmental stresses, demographic shifts, political insecurity, and increasing colonial intervention. The years of conquest in the decades before 1900 had upended daily life and provoked widespread migration, leaving many fields untended. Not long after the French began to establish colonial rule, the devastating 1913–14 famine left thousands in dire conditions. As the first decades of the twentieth century unfolded, the French would pressure those same rural populations to produce for new colonial markets. In the midst of rapid and dramatic change, women adapted their cultivation and preparation techniques to ensure their pots would continue to produce food.

GENDER AND EMBODIED LABOR

In the second decade of the twentieth century, another story circulated in the French Soudan featuring a young woman endowed with the fantastical skill to cook quickly. In the story she prepares an ordinary meal, one the audience understood required hours of labor. Yet her meal is ready seemingly in the blink of an eye. It is a testament to vernacular thinking about women's expertise. In the same tale, the young woman is compared to her husband and his younger brother, who also accomplish ordinary tasks with astonishing speed. The husband prepares meat from an antelope before the ball from his musket even reaches the animal. His younger brother humorously slips in the mud on the way to the fields. At the same time he saves a basket of seeds from overturning, which he had been carrying on his head. His accomplishment is to quickly change out of new and unsoiled clothes into older ones better suited to the tasks of farming.[6] Taken together, the three feats underscore the continued centrality of food production: grain cultivation in the fields, the provision of meat through hunting, and women's cooking. The story also reflects an idealized division of labor and endows both men and women with extraordinary abilities related to their gendered food labors.

In other stories dating to the early twentieth century, food is similarly produced as if by magic, but the corpus of regional tales also features, less idealistically, greedy characters hoarding food and entire villages suffering from famine. Such tales resembled actual histories of food shortages, hunger, and famine from the period. In many of these stories food signified moral lessons on the importance of hard work and generosity, and specifically the sharing of meals.[7] Women cooks, like the fantastically fast woman in the tale, produced the symbolic value of food in regional culture, but throughout the twentieth century their labor was also critical to the material production of rural subsistence in ways unseen in this tale and others.

The story about the three surprisingly fast family members, in particular, suggests the embodied nature of food production in rural Mali. Each of the three characters displays the remarkable ability to quicken ordinary physical movements associated with farming or the production of the evening's meal. However, the younger brother's comical fall suggests that he is still learning how to be a farmer. Moreover, the quickly averted loss of seeds gestures toward real risks to the harvest. Embodied knowledge was critical to rural survival but required cultivation. By contrast, the young woman in the story had mastered her embodied cooking tasks. Once the men return from the fields and the chance hunt, she tells the younger brother to bring her some millet from the granary so she can begin cooking. He brings her two baskets of grain, and when he arrives with the third basket full, she announces the meal is ready. She even chastises her brother-in-law for almost dropping unprepared grains onto the finished meal.[8] We are not told about the young woman's actions in detail, but the audience would have been well aware of the time-consuming and fatiguing work that she rapidly completes. In short, like the Generous Cooking Pot, she is a fantastical cook.

Women's work producing food was physical, and in this story it is valorized alongside the men's labors. In the conclusion to the story, the narrator asks the audience (in its original telling the story was meant to be interactive) who of the three characters is the fastest worker? The men's work in the fields and in providing meat was valuable. All the same it is noteworthy that a female cook is potentially as heroic as her husband, the hunter, an iconic figure in the regional cultural imagination.[9] And perhaps both are especially impressive when compared to the clumsy younger brother. Yet women in the audience likely chose the cook as the fastest character. Ultimately, women's expertise in performing quotidian physical labor is lauded in the popular tale.

Women's cooking in the midst of shortages was especially astounding. For many in the region, the mythical figure Muso Koroni embodies the labor and pain of women who in times of ecological crisis produced something to eat. Muso Koroni is a female deity credited with introducing cultivation techniques to human society. However, it is only through her pain that she transmits these techniques. Muso Koroni and her male companion deity, Pemba, create much of the material world, but she instills chaos, which angers Pemba. In response Muso Koroni takes flight, only to be chased by a third god, Faro, who floods the earth as she runs. Faro's actions restore order, but Muso Koroni is left to wander and suffer from hunger. To survive, she learns to collect food in the wild and to plant seeds for cultivation.[10] The myth shares many elements with historical research on how women's gathering of seeds led to seed cultivation. When women collected seeds and nuts, they carried them and planted some of them, sometimes deliberately, sometimes accidentally.[11] In the French Soudan, women evidently retained and continued to pass down the knowledge of this seasonal, varied, and opportunistic aspect of food production.[12] Into the twentieth century, women continued to depend upon and shape a food-gathering capacity in their daily labor. While Muso Koroni's pain and physical labor was transmitted to human society, productive human activity that creates bodily pain and fatigue is understood as a creative and noble endeavor.

The Muso Koroni myth speaks to regional ideologies about gender as well as labor. In fact, she is credited with innovating the rites that inscribed gender on human bodies (even though elements of both genders were understood to exist in male and female bodies). This original gendering of the body was associated with human procreation but also with the ideal of complementarity in the organization of human society.[13] Gendering the body produced specifically gendered labor tasks. For example, women's food production specifically included the fatiguing labor of collecting wild foods (not unlike the Muso Koroni myth). At the same time, men shared the legacy of producing subsistence through physical hardship. In fact, regional agricultural rites celebrate the tired arms and bodies of male farmers. For example, Ciwara dancers encouraged farmers to work hard during the agricultural season, and their bent-over dance positions purposefully fatigued the dancers' bodies, who in so doing shared in the overall physical labor of the harvest (see figure 1.1).[14]

FIGURE 1.1. Pair of masked Ciwara dancers by Rev. Père. Joseph Dubernet. From Joseph Henry, *L'âme d'un peuple africain: Les Bambara leur vie psychique, éthique, sociale, religieuse*, vol. 1 (Munster, Aschendorff, 1910), 144 bis.

Women also worked in the fields, but their ritually celebrated labor drew together agriculture, fertility, and childbearing. In particular, Ciwara agricultural rites featured a male and female antelope mask in ceremonial performances. The female antelope was distinguished by the representation of a baby on its back, mimicking the actions of women who carry infants in the same manner (worn by the dancer in the background of figure 1.1). The mask connected agricultural fertility to human fecundity. In this regional tradition, childbirth was akin to farming labor—both were productive and procreative but also painful. Childbirth, in particular, was understood to ennoble women.[15] It also centered women in the work of human survival and social reproduction.

Returning to the Muso Koroni myth and the figure of the surprisingly fast cook, when drawn together, an important tension emerges: women's critical labor was widely understood to be physically taxing but was ideally performed as if women were endowed with extraordinary physical powers. These popular representations of women evoke the early twentieth-century cultural discourses shaping regional understandings of women's social roles and labor. Strikingly, both the story and myth emphasize women's bodies. In the myth, Muso Koroni's bodily experience of hunger and fatigue is highlighted. Moreover, the changing material reality of the earth (its flooding and the presence of wild resources) shapes her wanderings. In the tale about the wife who is a fast cook, her ability to prepare a meal so quickly highlights women's embodied expertise. Both representations signal that women's experiences are rooted in the material—the material world of floods, hunger, and fatigue—but also in control over the body. Women mediate these material challenges with technologies like pots, but that labor also requires specific corporeal techniques.

To get a sense of this daily embodied and sensorial labor, imagine a woman winnowing grains by using the wind. She would have stood tossing grains up in the air from a calabash. She then would judge when the grains were clean by touch or sight. During the dry season, the same woman would recognize specific wild grains in the bush by sight or she would gauge the quality of seasonal fruits by touch. These kinds of sensory skills also served women who examined foodstuffs in the market. Women in Mali used their *musow minanw* (women's tools or things) to enhance their perception of the readiness of food, such as when they pounded grain or gauged that the meal was ready in the pot. The right feeling of ground millet or husked rice started with the feel of the pestle in a woman's hands. Similarly, the texture and consistency of the staple food *toh* (stiff millet porridge) were measured by the pressure against the wooden spoon in the cooking pot. This work was manual, but it required more than just physical exertion; it was also skilled sensorial labor. Moreover, the repetition of familiar food-preparation actions (and the addition of new ones) took on great significance when food was scarce.

As feminist scholars have noted, studies of the body from the past several decades, while generative, have focused on discourses *about* the body (its symbolic value and social construction) to the exclusion of "lived material bodies and evolving corporeal experiences."[16] The absence of the corporeal body in this line of feminist analysis was one means of resisting the persistent association between women and nature that also connected men with reason. In such dualistic reasoning, women were constrained in their experiences by biological determinism and often rendered inferior to men. An earlier school of ecofeminists had, by contrast, romanticized women's relationship to nature as superior to men's, thus rendering women as altruistic protectors of nature and society.[17] The contours of these debates are well worn but highlight the striking materiality of gender thinking in Mali. As Barbara Hoffman has demonstrated, men in Mali have historically associated women with nature—or the untamed and chaotic wilderness. In teasing out this masculinist perspective she cites the following proverb: "Woman is fresh mud; however, you shape her, she will dry that way."[18] While depicting a broadly critical view of women, the proverb reveals an understanding of gender rooted in the material world. Ironically, across the region, women (not men) are traditionally potters. Who then is shaping women's nature?

Women's lived experiences throughout the twentieth century demonstrate that they, like female potters, learn specific environmental and technical knowledge that they consider valuable and significant for the ways they relate *as women* to society and to their material worlds.[19] While gender relations in Mali as studied through proverbs, masked performance, oral tradition, stories, and political structure have been interpreted as evidence of historical male dominance, women's material and corporeal experiences speak to their key role in producing society.[20] Indeed, it would not be difficult to imagine women interpreting the story of the fast cook as representative of women's skill and social value. As Hoffman notes: "[Women's] acceptance and submission to a public ideology of subordination gives them cultural space in which to cultivate substantial quantities of actual power and effective authority."[21] Obioma Nnaemeka similarly theorizes that it is "the power of African women to work with patriarchal/cultural structures that are liberating and ennobling while challenging those that are limiting and debilitating."[22] In Mali women have cultivated expertise and social prestige through their specifically gendered food production labors. However, they have also challenged overly taxing demands on their labors.

MANAGING FERTILITY AND CRISIS

The specific association of women with human and agricultural fecundity is rooted in regional understandings of the environment. First, women and men approached the environment as essentially productive. At the same time, human intervention was required to realize its bounty. Pemba, one of the deities previously mentioned, introduced mythical *fonio* seeds and their life-giving potential to the earth. Pemba then transformed into a *balanzan* tree during a turbulent reign marked by tempestuous winds and rains. Initially, the earliest people ate shea nuts that fell from the sky, but when the nuts ceased to fall, the people suffered from famine.[23] The difficulties faced by these mythical people testified to the dangerous power of the elements, a common theme in the oral traditions of farming communities. Ultimately, the deity tree withered and dried up. In a subsequent episode, Muso Koroni, who was the mythical wife of Pemba, and the successor deity Faro, who reorganizes the world, introduce cultivation techniques. This next period of creation was not an idyllic response to Pemba's reign but one in which humans took responsibility for their own agricultural production. It is in another important episode that Muso Koroni flees Faro and suffers in her search for sustenance. In response Faro transforms into the Niger River, which floods its banks in pursuit of Muso Koroni. Human settlement then expanded along the river, creating the first fishing and farming communities.[24] The myth presents an environment that is fertile but subject to periodic disorder and destruction. Indeed, the desiccation of the balanzan tree also suggests sustained drought. In the midst of such crisis Muso Koroni wanders gathering and planting food, a survival strategy she teaches to people. The balanzan tree, ubiquitous in the Sahelian landscape, is a reminder of this origin story and the work required to create and sustain human life. It also symbolizes the unpredictable and potentially dangerous aspects of the region's climate. Yet when the land is dry after the harvest, the balanzan tree blooms.[25]

Ecological crisis has long been a persistent yet unpredictable element of the wider western Sahel.[26] Its inhabitants have faced periods of severe recurrent drought punctuated by years of increased rains.[27] In assessing the long social history of climate change, archaeologists Roderick McIntosh and Téréba Togola argue that for centuries residents of the Mande cultural region have inscribed the social crises and resource pressures associated with such climate volatility into social memory.[28] Through myths and other oral traditions, Mande societies have testified to dramatic environmental transformations faced by early western Sahelian societies, and they have also recorded social responses to those changes.[29] In these myths, legends, popular songs, and stories, ecological crises are often supernatural in origin, or they result from the violation of a taboo. For example, in one telling of the Soninke legend of Wagadu, the killing of the sacred snake Bida results in prolonged drought and large-scale migration to the south.[30] In other cases, humans act on the environment, drawing on their own sacred powers. For example, the first blacksmith in the Mande creation myth strikes a rock with his hammer to call for rain.[31] In another example, Fanta Ba, a mythical Bozo hunter, opens lands along the Niger River for settlement.[32] Broadly speaking, regional understandings of the environment were decidedly *not* deterministic, even if powerful forces brought potentially devastating changes in the environment: rather, people in the oral tradition who possessed specialized skills and powers counteracted negative forces and even transformed their environments.

By contrast, European observers tended to frame the region's environment as static, either abundantly productive or barren of any resources, and not as a product of human action.[33] For much of the late eighteenth and early nineteenth centuries, European travelers perceived the region to be a land of great riches.[34] In 1800 the Middle Niger region was a vast expanse of savanna plains punctuated by the Niger River (see map 1.1). Upon entering the region from the south near the Sotuba rapids (just north of

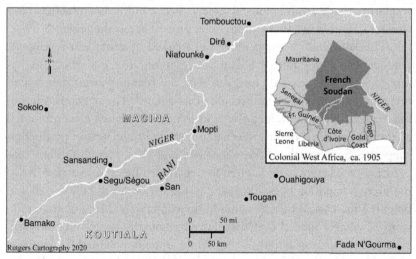

MAP 1.1. The Middle Niger region and nearby Volta territory in the French Soudan, ca. 1905.

Bamako), the river ran calm for river traffic to the north and widened its banks. During this period, river transport expanded to support commerce between the region's producers and their neighbors to the north and south.[35] In this same period, European travelers such as the Scotsman Mungo Park explored the river. Significantly, command over the Niger would later become a hallmark of French rule, especially for transport and irrigation.[36] For travelers coming from the northern Sahel, baobab and shea trees, managed by women, marked the farming landscape as distinct from the more arid north.[37] The region was predominantly agricultural and culturally Bamana, but even small towns could be cosmopolitan, hosting traders and religious pilgrims.[38] In addition, Fulbe pastoralists raised cattle in the Macina, and they controlled much of the floodplain rice cultivation in the inland delta. Culturally diverse fishing towns also dotted the river, and during the dry season northern herders traveled south to graze their cattle and exchange salt for millet or other grains.[39] The larger agricultural picture of this period was one of a dynamic seasonal trade in grains, fish, meat, and milk, as well as in prepared oils, spices, and leafy or vegetable-based sauce ingredients (the latter were produced by women). In 1799, Park remarked upon the tremendous agricultural wealth of the region. While traveling to the Bamana capital at Segu, he wrote: "Cultivation is carried on here on a very extensive scale; and, as the natives themselves express it, 'hunger is never known.'"[40] Certainly some towns fared better than others, but as Park himself noted, regional residents took pride in local agriculture and food production. Traveling in the 1820s, René Caillié similarly recorded rich botanical variety, diverse cultivation practices, and women's significance to the production process across the region.[41]

The latter half of the century was marked by warfare and religious conflict, which took a heavy toll on rural communities. Following the capture of Segu by Umar Tal's forces in 1861, the Islamic warrior state raided outlying towns for slaves, grains, and other goods.[42] The previous rulers of the Segu Empire had raided the countryside but nevertheless promoted a strong regional economy specializing in grain and textile production supported by slave labor.[43] Between 1863 and 1865, French military officer Eugène Mage was dispatched to negotiate a treaty with Umar Tal as the French had long been interested in the region for its fabled gold and potential markets. While traveling to the capital city, Mage observed the ruins of villages sacked by Umar. He subsequently wrote unenthusiastically to his superiors about the region's economic potential.[44] Mage was disappointed by the difficulties of travel and depressed trade under Umar's rule. At the same time, he made note of towns rich in grains and other agricultural products, especially those offered to him by women.[45] He noted women in market towns selling prepared snacks such as grilled peanuts, fried millet, shea butter pastries called "momies," and balls of couscous mixed with honey called "bouraka."[46] In the same markets women sold shea butter, millet, maize, tamarinds, spices, local *niébé* beans, peanuts, peppers, and specialty grains. In towns along the river, fisherwomen sold fresh and smoked fish.[47] The diversity of market foods described by Mage testified to the richness of local diets and the resilience of local food production systems, even in a politically unstable environment.

In the last two decades of the nineteenth century, the region saw a growing French military presence. French *tirailleurs* marched against the Umarians and other resistant states.[48] Moving across the countryside, troops from all armies captured both people and food, including livestock, fish, vegetables, and especially grain stores.[49] Yet, in 1888, Commander Frey wrote that the *only* resources in the western Soudan were slaves (a labor force and economic good).[50] It was a stark assessment and contrasted with earlier observations. While the many wars of the late nineteenth century increased slaving and dramatically reduced production for the markets, Frey's suggestion that the region lacked resources was hardly true for local inhabitants. Rather the resources available to them were constrained by the volatile political environment. In fact, French conquest relied upon the region's agricultural productivity. Nevertheless, in the following year, the French officer Jean Gilbert Jaime, while on a mission to explore the Niger River, wrote about the region: "The country produces nothing or almost nothing, and the climate is atrociously unhealthy. The Soudan is without exception, our worst colony. Despite the expense we will hold it in the hopes of finding, beyond its unproductive plains, other regions rich in resources to make up for our losses."[51] Famine hit the region in the years 1899, 1901, and 1902 just as the colonial government was establishing its rule, and rural populations were struggling to recover from several decades of regional warfare, drought, and declining food production. These natural disasters further contributed to dramatically shifting the European image of the French Soudan as a region economically stunted by an infertile environment.[52]

By the beginning of the twentieth century, the region was already undergoing dramatic socioeconomic change. It was a period of upheaval that would continue into the first decades of colonial rule. In 1903 the French formally prohibited the legal recognition of slave status in the French Soudan.[53] Soon thereafter, a massive exodus of slaves in 1905 from the Marka town Banamba stoked French fears of social disorder and declining agricultural production.[54] The exodus of large numbers of fleeing slaves from the Middle Niger region lasted almost a decade and had a dramatic impact upon the region's food supply and economy.[55] For example, in 1907 in Sansanding, where the French would later plan to build an irrigation dam (see map 2.1), the local administrator recorded the flight of at least 985 slaves. The runaways fled during the months of April and May, just before the onset of major field preparations. In a report on the situation, the French official in Ségou openly expressed concern for the loss of agricultural workers and potential future loss of labor.[56] A year later, large numbers of slaves left the northern edge of the agricultural region at Sokolo and nearby Gumbu.[57] At the desert border, raiding parties further disrupted local and regional food supplies.[58]

Women's labor, especially the labor of young or servile women, was already essential for rural food security. In the first three decades of the century, all women bore an increasing share of the field labor burden. Enslaved men and women, who had previously performed much of the field labor, had to negotiate with their masters to establish new roles in society. Many enslaved men fled. Women often remained behind, especially those who had children by their masters or were married to a local man. During these years, female pawnship also periodically resurged. These women provided labor as part of the repayment of a debt, most often contracted by the senior man in a woman's household. Former female slaves, female pawns, and many young wives all provided more labor in the fields (in the absence of many men) and in domestic tasks.[59] For example, in 1905, Mademba Sy, the French-appointed ruler of Sansdanding, reported that Birama Koita (who was likely enslaved) had abandoned his wife and their three daughters in November 1903.[60] This record speaks to the heightened movement of people (especially men) during this period and the disruption of households. The upheavals of the late nineteenth and early twentieth centuries prompted the reorganization of agricultural labor, and the insecurity of male agricultural labor in these years placed greater pressure on women. Yet many women resisted undue agricultural burdens. As was the case for departing slaves, women who abandoned their husbands did so during the months when their labor was most needed in the fields.[61] Certainly, the political, economic, and social upheavals of these years impacted daily life and the provision of daily meals in individual towns and households.

Against this backdrop, many of the first colonial administrators believed the French Soudan was plagued by underproduction and chronic food shortages, which spurred them to invest in agricultural research in the colony. The early colonial government dispatched a host of scientists to study potential agricultural exports. Shortly thereafter, they established experimental farms, built agricultural schools, and intervened directly in production. These colonial researchers tended to study rural production by looking at one particular crop (or another potentially profitable resource) at a time. When it came to food production, researchers and officials were primarily concerned with increasing the yields of local cereal crops like millet and rice (see chapter 2). For the most part, they misunderstood the complex food production systems managed by women. Women's agricultural knowledge was not vertical (as in the study of millet or rice in isolation), but horizontal, drawing on their experience with multiple crops and wild resources. More importantly, women (and men) did not see the landscape—even in years of hardship—as unproductive. Rather, they saw themselves as agents of environmental transformation.

FAMINE, FRENCH INTERVENTION, AND LOCAL INTERPRETATION, 1913–14

French West Africa was again hit by food shortages between 1913 and 1914. By this time much colonial discourse about the economic potential of the Soudan was framed by the threat of famine. As French officials increasingly became aware of the agricultural crisis and famine conditions, they moved to intervene directly in grain markets and sponsor additional research on agricultural production in the colony. The 1913 rainy season had been particularly poor, and millet harvests across the region were extremely disappointing. Even rice cultivation along the Niger River in the floodplains of the Macina suffered that year because of low floodwaters. By early February 1914, the colonial government was preparing to send rice and millet to especially hard-hit northern districts.[62] Despite the administration's efforts, the administrator for the northern Tombouctou region reported 3,428 deaths in his district as of March 1914.[63] Grain shortages were exacerbated by inflated prices in the northern areas and elsewhere in the colony. Only Bamako and other southern districts had sufficient grain surpluses, but the transportation of goods from these regions to the north along the river was delayed by low water levels.[64] During the famine, the administration registered thousands of deaths, and many who survived were displaced after having to leave their homes in search of food.[65] The disaster was devastating and widespread.

Still, ritual specialists in farming communities called for good harvests through ceremony and prayer. For example, in 1913 the community of Banankourou (near Ségou) called for the performance of a special agricultural rite. The Catholic missionary sisters who observed the ceremony noted its exceptionality and recorded that it was ordinarily held only once every one hundred years. Part of the ceremony involved a parade of women circling the village three times carrying cooking tools over their heads. The women's actions called for rain and enough of a harvest to fill their mortars and pots for the months to come. In the midst of ongoing drought and food shortages across the region, the same community held another ceremony to bless their seeds in 1914.[66]

Drought, food shortages, and famine forced the colonial government to respond to the crisis, albeit somewhat belatedly. In 1914 the administration ordered the construction of government-controlled reserve granaries. The policy move was directly aimed at improving the food supply. However, the reserves in these granaries were not to be touched except by order of the local administrator.[67] In practice, the policy move failed. For one, farmers already maintained personal granaries for annual consumption, as well as the next season's seeds and a security reserve. A granary was customarily built and controlled by the owner of the various grains stored inside. Because of their importance for daily meal production, the household head controlled the granaries that stored crops from the fields that everyone in the household cultivated. Every day this senior man distributed millet, rice, or other grains from one of the granaries for the day's consumption to the woman cooking. Anyone in the household, including women, could also build their own granaries to store products cultivated in individual plots: women often stored peanuts, for example. Household or family compounds could house ten or more granaries of varying size (see figure 1.2).[68]

When administrators ordered the construction of communal granaries in each settlement, they also ordered household heads to contribute to the new granaries a portion of the harvest that ordinarily would have gone into the household grain stores controlled by the senior man. The new granaries were not consistently maintained over the next decade. When farmers received grains from the new granaries called *magasani*—as they came to be called, after the French word *magasin* for "storehouse"—they had to purchase them.[69] Moreover, most of the new granaries were built using conscripted labor, a difficult burden for hunger-stricken regions.[70] Ultimately, the official granaries poorly replicated measures already taken by farmers to store surplus harvest. More to the point, individual households lost control over grains they contributed to the magasani. Access to their food stores required payment or a promise of payment in kind.

FIGURE 1.2. Granary for rice in Baguinèda with man holding a plow. Courtesy of the Archives Nationales de la France d'Outre Mer, ANOM FR 8Fi 417/31 Office du Niger Aménagement rurale, 1935 to 1954.

In the course of carrying out the official order to build granaries, the administrator in Fada N'Gourma noted that many cultivators hid their grains in "the bush" for fear of theft,[71] which prevented the administrator from successfully maintaining granaries. The same administrator also criticized farmers under his jurisdiction for selling their surplus "basket-by-basket," rather than saving the grains and profiting from one large sale (presumably to European agents or merchants). However, as he observed, farmers participated in an active and long-standing domestic grain trade with Fulbe herders and other groups. In years of severe shortages, many of their trade partners may have wanted to (or could only) purchase "basket-by-basket." For the administrator, the limited sale of grains was a particular concern in the years immediately following the famine because of the outbreak of the First World War in 1914. At that time, French West African colonies were called upon to provide France with foodstuffs and oils for the war effort.[72]

In general, administrative policy and the colonial market responded poorly to the ongoing food crisis, a fact that Lieutenant-Governor Clozel admitted to in his 1914 report to the governor-general in Dakar. Still,

Clozel accused farmers of courting disaster through poor planning and apathy. He failed to mention that the 1913–14 famine fell on the heels of widespread slave departures and substantial population movement. Moreover, 1911 and 1912 had also been disappointing agricultural years, making it unlikely that many farmers would have been able to constitute substantial surpluses ahead of the 1913–14 drought. Ultimately, Clozel argued for greater colonial intervention and an expansion of the transport infrastructure. In the 1914 annual economic report for the colony he proposed the expansion of the railroad system. Clozel reasoned: "Only by improving the economic infrastructure, and completing the proposed railway system, will export crop production gradually develop and protect, in poor years, against insufficient food supplies with the rapid and inexpensive transportation of products between regions."[73] Clozel's argument demonstrates the great emphasis administrators placed on large-scale technologies and infrastructure. Increased production and improved transport for export did not directly address food cultivation, yet Clozel assumed improved transportation networks would facilitate a greater domestic market in the case of famine. Following this reasoning, dramatic food shortages became an argument in favor of greater export cultivation rather than increased food production.

Despite ongoing administrative concerns for the colony's food supply, efforts at agricultural reform prioritized metropolitan needs. In 1913 the French West African government reorganized the Agricultural Service to concentrate on the production of goods for the colonial export market. The reforms especially emphasized cotton cultivation, which would be the primary export crop for the eventual Office du Niger scheme. The 1913 reorganization of the Agricultural Service also instructed staff in the Macina region to research possibilities for irrigating the inland delta of the Niger River and for promoting rice cultivation, especially white varieties of the grain. Rice farmers in the Macina grew rice that was typically described as red in color and not as marketable internationally as white rice. The reformed service additionally called for more research aimed at diversifying local agricultural production for export, including the establishment of fruit tree plantations. Concern for fuel reserves also prompted preliminary forest conservation efforts that limited rural residents' access to essential resources. These reforms came in the wake of widespread local hardship, but the policies paid little attention to the needs of farmers or other rural residents. Cereal crops like millet, maize, and fonio received little mention precisely because they were produced for the domestic market.[74]

One exception to these policies was the interest in shea nut collection and their potential industrial uses. In earlier years, European travelers observed that women collected and processed shea nuts for domestic consumption. For the most part, shea nuts were locally associated with women, female labor, and women's economies.[75] Increasingly, European agents purchased dried shea nuts, whole or shelled, but not the women's prepared butters or oils.[76] Denise Savineau noted that by 1938 women in French West Africa made a higher return on butter in local markets than on unprocessed nuts. Butter was also easier to transport. These two factors accounted for women's interest in transforming their surplus shea nut stores into butter for the local sale. However, Savineau suggested that men had begun to sell nuts to colonial agents, which undercut women in the shea market.[77]

Women tended to sell any surplus production of processed household and community goods on the domestic market. The colonial market, by contrast, promoted the sale of raw goods rather than processed or manufactured products (shea butter, beer, cotton thread, etc.). Significantly, workers in the processed or manufactured domain of the domestic agricultural economy were predominately women. Men tended to be more active in the sale of raw goods in both international and domestic markets (in some cases they were required to sell raw products for colonial export). The sale of peanuts (a crop associated with women) is a notable exception. Women cultivated Bambara groundnuts and peanuts, and then they processed them into cooking oil and preservable sauce bases.[78] Peanuts were associated with women's labor because they fit the gendered labor pattern of women producing ingredients for the sauce, while men cultivated the staple grain (see "Colonial Intervention and Gendered Labor Dynamics" in this chapter). However, it must be noted that men also began farming peanuts for export in these years—they sold the raw unprocessed peanuts on the market, which was a striking departure from the customary pattern. Men were even selling raw shea nuts to European purchasers. Indeed, women's earnings were challenged by the men's expansion into peanut farming and shea collection for the colonial export market. Domestic consumption and a fortiori women's production and manufacture were not a priority for the new colonial Agricultural Service. It was designed, after all, to support the new colonial export trade; local food production drew the attention of officials only in moments of crisis.

Under several years of French rule, the agricultural market was in flux, yet it was the 1913–14 famine that would shape the way rural residents thought about social relations and food for much of the twentieth century. Such crises were understood not necessarily as ecological crises but rather social ones. For example, in one folk story, published in 1923, the town of Dougounifin is hit by famine after an orphan is chased from the village. While rain falls on the fields of neighboring towns, Dougounifin suffers extreme drought. It is only when the town vows to care for the orphan that rain begins to fall again. The story might be interpreted as a morality tale about the material consequences of greed. It also suggests that ecological crises are communal events and require collective action. Strikingly, the colonial interpreter Moussa Travélé, who collected this tale and published it, notes that it is a women's tale. In fact, it is one of the few stories among those published by Travélé that he specifically associates with female tellers (and perhaps a female audience). While he does not elaborate on what makes it a woman's story, his translated title, "The Orphan (a Women's Tale)," implies connections between women, the communal care of children, agricultural success, and wider social well-being.[79]

Other stories recorded in local memory highlight the specific role of women's labor in surviving famine. In Koue-Bamana, a town that dates to the Segu Empire (ca. 1712–1861), old women tell a story about a great harvest of wild grains during a famine. In the story, which the women say happened in the distant past, the whole town was suffering from severe hunger. It became necessary for the women to find something for people in town to eat: a group of women came together and went to collect wild rice called *jéba*.[80] They started out looking for termite mounds because jéba could be found inside them. While out in the bush, the women found a mound that contained so much jéba that they could not collect it all. The women discussed what to do. They feared that if they did not take all the jéba with them at once, the rest would be gone upon their return. First, they put as much jéba as possible in the containers they brought.[81] Then they waited for nightfall. As it got darker, they removed the cloths around their waists and wrapped the rest of the jéba in this material. It was a modest technological solution to their problem. Each woman grabbed a calabash or basket of jéba to carry on her head while she also held her wrap full of grain with one hand, leaving the other free. Finally, the women started walking home, carrying all the precious jéba.

When the women did not return before nightfall, the people in town believed they were lost. The townsfolk quickly organized a search party. The men in the search party left the town beating alarm drums and shooting rounds into the air from their guns. After a while, the searchers found the women kneeling naked by their overflowing containers and wraps.

When the women heard the alarms, they knew the town had sent men to look for them. They quickly kneeled to cover their nakedness. And that's how the men found them. In the end, both the women and the men from the search party made their way home. When the women returned home they were still naked but had brought a wealth of food with them. They had found so much jéba that the famine was declared to be over. Everyone then began to celebrate.[82]

There are several ways to interpret this story. Of course, the women are heroes of sorts. Their trip to gather jéba saved Koue-Bamana from famine. There is also an element of sacrifice: the women risked danger and shame to collect all the grains they found. If the story took place in the first decade or so of the twentieth century or earlier, women who ventured outside their towns risked being captured by raiders. In fact, the urgency of the search party suggests just such dangers. The story is also instructive on how to protect a community and where to find food in times of need. For example, from the story we learn that jéba can be found in termite mounds. If Koue-Bamana was suffering from food shortages, nearby towns were probably also in need of food. This would have justified fears that either animals or other women may have come along and collected what the Koue-Bamana women left at the mound. Perhaps this was why the women could not leave the jéba and expect to find it later. The story was also meant to be entertaining. The elderly women who told the story laughed as they recounted the women's nakedness and how the male search party found them with nothing covering their bodies but the jéba. Like many regional stories, this tale links food and women's sexuality but also contains bawdy humor and depicts the resolution of a social crisis: in this case, there was suddenly so much food in the village that no one cared that the women were naked when they brought it home. The laughter inspired by the story may also serve to distance memories of hunger in the past from the moment of storytelling. While the storytellers did not specify when this episode took place, it is possibly a collective memory from the devastating famine of 1913–14. It may also be a composite memory of several episodes of hunger or famine. Regardless, the story is about bounty in the wilds and how women's collective efforts saved an entire town from famine. Hunger propelled women from town to leave the safety of the settlement, and their sacrifice was rewarded.

The story of famine in Koue-Bamana is complex. Embedded within it are signs of the extreme fatigue women must have endured during times of famine. Already suffering from hunger, the group of women walked deep into the bush and then carried a large load of grains home. Indeed, the hunger of town residents is presented in the context of the bounty of wild grains that saves the town. It is a foil against which hunger and possible starvation are presented as real and historical threats to the community. Yet in the story, women and men also experienced the food shortage differently. Food production was a female domain and remained so even when going into the bush was risky.[83]

ENGINEERING THE DAILY MEAL

While cooking was often a fantastic endeavor in folk stories, quotidian food production in the first three decades of the twentieth century was a complicated and labor-intensive endeavor. Imagine such a day for one woman: it is midafternoon at the beginning of the rainy season in 1915, and farmers are busy in the fields. Everyone is happy for the rain, especially when they remember the terrible drought from only a few years ago. In a town nearing the northern edge of the former Segu Empire, a young woman named Bintu leaves the millet fields that she tends with her husband's family. Her body is tired because she just finished her part of the day's weeding. Rising to leave, she takes the small hoe she used for work and grabs a large calabash.[84] Bintu spends more time in the fields than many older women did as young women. She must work more because there are fewer workers per household than half a century earlier when slave labor was common, and new demands for grains and cotton by the French added to the household labor burden.[85] Bintu reaches the road and waves to a female friend in the distance who is headed to a neighboring field. Her friend is carrying a covered wooden bowl on her head; it is the afternoon meal for the men and women from her household still in the fields. Bintu was spared that task for the day because her sister-in-law was in charge of the day's cooking and transporting the meal. Because Bintu completed her day's work in the family fields, she is free to tend to her own peanut plot or to collect other cooking ingredients in the nearby bush.[86]

After walking for some distance, Bintu reaches a shea tree and stops to fill her calabash with nuts that have fallen to the ground. She then balances the heavy load on her head and turns toward town. Once home, she will add the day's collection to her store of shea nuts. After the grain harvest, she will pound, cook, and process the nuts into butter for the year with help from her female in-laws. Close to town she decides to put down her nuts and collect some *saba* fruits (see figure 1.3). There is a little room left in her calabash for a small branch full of fruit, but she still must tie up a few loose ones into an extra piece of cloth from around her waist. Bintu began her work at dawn, so it had been a long day of physical labor. She will enjoy the fruit along with her mother-in-law and the family's children

FIGURE 1.3. Saba fruits in a metal bucket, 2010. Photo by author.

when she arrives home. Even though it is the hungry season, when grain stores are low during the rains, women like Bintu not only prepare daily meals but also offer delicacies of the season.

Ensuring the Generous Cooking Pot on a daily basis was not easy for women like Bintu, but it was an idealized standard. Women transformed raw foodstuffs such as millet, shea nuts, and fruits into meals and appetizing treats. Year-long food production required expert botanical, agricultural, and technical knowledge. On another day in the rainy season, Bintu may have collected baobab leaves to dry and pound into a powder for sauce making, or she may have gathered wild rice for the next day's meal. Women created and managed a foodscape that was yearlong in its scope, but it was seasonal in its diversity and accompanying activities. (What I call the "foodscape" refers in part to the gendered production of raw foodstuffs and meals for household consumption.) Over the course of a year, women like Bintu and her friend carefully managed their labor time and the agricultural and technical resources available to them. In so doing, they endeavored to satisfy the food preferences and tastes of husbands, mothers-in-law, and others at home. The survival of rural households and families depended upon an adequate supply of cultivated cereals like millet and rice, the production of which involved a great deal of male and female labor. However, beans and other garden crops like eggplant and okra, nutritious tree leaves, nuts, fruits, and wild grains from the bush were equally important. The cultivation and gathering of foodstuffs were only two aspects of the early twentieth-century food production system. This foodscape was simultaneously a function of the environment, technology, and taste, and it is something that was created, protected, and exploited largely by women. Ideally, it was also endlessly generative, indeed, as generous as the hyena's magical pot. The foodscape was also imbued with symbolic meaning, but this meaning was not fixed.

Agricultural production and the gathering of foodstuffs was complemented by labor intensive food preparation. Women in every household managed daily cooking tasks and organized the seasonal manufacture of essential foodstuffs. Most often this work was carried out by young women or women of dependent status under the supervision of senior women. It was the primary responsibility of rural women to produce food for their households, but many also sold surpluses of raw and prepared foods in local markets. They used small hoes, clay pots, wooden mortars and pestles, stone grinders, and calabashes among other tools to cultivate, transform, and prepare food. Grain and other stores for the year came from both cultivated fields and the bush, or forested areas. In this way, women turned the diversity of the agricultural landscape into satisfying and familiar meals, as well as seasonal delicacies. This foodscape emerged from the regional environment, women's labor, local taste, and female technological mastery.

Women's work was expected to support households that were self-sufficient and had enough foodstuffs to provide for visitors and regular social and ceremonial obligations. This generosity was a well-established cultural ideal made manifest in public life through women's labor. This managed material and symbolic foodscape shifted over the twentieth century: the gradual end of slavery and resurgence of female pawnship was accompanied by increased production for export in the colonial market and the introduction of new agricultural and domestic technologies. In some cases, women had difficulty feeding the household and adapting to new resources and conditions. Over the twentieth century, economic, social, and ecological changes all reconfigured women's labor and the local diet. Yet throughout this period, women treated their environment as a productive resource. Gradually, women's exploitation of available cultivated and wild lands, their management of labor time, and their understanding of how to make filling, appetizing, nutritional, and socially meaningful food similarly adapted. Women shaped the foodscape and it, in turn, shaped women's food preparation strategies.

Women were responsible for providing their households with a filling and diverse diet throughout the year. Millet and rice were staples of regional cuisines, and men and women worked hard to constitute house-hold stores of these grains. Staple cereals were cooked in a variety of ways and constituted the base for each meal. Women in Bamana farm commu-nities regularly prepared the main dish, toh (a stiff pudding eaten by hand), from pounded and cooked millet. Rice was popular in towns located along the Niger River's floodplain. *Basi* (a fine couscous made from millet) was a seasonal delicacy; it was prepared only when the leaves needed for the ac-companying sauce were found in rainy-season marshes.[87] Other seasonal grain delicacies included jéba (wild rice) and fonio (a small grain that cooks like rice or couscous).[88] Women also used millet and other grains to prepare hot porridges like *moni* and *seri. Nyo,* as millet is known in the Bamana language, is actually a broad term for a wide variety of millets with distinct growing conditions, flavors, and colors; similarly, farmers cul-tivated many kinds of rice. Each millet and rice type added variety to the toh and other dishes, all of which were accompanied by sauces made with vegetables, leafy greens, and sometimes meat.

Even in years of average cereal harvests women created grain stores from sources gathered outside the fields. They collected wild roots and fruits on a seasonal basis year-round. Women also collected stores of dried baobab leaves, a common spice called *soumbala,* and shea oil or butter, which gave sauces flavor and nutritional value (see figure 1.4).[89] Women

FIGURE 1.4. Hawa Fomba with nere seeds and soumbala, 2010. Photo by author.

Making the Generous Cooking Pot, ca. 1890–1920 ⌐ 47

used many of these ingredients to make snacks (e.g., fried millet and bean pastries) or fermented beverages such as millet beer. Regional cuisines varied widely out of necessity. However, women's cooking practices also responded to local tastes.

Wild grains were a seasonal food that people enjoyed when grain stores were low. The months of intensive field labor may have been lean, but the foods eaten and manufactured during this season were essential to providing dietary diversity. By the mid-1940s, ethnographer Henri Labouret and other French scientists argued that seasonal variation in African diets—especially during the rainy season when people ate less food on average—was a sign of dietary trouble and malnutrition.[90] As Diana Wylie argued for early twentieth-century South Africa, seasonal dietary change was not an indication of a poor diet. Rather, it represented an adaptation to the seasonality of resources. Food crises only really occurred when something disrupted the regular cycle of food production. But it was chronic lack of food that posed a health danger, not seasonal variations.[91] The situation was similar in the French Soudan. Nevertheless, the seasonality of food production in the colony would continue to cause concern among colonial scientists who feared that nutritious foods, especially proteins, were not consumed regularly. Labouret credited West African women's inventive food production techniques, but he still regarded the amount and quality of their food production as insufficient. He further believed the "poor environment" of the West Africa was at fault because it lacked the natural resources to support a healthy diet.[92] Women, quite clearly, did not view their environment in the same way.

As suggested in the Koue-Bamana narrative, collecting wild grains was often a group endeavor. When individual families suffered from grain shortages, the woman in charge of cooking for the day would procure millet in town. In such a case, the cook had perhaps been unable to find any wild grains recently or did not have time to collect them. Typically, the cook began heating water over the fire. Then she would go to a neighboring cook's hearth. Rather than say she had no grain, she would ask whether there was anything that needed pounding. After she pounded millet for her neighbor, that woman would give her a little something to cook that day. It was only when shortages were widespread that women went on a collective search for grains.[93] In both cases, women were charged with supplementing the grain harvest when it ran out. A poor harvest was followed by intensive women's labor.

Women obtained much of a household's essential food products from the bush. These wild products were processed into butters, oils, dried sauce condiments, fermented spices, and beverages. Indeed, the early twentieth-century French botanist Guy Roberty observed that the landscape of the Soudan was particularly rich in oil-producing trees and plants of potential interest to metropolitan industries. He called attention to sesame, shea nuts, and peanuts, all of which women cultivated in fields or harvested from the bush and processed for home use.[94] Women also collected fruits from the *nere* tree beginning in April, the hot season. As suggested by early travelers' observations of market women, each woman collected ingredients for her own cooking and use. If a woman's store of shea butter or soumbala spice exceeded what she needed to cook, she was free to sell her surplus. Production of cooking butter made from shea nuts and of the spice soumbala was labor intensive. For soumbala production, Labouret reported that nere seeds were dried, washed, boiled, pounded, and finally fermented for ten days. He also recorded nine steps in the process of making shea butter.[95] Clearly, it was a complex undertaking.

Women needed to use several specific tools to cook and manufacture ingredients like shea butter and soumbala. Basic kitchen items would have included at least one heavy wooden mortar and pestle set to pound millet. In the cooking process, the fish or meat was pounded together with some of the seasonings while salt was pounded with other ingredients.[96] Ideally, a cook would have had more than one mortar and pestle: one for millet and one for sauce ingredients. A wife would also need calabashes to hold stores of sauce ingredients, a *baarakolo* for collecting water[97] and at least one large clay container to store drinking water. A cook needed at least two earthen cooking pots and two kinds of wooden stirring spoons. If she was going to cook rice without pounding it, she also needed a clay steaming bowl. Every household needed at least one large wooden serving bowl for eating and one small one for the sauce. They would also need at least one drinking gourd, a woven cover for the bowls,[98] and serving spoons. In addition to kitchen items, women would have owned spools for threading cotton, one or more stools, bedding, some lengths of cloth, and possibly stores of medicinal plants.[99] Women's agricultural labor in the early part of the twentieth century followed seasonal patterns that integrated field labor, garden cultivation, the collection of wild foodstuffs and medicines, raw food processing, and cotton spinning or other industry.[100]

GENEROSITY AND DOMESTIC GENDER POLITICS

Women's labor supported the long-standing rural cultural values of generosity and hospitality. For example, in the late eighteenth century, Mungo Park frequently remarked upon the hospitality he received in the western

Soudan. While he was robbed en route to Segu and continued much of his journey in a destitute state, Park was able to rely upon local hosts to provide him with food and shelter. In many towns he was even invited to stay overnight. In the following century, René Caillié, Eugène Mage, and Louis Binger (who traveled in the Mossi territories where many Office migrants would later migrate from) also came to understand that the provision of water and food to visitors and travelers was a cultural norm that could be depended upon in the region.[101]

Women in early twentieth-century households were similarly responsible for showing hospitality to strangers. As countless European travelers observed, when a man offered food to guests, he displayed the agricultural plenty and wealth of his household and of his town. It was a woman's responsibility to actually prepare the food and carry it to the visitor.[102] Even when young married women (the household cooks) were away during the day elder women offered visitors *dege*. This cold porridge was made from spiced grain powders and milk (or water) mixed immediately before consumption. The dege powder was ground ahead of time for an older woman by her daughter-in-law and, thereby, ready for her to prepare quickly in the case of unexpected visitors.[103] Strangers passing through a settlement also expected to be offered water even if they did not eat. Most often young wives filled and carried the common drinking cup with water and presented it to the visitor on bended knee, a posture of respect and perhaps also of servitude.[104] The gesture also reveals women's central importance in assuring and sharing sustenance.

Generosity was a valued cultural ideal but one that was not always upheld. Once when Mungo Park was refused food by his male host in Doolinkeaboo, he approached a female slave in charge of preparing meals. He thought she would be more willing to help as he understood that women were in charge of offering hospitality to guests. He wrote: "I even begged some corn from one of his female slaves, as she was washing it at the well, and had the mortification to be refused." Park discovered that some women exerted greater control than others over the food stores. He concluded: "When the Dooty was gone to the fields, his wife sent me a handful of meal, which I mixed with water, and drank for breakfast."[105] One reason the female slave might have refused Park's request was that she was under the watch of other women. It was unlikely that she would have openly disobeyed the ruling of the household head. However, a woman more senior than the slave could overrule her husband. Women's control over food resources depended on hierarchies in the household. Moreover, as a senior woman, the wife in Doolinkeaboo had a measure of control over who ate from household stores or from her own personal stock of grain.

Another folktale that circulated in the early twentieth century was critical of men like Park's reluctant host, who violated the norms of hospitality. It also underscored the social value of women who reinforce the ideals of food sharing. In the story, a man with an excessive appetite leaves his village to set up residence in the bush with his wife. In the French translation the tale is called "Le gourmand" or "glutton." Another man, hearing the news of the gourmand's move to the bush, resolves to eat with him. Shortly thereafter this man passes by the gourmand's house in the bush. He tells the gourmand that he is traveling to the next village, and the gourmand suggests that he should continue because the village is not far. He adds that there is no food or water at his home. The guest persists in his resolve to stay, replying that if there is nothing to eat or drink, he will fast for the night. Meanwhile, the gourmand's wife prepares dinner as normal. When she informs her husband that the meal is ready, he tells her not to bring the food and that he will pray first before eating, hoping his guest would fall asleep before mealtime. Nevertheless a little while later, the guest calls out from under his covers that he is extraordinarily comfortable and could stay with the gourmand for much longer than one night. In the end, the gourmand tells his wife to bring the meal, which they all share. Despite the gourmand's greediness and his decision to isolate himself from his village, he is driven by his own hunger to share the meal.[106]

Significantly, the name of the wife in this version of the story is Fatima. It is also the name of Mohammed's daughter, who is widely regarded by Muslims as a female role model (and it would have been the case across much of the Soudan at the time). Fatima is venerated in particular for her labor in caring for the prophet as well as for her own husband and children. Fatima's character is in stark contrast to her husband's. She prepares the meal for her husband and his guest without question and, in so doing, serves to emphasize the moral lesson of the tale.

While Fatima's character embodies the cultural value of generosity through food, women's historical lives were more complex. Female hierarchies significantly affected the organization of household and food preparation labor. In the early twentieth century, daily cooking responsibilities rotated among the junior women in each household (cowives and sisters-in-law). One woman cooked the toh and the sauce for two days, while another woman assumed the responsibilities for the next couple of days and so on.[107] The woman cooking was in charge of procuring the ingredients for the sauce. She was also assigned the task of transporting the afternoon meal to household members working in the fields. After the cook prepared and transported the midday meal, she still had to weed, plant, or do other work in the household fields. That night she could expect the visit of her husband, linking her sauce to her sexuality but also potentially stirring the jealousy of her cowives.[108]

Through food preparation women could negotiate and improve their own material well-being. A woman only presented the meal on the days when she was responsible for the cooking. This practice ensured that everyone knew who cooked well, especially during shortages. By extension, everyone also knew who did not cook well, even when there were enough resources to provide a quality meal. As Paul Stoller has observed, mealtime offered women who were displeased with the domestic politics of the household an opportunity to materially express their displeasure.[109] At the same time, food preparation and presentation provided a stage for expressing social hierarchy, skill, and sexual allure.

A well-prepared and presented dish held distinctive cultural value. Food aesthetics entailed the appreciation of the meal, the taste, and the texture of the food.[110] For example, baobab leaves and other ingredients produced sauces with distinct colors (dark green, or sometimes red) that made the quality of the food visible. Main grain staples like toh, basi, and fonio also had distinct textures. Well prepared toh had an outer crust that was hard or sometimes crispy. The crust made it hard to get the toh into your hand to eat, but it also helped to preserve portions of the meal. In fact, the administrator-ethnographer Henri Labouret suggested that prior to colonial rule, women did not cook every day.[111] Thus, saving some of the meal would have been essential to food preparation. A cook added *gnumun* (from tree leaves called *nyemafuura*) to the toh to help give it a hard crust. This way "the skin does not spoil."[112] To make gnumun, women pounded the leaves, added the powder to water, and filtered the mixture that was finally added to the toh.[113] The resulting hard crust was a textural element that contributed to the quality of the meal.

On the days that a woman did not cook, she pounded millet in the morning for the woman making the toh. The more women there were in the household, the more social the task of pounding millet became (which also meant a lighter workload). Other regular tasks for women included sweeping inside the home and the courtyard, collecting water, and doing household washing (culinary items as well as clothing). Much of this labor fell to unmarried girls and young married women without daughters-in-law. Senior women oversaw this work and frequently commented on the quality of cooking or other completed tasks.[114] Cooking as well as other domestic and agricultural labor was well ordered; it also followed the generational and other social hierarchies between women.

Social discord was as much a threat to agricultural production as ecological crises, and women addressed these risks through their songs. One song that anthropologist Pascal Couloubaly identified among a repertoire of early twentieth-century Bamana women's songs decries household tensions between women.[115] In the song the tensions are specifically over the distribution of food resources. One of the opening lines asks the audience, "Who tolerates the injury of the household's mother?" The senior woman in question is often a mother or grandmother to many of the household's adult men and also mother-in-law to many of the family's young women. She would have been responsible for rationing grains and other basic foodstuffs to those younger women who cooked for the household. The injury revealed in the next verse is the theft of some of those rations by women in the household. Another verse suggests that the older woman failed to equally distribute grains, salt, or other foodstuffs among the women tasked with cooking. The implication is that women complain of a lack of resources when it is their turn to prepare meals and may even take more than their allotted share.[116]

This state of affairs is lamented. The final verses decry the injury such conflict causes the senior male household head, whose title might be better translated as "household caretaker" or "hearth caretaker." He customarily provides the household's mother with the stocks of grain and salt. While the song airs the potential grievances of young wives who feel disadvantaged in the household, it ultimately suggests that they should respect the authority of their household's mother. At the same time, the song reflects the ideal role the household played in the foodscape, ongoing concerns over food resources, and the potential problems of accessing household resources. In a strikingly similar historical example, the Catholic sisters stationed at Toma in 1928 recorded a domestic quarrel over salt. A woman they called "Julie" asked for salt from her father, but her request was refused. In retaliation, she hit him over the head with a "baton" (likely her pestle).[117]

Quite clearly, food production and preparation had sociopolitical implications beyond the framework colonial officials set forth in the early decades of the century. For one, the dynamics of food preparation and distribution shed light on domestic power relations between men and women. Senior men made decisions regarding labor in the fields and access to land. Men also exerted some control over the distribution of grains and other foodstuffs. However, women made decisions about the preparation of food on a daily basis: they exerted varying degrees of control over grains, prepared ingredients, and distributed food. Some women also earned profits from their food production and preparation labors. Thus, the foodscape was a domain in which women could establish the standing of the household and exert influence over domestic affairs.

TASTE, CELEBRATION, AND FOOD SECURITY

For women, local taste was long central to food production and well-being. They paid the same attention to taste in their daily preparation labors as they did in preparing for yearly celebrations. In fact, many of the foods women collected for the lean season were extremely pleasing to eat; there were also many kinds of nonfield grains and foods to be collected. Women made meals with grains from the bush, as they would with grains that came from the fields.[118] For example, women cooked jéba with fish and salt to make an appetizing dish.[119] Another wild grain called *fini* is still sold in the Niono market; it looks and cooks like rice and is much appreciated locally.[120] Wild rice varieties were found as far north as Sokolo during the rainy season.[121] Further south in Koutiala, wild rice was called *ja so*.[122] In Boky-Were, women cultivated a grain similar to jéba called *jé*. Boky-Were residents also harvested a water weed from the Niger called *gokun*, which people even sold because it was very popular to eat.[123] In the rainy season children looked forward to eating many wild fruits, as well as a kind of weed called *bré* that looked like fonio.[124] While field grains were harder to come by during the lean months of the rainy season, fruits and wild grasses provided important nutritional supplements; they also greatly diversified diets and assured year-long food security.

Celebrating taste and moments of plenty was also an important element of locally understood food security. In agricultural towns, men and women organized town-wide festivals (*san yelema nyenaje*) to mark the success of the agricultural year.[125] To enliven these events women brewed millet beer and prepared other fermented beverages. Eating as a social and sensory experience always mattered but was heightened during these agricultural festivals. Every year men and women hoped for a good harvest because it meant that the household would have enough food to last the entire year (or longer). Farmers also saved some grains from the harvest for the next year's seeds and sold a portion to pay taxes imposed by the colonial government. To do all this, farmers (men and women) looked for ways to increase overall production. Yet the harvest was more than a quantitative measure. Good harvests enabled senior men to finance junior men's marriages or to participate in other important social exchanges. In addition, the harvest festivals were moments to meaningfully display those social ties and ritually share the abundance. All comers would eat and drink to excess. Accordingly, the yearly rites were also opportunities for women to display their skills in food and beer production. They produced and displayed in excess the values of rural hospitality.

The beer harvest particularly was associated with maintaining strong social ties. Often specific household fields were planted solely with millet intended for beer brewing. Women made large amounts of *dolo* and *di-dolo* (millet beer and honey beer, respectively) for the celebrations. They also brewed for sale. Beer was certainly an economic good, but it was also an important element of social life. Beer along with food was offered to men following collective work parties. It was also consumed during religious ceremonies and of course during agricultural festivals. The excessive consumption and enjoyment of beer were embodied means of measuring the harvest, maintaining rural social relations, and ritually ensuring future abundance.

In the first half of the twentieth century, beer still featured prominently in the communal harvest parties. In good years, neighboring towns were invited to share in the abundance of grains. Celebrating a plentiful harvest may have been one way of sharing good agricultural luck with neighbors whose harvests had not been as productive. For example, in Koue-Bamana women described what harvest parties were like before the mid-twentieth century. They explained that the party was held after the harvest to show that it had been a good season. It was a playful atmosphere, and everyone ate as much as they wanted: "Music was played, toh was cooked and fish was served." The town also slaughtered goats and sheep, and the women made a lot of beer.[126] Food preparations were all organized by the oldest woman; she instructed each woman what to do as they prepared the toh, the sauces, the meat, and the beer. Women in Kouyan-N'Péguèna also recounted that for the festivities, one young man from town was selected to serve as the master of the event. The public place was prepared for him like a palace, and he was brought in on a horse to the sound of drums and rifles.[127]

The *dugutigi* of Molodo-Bamana, a town dating to the Segu Empire, described similar harvest parties.[128] Drums were played, and there was toh to eat, as well as goat and cow meat. According to the dugutigi, the men gave the millet to the women to make food and beer. Two types of beer were served: millet beer and honey beer. Men and women worked together to make the party a success. In addition, all of the old women worked in one house together. During the party, the beer drinking was in one place and eating in another. Only old men and some young men and boys drank. In particular, the old men compared the quality of each woman's brew, always complimenting each one but sometimes making suggestions for enhancing the beer's flavor.[129] In Boky-Were, another old settlement, men made masks and went into the bush to dance, while women clapped and sang. People from other towns also came and were given food. For parties here, women made *ngunajin,* a fruit drink that according to the dugutigi made people "crazy."[130] Imbibing alcohol in both towns was central to their annual festivities, but the specific taste of each town's brew was important as well.

All women were integral to these public agricultural rites and the social gatherings celebrating the harvest. In many villages women also performed songs to animate the events.[131] The celebrations brought entire villages and even neighboring villages together, making them important regional social events. Women in particular had a stake in the economy of the harvest, largely by transforming its bounty. Moreover, they fared better economically in years when surplus shea butter, beer, and other products could be sold. Significantly, the quality and quantity of beer and other fermented drinks testified to the success of the year's agricultural season. This was a qualitative measurement of production based on women's production. In either good or poor harvest years women's surplus manufacture of beer or production of wild foods was a barometer of well-being for the community. It was measured by the embodied sense of satiety and fulfillment. Yet whether harvests were good or bad, women were expected to ensure the pleasurable consumption of beer, prepared dishes, and other foods.

Beer brewing had a long history in the region before the arrival of the French. Oral traditions from the Segu Empire suggest that the mother of the empire's founder, Mamari Kulubali, joined her son at the capital, Segu-Sekora, where she brewed millet beer for his followers. Mamari was not an attentive farmer. In fact, the stories of young Mamari recount that he failed to stop a water spirit from eating his mother's eggplant crop. Mamari later caught the thieving spirit and received some fonio grains in compensation. However, he allowed birds to eat the crop after it was planted.[132] Significantly, Mamari's mother was always more attuned to agriculture, and her beer would become important to politics in the Segu Empire.

Specifically, Mamari shared his mother's beer with his warriors. Each paid a small amount, some of which went to Mamari's mother. The price for this millet beer was called *djisongo* and is referenced as the origin of taxation in the empire.[133] Young men's associations, which were the institutional basis for the empire's warrior class, had long been major consumers of millet beer.[134] These groups of young men provided much labor during the harvest. Later, during the era of the Segu Empire, young male warriors acquired millet beer not through agricultural labor but by raiding millet stores and collecting taxes for the state. This is perhaps the state-rural community dynamic that prompted the hyena in "The Generous Cooking Pot" to retrieve the magical pot from the ruler: it ensured rural communities control over their own production.

Nevertheless, in the Segu accounts beer symbolizes how Mamari shared the bounty of the empire with his warriors. It is also representative of the material wealth of the state and political unity among its warriors.[135] Therefore, the women who produced beer played a major symbolic role in the politics of the region as well as the agricultural economy. Ethnographers have long noted the strong ritual and social value of alcohol across the continent. Many fermented beverages were made from vital staple grains (often millet or sorghum) or other foods. The preparation of alcohol altered the physical state of the grains, and the resulting drink influenced people's behavior.[136] In precolonial Africa, broadly speaking, alcohol was not meant for everyday consumption. Rather, it was tied to the agricultural season, periods of heavy labor, and the end of the growing season. Beer could be crucial to recruiting male labor for the time-sensitive harvest. Once the harvest was in, drinking marked the end of heavy labor and the beginning of a period of rest. This was especially true for men.[137] The women who brewed beer, therefore, controlled a powerful and desired substance.[138]

Mungo Park, who traveled in the Segu Empire, liberally partook of local brews. Notably, Park found the flavor and quality of regional beers pleasant to his palate. In the course of his travels he was regularly offered beer to accompany other local amusements. He wrote favorably about one such occasion: "About four o'clock we stopped at a small village, where one of the Negroes met with an acquaintance, who invited us to a sort of public entertainment, which was conducted with more than common propriety. A dish, made of sourmilk and meal, called *Sinkatoo*, and beer made from their corn, was distributed with great liberality." In this description he also explained that beer was not only for men. Women were visible consumers of the drink (in contrast with many memories of early twentieth-century harvest parties). Park continued, "The women were admitted into the society; a circumstance I had never before observed in Africa." He further highlighted the social nature of drinking: "There was no compulsion; everyone was at liberty to drink as he pleased; they nodded to each other when about to drink, and on setting down the calabash, commonly said *berka* (thank you). Both men and women appeared to be somewhat intoxicated, but they were far from being quarrelsome."[139] Sociability around millet beer consumption was quite pleasant in Park's experience and was accompanied by the consumption of food.

Elsewhere, Park wrote about a special house for the consumption of beer: "As most of the people here [Moorja] are Mahomedans, it is not allowed to the Kafirs [non-Muslims] to drink beer, which they call *Neodollo* (corn spirit) [*sic*] except in certain houses. In one of these I saw about twenty people sitting round large vessels of this beer, with the greatest conviviality; many of them in a state of intoxication." Here, beer drinking was a marker of non-Muslim status, but it was also institutionalized to an extent through the creation of beer houses. Park's continued description of drinking in Moorja made plain the association between a bountiful harvest and the enjoyment of beer: "As corn is plentiful, the inhabitants are very liberal to strangers: I believe we had as much corn and milk sent us by different people, as would have been sufficient for three times our number; and though we remained here two days, we experienced no diminution of their hospitality."[140] In this way, plentiful millet beer was a marker of a successful previous harvest, peaceful social exchange, and leisure.

In 1805 Park returned to West Africa and toured the Sansanding market. He observed that a separate area of the market was reserved for beer sales.[141] By this time, Sansanding was a major trading and grain production center founded by Muslim Marka merchants and plantation owners. Even though beer drinking was widely considered a "Bambara" (non-Muslim) pursuit, it was tolerated in predominantly Muslim towns.[142] Harvest festivals were sometimes Islamized, and some Muslims drank millet beer. In fact, Mamari's mother, who is said to have sold dolo, is also remembered as a devout Muslim.[143] Islam spread in the region during the nineteenth century, contributing to the emergence of a series of Muslim states. Yet beer production and consumption remained widely culturally important.

Beer carried strong associations with both material well-being and politics. It was also a valued commercial good. In 1902, almost a century after Park's visit, dolo sales were ongoing in Sansanding. In a case that came before the French-appointed ruler Mademba Sy, a stranger was imprisoned for cheating a millet beer seller out of his product and fooling the same seller into purchasing meat for him.[144] This case was clearly important enough for Sy to write to the French regent in Ségou about the theft. Indeed, millet beer was serious business across a wide region. About a decade earlier in Ouagadougou, Louis Binger had observed that dolo was one of the most important items for sale in regional Mossi markets.[145]

The availability of millet beer continued to be a sign of a productive harvest and a strong economy. In the first decades of the twentieth century, Catholic fathers observed the ongoing presence of a specific dolo "house" near their mission in Ségou, and it was frequented by African soldiers in the colonial forces.[146] In the late 1920s, Emile Bélime described the local importance of dolo as follows: "Millet also serves the production of dolo, a sort of beer, very much enjoyed by the Soudanese. When common grain stocks are filled, the fields owned and cultivated by the woman are then used for this purpose [beer brewing]. She sells it at high prices, even to her husband. This fact, reported by Costes [a French agricultural researcher], shows that dolo has become, for the Soudanese, a drink of first necessity."[147] According to this account, some women cultivated millet for beer in personal fields, as well as producing millet beer from the surpluses of household fields. In years of productive harvests, women retained more profit from their own fields. To Bélime, beer was strongly associated with women, and it appeared that beer was a staple of rural life. He seemed to also think women earned a great deal from beer brewing.

Despite such observations of rural life, French policies from the first three decades of the twentieth century discouraged women's beer-brewing activities. As early as the 1890s, the French government was under popular pressure within France to combat drinking in the colonies.[148] However, it was not until the second decade of the twentieth century that the colonial government seriously attempted to control beer production in the French Soudan. During the First World War millet beer production was banned. Acting Lieutenant-Governor Raphaël Antonetti outlawed dolo with the goal of reserving grain production surpluses for state provisioning efforts.[149] Not surprisingly, the ban on beer production met with great resistance.

When the government again considered a ban during the difficult harvest years in the 1930s, local administrators reported to their supervisors that such measures had failed in the past. Yet many French officials tended to support a ban because they believed that farmers wasted food stores to make the alcohol. The two opinions prompted debate. In 1934 officials in Bamako voiced concerns that millet beer consumption was detrimental to the food supply. In response, the administrator for Ouahigouya[150] wrote that "the situation is even more alarming because district inhabitants, who are heavy dolo drinkers, cultivate millet almost exclusively. Shortages of this grain could not, as in other areas, be balanced by the harvest of other grains that ripen at different times, or by root crops." Yet the same administrator from Ouahigouya doubted the effectiveness of any such ban or his ability to enforce one citing past experience.[151] Around the same time, François Sorel, who studied food in French West Africa as a doctor for the colonial army, suggested that millet beer be reserved only for special occasions because farmers barely produced enough millet to eat. According to Sorel, they could not afford to use their millet solely for beer making.[152] Clearly, these officials were missing why beer was as important locally as the millet food crop. It symbolized strong ties between neighbors, the ritual fertility of the land, economic strength, and even political order. Beer also embodied leisure, the sense of fullness and health, and cultural identity. Consuming beer in times of shortage, then, was not only related to present food supplies but also about assuring the future harvest and community renewal.

COLONIAL INTERVENTION AND GENDERED LABOR DYNAMICS

Administrative interventions in these early years of colonial rule paid little attention to such celebrations or local consumption, but they did promote agricultural abundance. Despite the assessment of the region as poor in terms of food production, in the first two decades of the twentieth century, French administrators and merchants investigated the possibilities of promoting the cultivation of some wild grains for regional trade or export. In 1911 the European rice merchant M. Simon, based in Mopti, was interested in farming the wild rice that grew along the flood plains of the Niger River. He observed residents in the Mopti area eating this rice during poor harvests. He further remarked that the grains were resistant to bird and insect attacks. In a letter to the administration in Bamako, he explained: "The only care required [for wild rice] is weeding. It already victoriously resists them. In fact, the harvest is guaranteed. It provides an excellent food grain, if not better than the other rice. . . . I asked widely why most farmers refuse to cultivate this rice. No one could give me an adequate answer for this attitude. No one has tried the exclusive cultivation of this rice."[153] Simon added that he thought the production figures for harvests of this grain could be considerable. What Simon proposed could be termed the "domestication" of wild rice. He assumed that agriculture ought to be a profitable endeavor; whereas this rice grew wild without any labor (except the harvesting). Locally, it was a reliable food supplement to the cultivated fields. The wild grain was also a woman's product, which male farmers might not have wanted to plant because the harvest might be understood as belonging to the women who worked the fields.

During the period 1912–14, the colonial government also briefly investigated the industrial possibilities for cultivating a wild sesame called *bénéfing*. Primarily, researchers hoped the grain would prove useful for the production of paints and varnishes in French industries. Ultimately, local administrators assessed that bénéfing was not likely to grow in large quantities; it was also difficult to harvest.[154] On this subject, the administrator for the Bamako district wrote to the lieutenant-governor of the French Soudan: "Bénéfing grows spontaneously in the Bamako district, especially in swamps, marshes and humid areas bordering the Niger. Unfortunately, the plant is nowhere abundant enough to allow for profitable exploitation. The harvest would be all the more difficult because the grain is light and thin."[155] In that same area, women made efficient use of bénéfing. They collected the grains and pounded them; they then added bénéfing to their stores of dried beans, since the powder acted as a natural preservative.[156] Women probably did not collect bénéfing on a large scale but gathered just enough to produce the powder they needed to help preserve stored grains. Thus, women's botanical knowledge not only allowed them to address local dietary needs and taste but also enabled them to make the most of the household's major harvest.

At the same time that French agricultural officials and experts were promoting increased export production, a few researchers documented regional agricultural practices that were closely attuned to local consumption preferences. In fact, researchers recorded a great diversity in the grains and other foods that made up regional diets. Even political administrators made note of a wide variety of food products with potential industrial uses. For example, the French Agricultural Service sponsored a few studies of millet and sorghum, as they were the primary cereals cultivated in the colony. They found that farmers grew several types of millet and sorghum according to local dietary needs. The French botanist Guy Roberty, who was active in the Soudan in the late 1920s and 1930s, recorded a great diversity of millet types.[157] He rigorously noted local names and character-istics for each type of millet. One type, *sanio*, had a long cultivation cycle but was reportedly useful in several kinds of cooking preparations. Another type, *shallu*, according to Roberty's informants was particularly appetizing. There were also several quick-growing varieties such as *kindé* and *souna*, the latter only cultivated in times of food shortages because it acted as a laxative, which was most often undesirable. Roberty also recorded *ba-buegne*, a variety cultivated by Somono and Bozo fisherman that was resis-tant to flooding and grain theft by birds.[158] This variety was distinctly red in color. Another red millet variety was grown for dye near the town of Ségou but not eaten because it, too, had laxative properties. Bamana farmers and some herders also grew *hassa-kala*, a sweet millet.[159]

In 1934, the Agricultural Service commissioned a report on millet seed selections to find the most productive varieties. The unnamed researcher who wrote the report found that the prominent millet variety, commonly called *kéniké* or *bimberi*, was actually divided into several types by local farmers. *Bimberi fin*, or dark bimberi, kept well—as long as six years. An-other variety called *soumpé* was cooked and eaten like rice. *Counsouli* was another type resistant to bird attacks and *soubaku* was a variety whose flow-ering portion of the plant was to the researcher's eye noticeably long. The 1934 report also indicated that it was common practice to plant fields with several varieties intermingled. Rather than select one type of seed, local farmers selected and planted several varieties for their resistance to bird theft, flooding, or drought, as well as for taste and cooking ease. Women probably helped to select or directly selected the seeds for varieties that were best for cooking.[160]

Roberty and several other researchers carefully noted their observa-tions about the various desirable qualities of different kinds of millet. Yet, French policy stubbornly emphasized greater crop yields over the other desirable characteristics of millet varieties. In fact, the study of millet from 1934 was meant to identify and help standardize the most productive mil-let variety. Moreover, official interest in food production was exclusively focused upon grain production in the fields. A French masculine bias was one reason that colonial observers concentrated on grain production in the fields, which was also largely associated locally with men. Even though women's labor was essential to grain production in the fields, women were associated culturally with sauce production.[161] This gendered dynamic is expressed in the Bamana saying: "It is for the husband to supply the toh and for the woman to supply the sauce."[162] This specifically gendered female labor was often tied up with being an accomplished woman in Bamana society.[163] Locally, food production was a significantly gendered process, but both men and women were expected to contribute to the food supply. Women's labor contributions to grain production were underemphasized in a local sense as well, despite its importance. One difference in these perceptions was that French observers tended to exclude women from the food *production* process altogether, whereas locally women were understood to be central. Administrators tended to only recognize women's food *preparation* roles.

The observations of colonial administrators reinforced the research of the natural scientists. For example, both administrators and agricultural officers similarly noted the predominance of women cultivating rice in the southern districts. These men also observed that women and children picked cotton and weeded fields. (They also noted that products like shea butter were manufactured by women.) However, they did not often seek to understand the dynamics of gender and age in agricultural labor. Most often policy directives indirectly pressured women to alter their labor time management. More than a few administrators were distressed by women's high degree of participation in fieldwork. For example, in 1916, one official commented that "the cercle [district] administrators, or their support staff, must observe field work by sight and note on the spot new potential fields, taking into account the agricultural calendar. They must not tolerate, during the agricultural labor season, the spectacle seen too often of women laboring in the fields while the men relax in the village smoking, joking, and drinking dolo."[164] This comment was recorded during the push to increase production for the war effort. Certainly, the description of men relaxing speaks to fears that farmers in the colony were not working hard enough to provide food for France. It is also likely that the administrator doubted that women farmers would bring about the necessary production increases.

Unwittingly, the writer of the above passage provided a portrait of women's central role in agriculture. Planting and weeding were predominantly female tasks and required a great deal of labor time during the growing season. In southern areas of the colony women alone cultivated rice in swampy lands.[165] The grain harvest generally began in August with the quick-growing varieties of maize, fonio, and millet; women collected the cut plants and carried them to granaries.[166] Women also threshed and cleaned the grains, appropriating small amounts of grain for their personal stores.[167] Similarly, women collected cut rice paddy and transported it from the fields for processing and storage.[168] Depending on the region, women also planted gardens in yards by their houses or at the edges of grain fields during the rainy season and the following cold months. Women grew plants like beans, Bambara groundnuts, and leafy plants like *n'tioko* and *da* used in sauce preparations.[169] Women also cultivated calabashes, which were essential household goods when dried and cut in half.

The majority of households in the region of the French Soudan where the Office would be established had the capacity to provide the basic subsistence needs for their members well into the 1920s.[170] Women helped to ensure this household resiliency by integrating their food-related labor with other productive activities. In particular, women picked cotton, cleaned it, and spun it into yarn. The cotton yarn was then woven into cloth for household needs and for sale. Cotton cultivation and cloth production complemented food production because the plant was sown in food-crop fields; cotton was also harvested after the major food-crop harvests. Women tended to the grain harvest first and only later picked and processed the raw cotton. From as early as the fifteenth century, travelers remarked upon this dry season activity in West Africa, as did twentieth-century colonial observers. However, the later French observers failed to understand its meaning for women's labor time.

In the late nineteenth century, early colonial administrators and entrepreneurs noted with great interest that cotton cloth production was a highly productive and lucrative domestic industry.[171] In economic terms, raw cotton was less remunerative than millet and peanuts, but woven cotton cloth was highly valued in the region. Bands of woven cloth were culturally meaningful gifts for important social events like weddings. At times, cotton cloth even served as a means of monetary exchange.[172] Persistent yet uneven French intervention in regional cotton production foreshadowed the disruption of women's labor organization and time for food production.

Beginning in the 1890s, French officials expressed great interest in profiting from cotton cultivation in West Africa. In 1898 General Louis de Trentinian, who was then lieutenant-governor of the French Soudan, sponsored a series of scientific missions to study local cotton production and the possibilities for its expansion and industrial use in France.[173] It was the general consensus that colonial intervention was needed to improve and increase cotton production. For example, E. Fossat, who took part in the 1898 series of missions, reported that farmers left the mature cotton on the plant for too long after it was ready to be picked. He further observed that the unpicked cotton suffered from exposure to the sun, resulting in discoloration. In one exception, Fossat reported favorably on the production of twenty to thirty tons of "good quality" cotton in Sansanding. It was a region then under the direction of a French-appointed Faama (king).[174] The Faama, Mademba Sy, was a former colonial interpreter and telegraph service employee who supported French rule. He coerced his subjects into cultivating cotton that he then sold for export.[175] Fossat's positive assessment of cotton production in Sansanding suggested to metropolitan officials the possibilities for an increase in quality cotton production (long fiber cotton with no discolorations) under French influence.

At the same time, Fossat's report neglected to mention the labor conditions for large-scale cotton production under Mademba Sy. The extent to which he was aware of forced cultivation under Sy is not clear from his research notes. Fossat also missed or glossed over the fact that women picked cotton but that they only did so after the major grain harvest. Harvesting earlier would have endangered the food supply. His single-minded focus on quantity and compatibility with French industry is one example of a broader tendency among colonial agricultural experts and administrators. For several decades, the French experimented with numerous varieties of cotton, distributed seeds, and tried unsuccessfully to capture cotton production.[176] The same officials overlooked the use of forced labor and what should have been obvious gender considerations for French agricultural projects.

The history of colonial cotton cultivation in the French Soudan and Africa more generally has been widely documented. During the first two decades of the twentieth century, the French colonial government instituted several measures in an effort to increase Soudanese cotton production for export. The administration experimented with higher payments to farmers and established research fields; the colonial government even imported plows for land preparation and ginning machines to process raw cotton and reduce transport costs. A group of French cotton lobbyists known as the Association Cotonnière Coloniale also sponsored research and distributed new cotton seeds to farmers in the colony. The early colonial administration had high aspirations for the role of cotton in the colonial export market even though European merchants sometimes proved to be unreliable purchasing partners. In 1912 Governor-General Clozel issued an order for compulsory cotton cultivation. This policy was upheld through forced cultivation in fields managed by local administrators called the "champs de commandant." Despite these efforts, the colonial administration failed to meet metropolitan expectations.[177] Some colonial observers advocated competitive pricing for farmers and others promoted greater controls over Soudanese cultivation methods and labor.[178] In the absence of a political consensus, the French sponsored agricultural research missions to study the possibilities for irrigated cotton cultivation.[179] Cotton production increased in the 1920s, but the resulting cotton harvests furnished the domestic market rather than the export market.[180]

French researchers turned to the study of local farming practices. They paid particular attention to the local practice of sheltered cotton cultivation. This method involved planting cotton seeds in grain or other food fields as a secondary crop. The French wanted to flip this relationship and make cotton the primary harvest. In a 1928 investigation of cotton cultivation in the colony, Emile Bélime wrote: "In the Soudan, farmers plant cotton, by scattering seeds, together with sorghum. The cotton benefits from the field preparation, fertilization, and weeding done for the grain crop; the only additional work required is the harvest. Also, this is done by women and children."[181] He was well aware of women's role in the production of cotton but took their participation in the fields for granted. Most of his contemporaries similarly recognized that intercropping was efficient. First, the practice saved labor time. This benefited men, who were not involved in picking cotton. The American agricultural scientist R. H. Forbes further observed that the shelter provided by millet and corn stalks prevented cotton plants from drying out. In fact, local sheltered cotton varieties proved extremely productive in colonial research trials.[182]

Women were integral to the production of raw cotton and yarn for cloth, and the intercropped cotton harvest was easily accommodated by women's agricultural labor calendars. Local cotton varieties had long cultivation cycles and were harvested after grain crops and peanuts. When the millet harvest came in, married women collected the grains cut by men and transported them to the granaries. Young men in work parties usually cut the grains, and girls brought water to the group.[183] After the grain harvest, the married women and often the children picked cotton. The time lag between the grain and cotton harvests allowed women to space out their work transporting, storing, and threshing cut millet, as well as their cotton harvesting labor.[184] During the subsequent dry season women cleaned and spun cotton into yarn to make clothes for family members or for sale. In fact, women kept one fifth of the cotton crop for their own use. In this way, the women's cotton industry accommodated with their other agricultural and domestic tasks.[185]

Even though the French failed to promote large-scale cotton production of any variety for export in the first three decades of the century, their agricultural interventions disrupted women's labor patterns. First, the cotton harvest was predicated on women's picking labor. Already in the first two decades of the century women were protesting their increasing field labor burden. By the 1920s women processed greater amounts of cotton because more cotton was being produced. Women earned money through their cotton labor, but the total amount of cotton to be picked and processed increased. The introduction of a variety of cotton that was harvested at the same time as the major grain crop also meant that women worked longer hours. In all this time, the colonial government and its researchers continued to overlook women's roles in cotton production. The ramifications of this oversight were evident in early cotton enterprises in the Soudan. In 1920 French businessman Marcel Hirsch opened a private irrigated cotton scheme in the northern Niger River Bend at Diré. The enterprise employed male wage workers, but it faced labor shortages at harvest time, among other problems.[186] In 1924 Director de Loppinot requested the administration's assistance in persuading the wives of workers to pick cotton. Lieutenant-Governor Terasson directed local officials to intervene. Terasson did so with assurances from de Loppinot that the women's harvesting work would not disrupt normal food preparations done by the same workers' wives. It is unclear how Diré's director came to make such claims. In response to Loppinot's request, Terasson suggested that women were more likely to accept this work if they at least worked alongside their own husbands.[187]

Pressing women to work in the Diré fields was a delicate matter. The administration was already wary of repeated requests for labor intervention at Diré, complaints over pay rates, and allegations of abuse.[188] Administrators were further wary of asking the wives of workers to labor in the fields with the men. Across the Soudan, men and women carried out distinctly gendered tasks in the fields and at separate times. As noted previously, cotton picking was a task done by women or children, which probably accounted for some of the enterprise's problems at harvest time: picking cotton was not a task for men. The enterprise also proposed intermingling the wives of some workers with other male workers, which seems to have been an offense to the husband and the wife. The final resolution was that the wives were pressured to pick cotton for .10 FR per kilogram harvested. Supervisors were also expected to place wives in the fields near their husbands.[189] Further south in the Middle Niger region, young girls customarily worked alongside their mothers, and sons worked alongside their fathers.[190] Gendered tasks were also performed at separate times. The wives of workers at Diré, situated in the northern region of the colony, did not necessarily share an analogous gender dynamic for agricultural labor as women farming in the Middle Niger region.[191] Nevertheless, the example is instructive. First, agricultural interventions that disrupted acceptable gender roles met with resistance from both men and women. Second, administrators recognized women's cooking work but underestimated the time required to produce meals. They also misunderstood how women negotiated, balancing this daily task with other work including field labor.

Regardless of the French failure to understand women's labor (or their technological savvy) in the agricultural economy, women's food production was a technical affair, and they were at the center of agricultural life. Indeed, women's use of their cooking tools made food production a daily public event. When French observers did take notice of women's work in this domain, they most often remarked upon the mortar and pestle. Travel writers at the turn of the century often waxed lyrical in describing women's daily tasks of pounding grains. For example, the soldier E. Thiriet wrote in his memoir:

> From the appearance of the sun's red disk on the horizon, and still long after the day falls, there is a muted sound, rhythmic and steady—music familiar to Africa. It is the sound of the pestle. It is a bit of wood about the width of an arm and 1.50 meters long. Women and captives use the pestle to pound and grind the corn and millet used for making couscous. It falls into deep mortars made of wood and stone and hits again mutedly. This sound can be heard wherever you find yourself, always and without end. To make the task amusing and break the monotony, women sometimes accompany their arm's work with a low monotonous chant or nostalgic song. They also join their rhythmic movements with neighboring women, beating fantastic songs. Also, they often make a show of their ability by throwing the pestle in the air and clapping one or more times before taking it as it hits the mortar.[192]

Here, daily life is marked by the continuous sound of women preparing food. There is the sound of the pestle hitting the mortar. While pounding, women sang, sometimes inviting other women to join in. Lifting the pestle and pounding it into the mortar was physical labor. As suggested by this passage, women sometimes transformed what could be fatiguing labor into an enjoyable task. They also took great pride in their skill.

Thiriet was relatively specific when describing the mortar and pestle. The pestle is wooden, about the size of an arm around and one-and-a-half meters long. It was used with a deep mortar made of wood or stone. Many mortars were large containers that could accommodate cooking for a household of thirty or more people. Certainly, women took great care in selecting quality items. For example, the pestle had to be made from a solid piece of wood. Any crack, and it was likely to break quickly.[193] Similarly, women used flat stones with shorter pieces of round wood to grind millet and other foodstuffs. The stone itself was likely chosen for its appropriate shape and reworked over time to improve its efficiency. These tools were maintained by women, much of it the personal property of married women. As it happened the mortar and pestle, much maligned by outsiders, heightened the centrality of food production and women's physical labors in daily life. The sounds of women's work punctuated the sensorial experience of daily life. The muted and repeated pounding along with women's songs and clapping created an everyday rural aesthetic. They were also auditory reminders of women's attentiveness to the production of appetizing and satisfying meals.

2 ⌁ Body Politics, Taste Matters, and the Creation of the Office du Niger, ca. 1920–44

THE OFFICE town of Nemabougou was established along the Macina Canal with the settlement of thirty-seven families in 1937. They were tasked with cultivating irrigated rice and were obligated to sell their crop to the project administration. The first agricultural season (1937–38) was difficult. Floods damaged some of the fields, three of the work oxen supplied by the Office died, and sixteen more oxen were either too ill to work or too difficult for the farmers to manage. A few men among the settlers were even new to agriculture as a livelihood (having worked previously as weavers), and another man made his living as an Islamic teacher before coming to the Office. Not surprisingly, the rice harvests were disappointing for many of the town's first settlers, who continued to rely on food rations from the project to survive. Living in Nemabougou was miserable. When the new settlers arrived, they found that their houses were only partially constructed, and the nearby canals brought so many mosquitoes that Abdoulaye Menta asked a visiting administrator for a sleeping net. Before the end of the first year, Moussa Bouaré and his entire family had fled the project. They were followed by the Islamic teacher Abdoulaye Farouta and his family.[1] Without much to animate daily life and little to eat, what choice did these men and their families have but to leave the Office?

After the first harvest, another settler, N'Golo Diarra, asked the colonial administration to help him bring stores of millet from his hometown of Konomoni to Nemabougou. What makes the request striking is that N'Golo was among a small number of settlers to see a good rice crop. Why, then, did he ask for millet from his hometown? N'Golo's request suggests not only a lack of secure grain stores in Nemabougou but also the importance of familiar food for new Office settlers. According to a household survey from 1938, N'Golo headed one of the largest families in town with twenty-two reported members. Together they cultivated rice on ten hectares of Office land; three more hectares were planted with millet and maize; one hectare was planted with peanuts; half a hectare was for cotton; and another partial hectare had diverse plantings. On paper N'Golo's family had many workers in the fields and plenty of rice. Indeed, he and his family had even arrived in Nemabougou with some food stores, which they consumed in the months before their first harvest. As a consequence, they only received a supplementary ration from the Office. However, as was the case for many Office households, the official statistics failed to capture the daily realities of life in Nemabougou for settlers like N'Golo and his family.[2] Even settlers who succeeded at cultivating rice were not growing enough of what they wanted to eat.

Nemabougou was not an exceptional case; food was a major concern for all settlers at the newly founded scheme. The same administrator who questioned N'Golo for an official inspection of the project recorded multiple complaints from other male household heads about food. For example, Youssef Tangara was in charge of a household that depended on the food ration provided by the Office, but he found that it was not enough. To make matters worse, the harvest in his Office fields was poor that first year. The administrator recorded that his family had not grown enough to feed themselves for the upcoming year. Tangara was not alone. Abdoulaye Menta and Amadou Menta similarly reported that they did not have enough grain to fill their food stores. Sine Tangara had the same complaint, but the administrator qualified it with a note that Sine did not work much in his Office fields, as he frequently went home to Boky-Were (located not far from several Office towns along the Macina Canal), where much of his family had stayed behind. Bogoba Coulibaly, who also was from Boky-Were, left his family in Nemabougou and went home to tend their cattle. The Office food ration was similarly described by other male residents as barely enough for subsistence, and clearly some men pursued strategies outside of the Office to meet their family's basic needs.

The problem with food in Nemabougou had to do with more than just the problems of the first irrigated harvest or the quantity of the ration. Food security was a basic survival need that settlers associated not simply with the amount of food available but also with taste. For example, according to the 1938 household survey, Mamadou Famenta, like N'Golo Diarra, cultivated a good amount of rice. Mamadou was also in charge of the largest family in Nemabougou, with twenty-six members including eight men, nine women, and nine children. The administrator even noted that Mamdou had no complaints about the ration. At the same time, he wrote that Mamdou was not accustomed to eating rice. His family cultivated thirteen hectares of the grain for sale and for food, but only one hectare of Office land was reserved for the preferred grains millet and corn.[3] This same preference for millet was displayed by other Office farmers settled along the Macina Canal, many of whom used the profits from their rice sales to purchase millet.[4] Taste mattered to the settlers displaced from their hometowns, not simply because they missed familiar dishes but because the disruption to their daily meals also signaled the loss of people. When household members fled, food production and preparation suffered.

Another man surveyed in Nemabougou tried to explain to the administrator how the breakup of his Office household that year had made it more difficult for him to work in his fields and consequently to eat. Lanciné Traoré and his immediate relations (wives and children) came to Nemabougou in 1937 with his brother and his brother's family. Sadly, during the first year, Lanciné's brother died. His death precipitated the departure of Lanciné's nephews from the household, who returned to the family's home in Diado. The following year the household survey for Lanciné's family listed nine people (two men, two women, and five young children) whereas it had been as large as twenty people a few months earlier. Following the harvest Lanciné sold some of the rice from his Office fields. But as he explained to the administrator, it was hard to work (and thereby eat) after his household lost so many members. Lanciné wanted to return to Diado, where his nephews would feed him.[5] This case brings to light the critical importance of household size, the gendered divisions of agricultural labor, and affective relations in rural life.

N'Golo Diarra, whose story opened this chapter, also complained that his son had left him. After his son's departure, N'Golo asked the investigating administrator to relieve him of the tax burden for the son who fled Nemabougou. Other men in the same town were also troubled by the flight of sons or brothers.[6] Office administrators certainly would have agreed with the men that household size directly influenced available field labor. However, in recording household statistics the official did not convey the challenging labor and social implications of breaking up families, not to mention the communal activities associated with food production. Indeed, the household surveys for the first year at Nemabougou portray a community in crisis.

While the 1938 household surveys primarily recorded the complaints of men in the first Office towns and noted the departures and the frequent absences of male relatives, the records masked another demographic problem related to food at the Office: the absence of women. An overwhelming number of the women reported living in Office towns had fled to their hometowns or the hometowns of their husbands, as had the wives of Lancine's nephews. Many women simply refused to live under the dire conditions of the newly established Office, and their departure was devastating for the daily provision of food. When Lanciné suggested that his nephews would feed him at home, he meant that when the women were part of the household, together the family produced enough to eat. Women especially supported and animated daily life through their cooking. Women were responsible for the cultivation of appetizing sauce ingredients and the daily preparation of food, and women's work in this capacity was highly valued. Their labors also supported the yearly harvest celebrations. When Lanciné's nephews left Nemabougou, they would also have left with their wives and any children. In other families, when the most senior men complained about younger men leaving, they often meant that many people (not just their sons) were leaving. Indeed, Office administrators recorded many women in the population statistics for the Office who were not present. Their collective flight prompted serious concerns for those remaining at the Office: How would settlers survive without women? Put simply, an absence of women meant an absence of rural life.

FOUNDING THE OFFICE DU NIGER

In 1919 the governor-general of French West Africa (the AOF) commissioned Emile Bélime to study the possibilities for irrigated cotton production in the Niger River Valley. Following a year-long mission (1919–20), Bélime envisioned a cotton production enterprise that would involve the construction of a massive and technically sophisticated dam at Sansanding, a smaller one at Sotuba, and three major canals to distribute water across more than a million hectares of land (see map 2.1). Over the next decade Bélime heavily promoted the project to government officials in Paris and

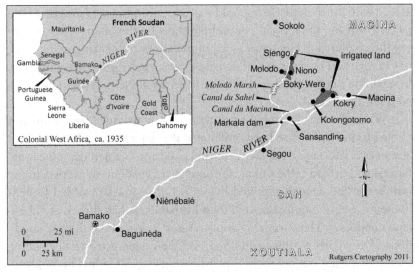

MAP 2.1. The Office du Niger and surrounding region, ca. 1932–47.

to the territorial government in Dakar.[7] He argued that his scheme would support the ailing French textile industry. He would later claim that the Office would improve the standard of living for African farmers who resettled at the project by teaching them "modern" farming methods and providing them with a better income from the sale of cotton.[8] Yet, as we shall see, farmers rarely earned enough to pay for the water and agricultural services of the Office. It was not a project initially designed or promoted to address local food needs, but as regional officials grew increasingly concerned about food supply, Bélime also promised the project would address colonial economic development and regional food concerns.

Bélime gained support for his plan from Albert Sarraut, whose colonial reform initiatives roughly coincided with the proposal for irrigation in the Niger River Valley. In 1921 Sarraut, the minister of colonies, proposed a program commonly referred to as *mise en valeur* to organize the vast colonial administration and to create value in France's overseas territories. His broadly proposed policy at least rhetorically promised mutual benefits to France and its colonies, and it emerged at a moment when France was struggling financially. The First World War had greatly weakened France economically and demographically.[9] Sarraut's plan never received formal approval. However, it was followed by a trend of investment in the agricultural services meant to strengthen the export economy in West Africa and that heavily favored technologically oriented interventions. Bélime ultimately framed his project so as to speak to the ideas of Sarraut and like-minded colonial officials. Even though Sarraut's legislation never passed, Bélime won financial backing in 1925 from the territorial government in Dakar.[10]

In 1924 Bélime significantly altered the design for his scheme to include both cotton and rice production. He made the change under pressure from Jules Carde, the new governor-general of AOF, who was concerned about recent droughts and the threat of famine. He was among a group of officials who saw irrigation as the solution to recurrent environmental concerns. The final plans for the scheme included a cotton-producing zone watered by the Sahel Canal and a separate rice-producing zone watered by the Macina Canal.[11] Rice produced from the scheme in the French Soudan would then be exported to farmers in Senegal.[12] In its redesign, the idea of the Office also resonated with the prevalent view of a Greater France composed of several interrelated regions.[13] The French Soudan would supply rice to Senegalese farmers who in turn supplied peanuts to the French oil-seed industry. Of course, the French Soudan would also supply cotton to the textile industry.

Even before Bélime had secured funding for the primary site of the project, the colonial government established two small-scale irrigation centers in anticipation of the larger project. In 1921 the experimental station called Niénébalé was founded. Farming at Niénébalé was achieved through pump irrigation. It was not until 1926 that the station recruited a small number of African families to settle in pilot villages and farm using irrigation techniques.[14] One year earlier the newly created Service temporaire des irrigations du Niger (STIN) had begun work on the first dam just north of the capital Bamako near the Sotuba rapids. Beginning in 1929 the Sotuba dam irrigated land for another group of pilot farming settlements collectively called Baguinèda.

In 1932 the French colonial government formally established the Office du Niger. However, construction did not begin until 1934.[15] France's economy and that of its colonies were again greatly impacted by the market crash in 1929, and the poor financial state of the colonial government led to postponed construction. The principal dam was not completed until 1947. Later, colonial officials attributed the long delay to the eruption of war in Europe in 1939.[16] Just prior to the beginning of construction on the dam, the project was relocated from Sansanding to a site called Markala (located eight kilometers to the southwest of Sansanding). The worksites were often dangerous, and many workers drowned or were electrocuted.

Workers also complained about the long hours, the liberal use of corporal punishment by French supervisors, and unsanitary food preparation.[17] The Macina irrigation canal was finished in 1935, and that same year the Office opened the first agricultural town, which was called Sangarébougou. This first group of settlers cultivated rice. Two years later both the Sahel Canal and the cotton sector were opened in 1937.

Around the same time that Bélime was elaborating plans for the eventual Office, the colony's Agricultural Service began to promote the extension of plow cultivation. Farmers across the region employed several types of hoe for land preparation and weeding. Early colonial agronomists bemoaned the fact that the plow had been unknown in the Soudan before French conquest. In the 1920s, French experts actively advocated replacing hoe cultivation with field labor aided by the oxen-drawn plow. They assumed that the plow use would bring about a shift from communal land tenure in the French Soudan to supposedly more advanced individual land ownership.[18] Proponents of the technology further argued that plow cultivation would increase crop yields, a goal that was consistent with the general push for increased production of commercial crops in the colony and, by extension, food crops. Initially, the colonial government distributed a limited number of plows to rural notables and agricultural students. Agricultural officers also organized educational tours and demonstrations to teach male farmers how to use the plow.[19]

Following these preliminary efforts, the administration planned to intensify plow cultivation across the colony but proposed to sell them on credit. The projected increase in demand for the equipment and the high cost for each plow (which included the cost of transport from Europe for the imported plows) persuaded agricultural officers to cease giving away plows for free. Indeed, men at the Office were expected to purchase a pair of oxen and a plow on credit.[20] For most of the twentieth century, French agronomists expected the plow to improve agriculture and, by extension, society as a whole. As historian Michael Adas has noted, Europeans tended to use the technologies they brought to their colonies as a measure for evolutionary development.[21] The energy that French authorities put into the promotion of the plow and other Western technologies on the assumption that the machines in and of themselves would trigger mise en valeur is consistent with Adas's argument.

In addition to the plow, irrigation technology would become one of the hallmarks of French development efforts. As early as 1916, the lieutenant-governor sent a circular to all cercles (districts) proposing an irrigation technique using small dams, dikes, and water gates to improve rice production.[22] That same year, the agricultural station at Koulikoro operated an irrigated rice cultivation project at Tienfala (to the northeast of Bamako). Notably, the workers were Somono (fisher) women who made use of the local swamps to grow rice using a system of dams and canals, and their work potentially inspired the all-colony circular.[23] While the Somono women likely did not perceive of their irrigation work as mastery over nature, it is a trope that describes the overall French attitude to technology at the Office. Yet, none of project's infrastructural plan was presented as artificial. Colonial agronomists and ethnographers argued that historically the whole region had long ago been fertile and well watered.[24] One early agricultural researcher Costes interviewed residents in the region about historical floodwaters and ancient swamps, and the resulting Sahel irrigation canal eventually followed an old waterway that his informants mentioned.[25] In Bélime's vision, technology promised a regeneration not only of this ancient waterway but also of the ancient population. The Office du Niger was going to grow people as a resource for the empire.

The idea of cultivating populations in the French Soudan was not novel. By the 1930s population science was an international concern and a regular part of colonial policy and planning. Western scientists and policy experts believed populations could be managed; however, they debated over the best policies to control population growth and to ensure the "quality" of people.[26] As early as 1909, the French Colonial Ministry requested accurate population statistics for the West African colonies. Initially, the goal was to assess the military conquest, but collecting statistics became a tool for exerting authority. In the colonial context population statistics directly impacted military recruitment, tax collection, and labor conscription and recruitment. It was soon foundational to governance.[27] The Office du Niger stood out in French West Africa as an effort to shape a population through technology. Bélime, however, was an irrigation engineer, not an expert in demography. Nevertheless, amateur population thinking was an important part of his pitch to government officials. Bélime argued that large-scale irrigation technology and plow cultivation would produce cotton, food, *and* people.

The Office did have an impact on population numbers: it provoked a demographic crisis and exacerbated food shortages. In fact, in these early years, male farmers directly equated the much-vaunted plow with food problems. After seeing the plow in action, many men concluded it ruined the soil, thereby harming future harvests. It also lacked the force of human labor, which they believed was vital to the production of energy-giving food. Some men even refused to eat food from grains cultivated by the plow.[28] They feared the Office was simply cultivating hunger.

COLONIAL BODY POLITICS: RECRUITMENT FOR THE OFFICE DU NIGER

As an agricultural project, and as an institution, the Office was dependent on African settlers. Bélime wrongly anticipated that the scheme would attract former slaves in the Macina region seeking land free from labor obligations to their former masters. He also expected other farmers in the surrounding region to seize the opportunity to farm with irrigation.[29] In reality, the Office was established through the forced migration of "families" mostly consisting of men but also of a smaller number of women and children. In fact, these initial settlers were locally called *tubabu jonw*, or "slaves of the whites."[30] They were joined by a very small number of voluntary migrants who had been previously employed by the colonial government as soldiers, wage workers, administrative guards, and interpreters. Male settlers and their families were expected to farm pre-assigned plots under the instruction of French staff. They paid a host of water and other fees to the institution out of the profits from the sale of their crops. A woman at the project did not have access to any Office fields for her own use, but she provided a great deal of labor in the fields assigned to her husband, male relative, or in the case of a female pawn, the man to whom she owed her labor.

The Office was meant to be a settlement scheme, not a wage labor enterprise. Bélime, the project's architect, argued that the goal of what he called "colonization" would best be accomplished by recruiting whole families and even entire villages. Top officials also favored this form of recruitment because they believed it preserved local custom and authority.[31] Populating the project, therefore, was represented as a matter of transplanting people. It was also a matter of colonial body politics. For the colonial state, controlling the labor and residence of populations was central to the Office as a development project. At the same time, the laboring body became a site of contestation as large numbers of settlers, especially women, fled the project.

The Niénébalé agricultural station, founded in 1926, was the earliest precursor to the grand settlement scheme, and planners for the larger project were encouraged by its much-lauded early "success." Families recruited for Niénébalé grew experimental cotton varieties, animal fodder, peanuts, and other food crops using pump irrigation. While initial cotton production results for Niénébalé were far from dramatic, officials touted the experimental station as a model. Much of this was due to its early recruitment record. For example, the governor-general of French West Africa wrote encouragingly to the governor of the French Soudan: "It is reassuring enough to look at the example of Niénébalé, where the population doubled in two years, to guarantee that the development of irrigated farming will provoke immigration from the neighboring populations in less favorable conditions."[32] Years before large-scale recruitment for the larger project even began, many like-minded officials were confident that populating its towns would pose few problems.

Niénébalé and the new villages associated with the first irrigation dam at Sotuba were all supposed to be testing grounds of sorts.[33] Families followed instructions regarding crop rotations, seed varieties, when to plant and weed, and when to harvest. They even lived in housing laid out on a grid near the station's fields.[34] The practice of settling farmers in a new town near the station was designed to not only disseminate new farming practices but also promote modern domestic life.[35] Bélime and other colonial officials were right in assuming that people were critical to the successes of the trial irrigation stations. However, the agricultural knowledge and technical expertise of these settlers were not immediately recognized. The same held true for vernacular demographic knowledge, which in this case refers to the practice of forming large households with multiple generations as a means of sharing the burdens of food production and social reproduction. More women in a household usually meant more prosperity.

The local emphasis on large households with many women is evident from the records for early settlement at Niénébalé. In 1926, the station recruited families to work in its irrigated fields. Both men and women were paid wages. At the end of the year every family asked to leave Niénébalé, but a little over half the families were persuaded (or even coerced) to stay through 1927. Beginning that year, family members no longer worked for wages; they were expected to cultivate their irrigated fields following a set agricultural program. After the harvest, they were expected to sell their harvest to the station. In 1926 there were fourteen families working for wages at the project: thirty-eight men, twenty-two women, and fourteen children. In 1927, nine families remained, with twenty-eight men, thirty women, and twenty-four children. While the number of families decreased, the total population increased from seventy-four to eighty-two people.[36] The number of male workers in the total population went down. Only the number of women and children increased. This demographic shift coincided with the change from wage work to independent family farming.[37] Those who stayed knew that permanent living at the project required more women.

Early population statistics from Niénébalé were used to support optimistic predictions for settling towns in the larger project along the Macina and Sahel canals. However, officials took for granted the gender shift in favor of women that took place at Niénébalé when they were planning and recruiting for the larger scheme. In 1934 Emile Bélime, serving as the project's first director, claimed that a total population of eight thousand between Niénébalé, Sotuba, and another station to the north at Diré had been easily achieved. He suggested that recruiters would find similar success for the larger project. Even though much of the major irrigation infrastructure was unfinished, in 1935 the first town—Sangarébougou—opened with 314 people in the rice sector Kokry. The first cotton-producing town was Kolony (also called "km 26"), and it opened near Niono in 1937 with 308 people.[38] Recruits for all the experimental stations had come from surrounding towns; it was a strategy that would prove difficult elsewhere.[39]

For the larger scheme, officials also sought recruits from among nearby populations. However, recruiters for the project met with persistent resistance. In a 1938 report, Office administrators admitted that local recruitment was not working. They nevertheless justified their actions: "In the beginning, it will obviously be necessary to recruit settlers from outside, but very quickly and increasingly, the growth only of a prosperous population by birth will sensibly add support, and become sufficient after a few decades."[40] Notably, they still expected that families would stay and grow once recruited. While none of the Office planners were population experts, colonial officials and local administrators had long been directly involved in reporting population numbers, labor recruitment, tax collection, and demographic analysis. However, as Raymond Gervais and Issiaka Mandé have argued, the methods of colonial census takers left much to be desired.[41] With the founding of the Office, the same administrators were responsible for recruiting families to send to the project. Most of these men recognized that recruitment often fell to politically or socially vulnerable individuals. Nevertheless, they generally supported labor conscription and some degree of coerced migration, including by force. The number of people sent to the Office steadily increased with the idea that each new town would start with around three hundred people. By 1938 the Office listed 4,264 residents in towns administered by the Kokry center. Niono (which only opened in 1937) had 1,428 listed residents in 1938 and 2,281 by 1939.[42] The vast majority of these men, women, and children were designated as "volunteers" by their village chief, whose orders were backed by the colonial guards, always in the company of the administrator.[43]

After a few years, administrators were actively seeking Office recruits in regions they deemed unproductive or that were designated as labor reservoirs. In 1939 the governor of the French Soudan reported to the territorial governor that "we should only carry out propaganda in poor, overpopulated, or isolated regions far from the major centers where agricultural outreach is enough to better the circumstances of farmers."[44] In other words, most of these groups were vulnerable to coercion. For example, in 1939 the Ségou commandant, Robert Léon, targeted sixteen hundred people in seven villages in the canton of Sagala for recruitment precisely because they were "poor" and "isolated." In the end, the Office did not have space for the Sagala families that year, and they were not relocated.[45] In 1940, officials planned to recruit three entire villages recently hit by an epidemic of sleeping sickness. Dr. Ethès from the colonial medical service was dispatched to investigate the implications of moving between 450 and 500 residents from towns hit by the disease to the cotton sector at Niono. Correspondence between the lieutenant-governor of the French Soudan, the doctor, and Office officials about the potential recruits actually suggested that the move would help improve their health. In reality, residents in both the cotton and rice sectors had experienced high rates of mortality in these years, rendering implausible the notion that a move to the Office would be more salubrious. In the end, the idea of recruiting the villages affected by sleeping sickness was discarded in favor of recruitment from more populous areas.[46]

In the same period, Office representatives were instructed not to spread propaganda in areas where farmers already owned plows, or participated in colonial outreach programs designed to teach "modern" farming methods. In fact, letters exchanged between the lieutenant-governor and the Ségou regional administrator recorded that plow owners vigorously protested recruitment. Settlement at the Office was advertised to potential farmers as a means of economic improvement *because* of the scheme's association with technology. At the Office farmers were expected to adopt plow cultivation (although in these years few plows were actually distributed) as well as irrigation techniques.[47] Planners also projected increasing mechanization for planting and harvesting in the near future. In fact, some agents told potential recruits that at the Office, machines would do the work for them.[48] Plow owners had already adopted some French techniques, but few among these already "modernized" farmers saw advantages to joining the scheme.

Administrators reexamined the labor pool when recruitment efforts were less than encouraging. In some cases, settlers who demonstrated unsatisfactory agricultural skills were evicted, as was the case for several Moor families with prior experience as herders. They were evicted from Sériwala between 1940 and 1941.[49] Demographic observers believed that Mossi populations (from regions in the Upper Volta previously administered by the French Soudan) had higher fertility rates than other groups. Some colonial administrators went so far as to argue that the Mossi population was too large for the land they lived on to support them. They were also widely represented in colonial literature as skilled agriculturalists.[50] As early as 1932, the lieutenant-governor of the French Soudan suggested Bélime look for Mossi recruits despite the high transportation costs (Upper Volta was farther from the project than any of the other regions targeted).[51] While Mossi recruitment did not peak until the 1940s, from the late 1930s Mossi regions were regarded as Office labor reserves.[52] In 1937 thirteen Mossi families were settled in the rice sector at a town named Ouahigouya, after their home region. Their numbers increased, and in 1939, the administrator in Ouahigouya sent eighty families to Kokry.[53]

Criticism of recruitment from within the ranks of the local administration pointed to the lack of planning. In 1944, almost a decade after the first town's founding in 1935, the officer then in charge of the Ségou region, Joseph Rocca-Serra, recalled, "In the beginning, it was accepted that the indigènes would rush to the irrigated land."[54] He reported that administrators resorted to conscription when local farmers resisted recruitment. Because of the lack of actual volunteers, most settlers were ironically labeled consentants, or those who have consented (following significant pressure from local chiefs).[55] During the war, recruitment intensified in neighboring regions and Upper Volta to include Ségou, San, Koutiala, Tougan, and Ouahigouya. The ongoing practice of designating thousands of unwilling migrants to join the Office deeply unsettled the Mossi region of Ouahigouya, provoking massive migration to neighboring Ivory Coast in 1941. At the end of the war in 1945 the Colonial Ministry ordered an investigation into the finances and social conditions of the Office. That mission was commonly called the Mission Reste after the governor of the Soudan. Inspector Lenoir was dispatched to investigate complaints of abuse by recruiters. He verified what many administrators already knew: even the recorded number of "volunteers" masked coercion and displacement. Lenoir discovered, for example, that for 1942, where 750 of the total 925 recruits were listed as volunteers, only 176 had even "consented." By the end of the war this pattern intensified for Ouahigouya and other regions.[56] The colonial body politics of recruitment for the Office was nothing like a natural "transplantation" of people but was widely experienced as violent forced removal.

HUNGER AT THE OFFICE: A CRISIS OF COLONIAL MAKING

Settler population figures played a major role in project rhetoric, but the agricultural engineers who designed the mechanical and agricultural landscape did little to plan for the nuts and bolts of settlement. Every year for the new agricultural campaign, administrators sent hundreds of families at a time to the Office. They routinely collected the names of male "consentants" and the number of family members from appointed local chiefs.[57] The very first towns were settled by families in the Macina and Ségou regions. In later years, recruits and their families traveled long distances from their hometowns, some stretches by foot, others by boat or truck. Women and children often arrived very ill; some died en route. Not surprisingly, the project regularly suffered population losses from waves of fleeing individuals and families.[58] Many of those selected fled even before making the journey to the Office, especially those from regions in Upper Volta.[59] The reality was nothing like the confident predictions of farmers flocking to well-watered fields and prosperity.

By the late 1930s, the Office du Niger was already subject to criticism for harsh labor practices and the poor material conditions of its villages. In 1937, the short-lived Popular Front government in France commissioned several inquiries into conditions in French West Africa. Mme. Denise Savineau toured the project as part of her investigation into the situation of women across French West Africa. She reported: "In the villages that surround the Office, no one wants to be a part of it. On the contrary such a neighbor is unwelcome." She observed flooding in these neighboring towns, forced labor to build the project roads, and strict control of movement in the area.[60] It is no wonder that neighboring towns refused incorporation. Savineau, like other critics, also observed the overwhelming presence of mosquitoes, high levels of malaria, and dysentery in Office villages. She pointed out that women and children in these towns disproportionately suffered and died from these illnesses.[61] In 1937, the year before

Savineau submitted her report, the health service at the Office reported high numbers of yellow fever cases.[62] In 1938 the official who surveyed households in villages along the Macina Canal also observed farmers afflicted with guinea worm.[63] The canals flooded the fields of surrounding villages and concentrated waterborne illnesses in Office towns. Rather than attract farmers, modern irrigation technology initially repelled neighboring populations and harmed the health of Office residents.

Savineau further observed that the Office pressured male farmers to exact taxing labor from their wives and children.[64] Women and children weeded and harvested the fields. Wives cooked and traveled long distances between their cooking fires and the Office cotton and rice fields to deliver the midday meal. They also carried food to family members working in food crop fields distantly located from town and from the irrigated fields. Some of the food fields were at such a distance that during the busy rainy season, some family members moved to temporary housing next to the fields to complete the work.[65] The Office partially prepared cash-crop fields for settlers, but when families arrived they had to cut down bushes and trees in forested areas to plant their personal millet and maize fields.[66] Yet Savineau did not observe women cultivating their own fields for food or profit at the Office as was typical elsewhere.[67] Life at the Office was harsh, and it was especially so for women.

Chronic food shortages were a glaring problem for the Office. Farmers were not allowed much land for food production as compared to the crop rotations allotted earlier settlers at Niénébalé. In the first years at Niénébalé, residents actually grew more food than cotton. There, in 1928, a Niénébalé family typically cultivated 115.45 hectares (ha) of millet, 12 ha of manioc, 6.55 ha of corn, 7 ha of potatoes, and .28 ha of rice. They also grew 112 ha of peanuts, as both a food and a cash crop. In addition, they grew watermelons and niébé beans in unregulated plots. That same year they grew only 32 ha of cotton.[68] In fact, agronomists at the station had to persistently pressure farmers to increase the area reserved for cotton planting. The variety and high percentage of food crops cultivated at the experimental station responded to real food needs and importantly allowed women to produce diverse and delicious meals.

Planners at the Office did not pay attention to what the earlier Niénébalé settlements revealed about the optimal balance of food and cash crops. When preparations for the new Office settlers were made, more effort and money was expended on preparing irrigated fields and purchasing equipment than on ensuring their food or other material needs. For example, in 1936, the Office planned to build five new villages near Kokry. A total budget of 1,235,000 francs (FR) was approved in July of that year, but as with other financial matters at the scheme, management of the budget was negligent, and line items did not always total the budgeted amount.[69] Funds were designated for the construction of the villages, basic housing for 135 families, and an instructor's house. Technology outweighed all other costs. The funds allocated for land clearing, mechanical labor in the fields, and extending the irrigation system were estimated to cost 510,000 FR. Settler housing was estimated at only 210,000 francs. The budget allotted 2,000 francs per family for the purchase of oxen, plows, rakes, and carts. Only 1,000 francs were set aside to purchase six months of food rations for each family—the smallest budgeted expense.[70] Then in 1938 the Gruber Canal ruptured near Niono, leaving many settlers in the cotton sector with no harvest to sell and little to eat. For all the expense of the irrigation technology, it was highly unreliable. The ensuing floods ruined cotton and food crops in all the villages along the canal. One village, Fouabougou, was moved to another location as a result. Officials compensated farmers for the lost crops, but widespread food shortages nevertheless resulted.[71]

Recruitment for the project intensified the material stresses upon people in areas already hit hard by environmental catastrophes. In the early 1930s, a series of droughts across the colony hurt farmers. Officials reported that the 1934–35 harvest would be poor because the rains had stopped early. In response, the governor requested reports on the food situation from every region. That same year, recruiters heavily targeted Ouahigouya partly because the region had suffered due to drought and locusts.[72] In fact, recruiters hoped to take advantage of poor harvests to recruit families with seemingly few other options. For example, Office agent Mr. Blanc told Sériwala residents that they would die from hunger if they remained. But if they moved to the Office, according to Blanc, they would have food.[73] Other Office agents like Mr. Mougenot offered meat to men if they agreed to sign up for the project. When faced with refusal in Siguiné, Mougenot retaliated by ordering the village to send twelve men to work on a nearby road and threatened to recall the oxen and plows being used in the village.[74]

On the long journey to the Office many recruits and their families already suffered from hunger. For example, in 1941 a large number of Mossi recruits died en route to Niono from hunger-related illnesses. Joseph Rocca-Serra led an investigation into the deaths and found that when the group of 1,170 people left the Upper Volta, they did not have enough provisions for the several-week-long trip. He blamed the *Naba*, the Mossi political leader, rather than recruiters, for failing to provide the families with the necessary supplies. Before reaching Niono, twenty-four people in the group were dead, and another fifty died shortly upon arrival at the Office. Women and children were the most affected. Office medical staff attended to many more who arrived ill or became sick in their new village.[75] Months later, widespread illness—especially dysentery—persisted in their village. In response, medical observers advised the Office against an abrupt change from millet to rice in their diet. In fact, this was a general medical suggestion for similar problems across the Office, one that shifted responsibility for the ill health of settlers away from coercive recruitment and toward cultural difference.[76] Despite the official rhetoric that heralded the Office as a solution to food shortages and malnutrition, many arriving settlers already associated the project with deprivation and hunger.

STANDARDIZING TASTE AND CULTIVATING HUNGER

Office staff often blamed farmers for the poor agricultural and economic performance of the scheme in its first decade. Though many settlers sabotaged irrigation works or deliberately refused to work, all residents worked hard to eat. For example, instructors noted that settlers spent more time maintaining their food crops in rain-fed fields than they did attending to their irrigated cash crops.[77] Women in particular worked in millet fields and, generally, without the much-vaunted plow.[78] Officials promised rations to first-year settlers, but their distribution was irregular and not sufficient to satisfy their consumption needs.[79] In January 1937 the Office even sought the eviction of several settlers in Kokry area villages for failure to attend to their irrigated fields. N'Golo Tangara was one of the settlers targeted for eviction that year. The instructor for his village observed that he spent too much time hunting.[80] Hunting may have been an occupation or pastime for Tangara, but it was also a way to provide meat for meals and a valued good to exchange for grains. Men and women also spent a great deal of time fishing.[81] There was simply no way for families to survive without supplementing their rations. Meat and fish offered much-needed protein to the diet but also familiar and pleasant flavors.

Severe food shortages across the Kokry area in 1938 brought attention to the multiple problems of food supply at the Office. By January of that year, colonial officials were investigating reports of the shortages. In addition, officials sent forty tons of rice and twenty-five tons of millet to be sold at fixed prices to Kokry farmers.[82] Local administrators were also ordered to oversee the construction of reserve food granaries, a policy dating to the years after the 1913–14 famine in the colony.[83] The Gruber Canal rupture had damaged crops in Niono, but technical problems were not blamed for the crisis in Kokry. The French investigators and officials charged with looking into the food crisis blamed the project's strict controls over the sales of rice and the excessive surveillance of Office settlers. Colonial investigators cited several other problems related to food supplies for all settlers. The head of the health service reported that existing rations were not adequate for the heavy labor expected of all residents. For one thing, the promised rations for new immigrants did not include meat.[84] Moreover, he reported that rations were habitually reduced or withheld as a form of work discipline.[85] Another administrator revealed that many families survived only by procuring food in their hometowns and bringing it to the Office.[86]

In August of the same year, the office of the governor of the French Soudan sent a message to the territorial government:

> I take advantage of this occasion to point out the inconvenience of too large an extension of rice cultivation on Office lands to the detriment of millet cultivation, which is the Soudanese staple. At Baguinèda for example, all the harvested rice is exported to Senegal and it is the farmers on dry land who provide millet for the colons. During the lean season just before harvest, in fact, then, it is hardly surprising that rather than being able to count on assistance from the Office regarding provisions for the population of indigènes, which seems rational, around 1,600 tons of millet must be sold to colons, tons that they consume, but do not produce.[87]

In short, the Office supplied a wider regional market but failed to feed its farmers. Instead, villages were importers of their preferred grains grown elsewhere. There was also no guarantee of the quality of imported food. In fact, the shipment of millet that was collected and sold that year in Kokry had been rejected by the Labor Service; its sale in Kokry led to yet another investigation.[88]

Interestingly, in 1938 several Office staff members (as well as other agricultural officials associated with the scheme) participated in a colony-wide inquiry into Soudanese living standards and diets. The concerns about food provisions at the Office reflected a larger colonial debate over poverty, hunger, and scientific approaches to diet. Vincent Bonnecase has demonstrated that a marked shift occurred between the 1920s and 1950s from a colonial concern over the threat of famine to malnutrition and quantifiable food rations.[89] The 1938 Guernut commission for the AOF, similar to other colonial surveys on food and diet in these years, aimed to quantify the basic ration for its subject populations.[90] The set of surveys for the French Soudan recorded household food expenses, the average number of meals per day, crop information and harvest records, as well comparisons between wealthy and less well-off households. But very little information was collected relating to nutrition.[91] Why would staff at the Office, where the food supply situation was admittedly bleak, be asked to conduct some of these surveys? It may be that they were meant to apply what they learned to better calibrate the Office household ration or food production allotments. Of course, they might also have been seen as knowledgeable about worker rations and the average diet because of their work at the Office. Perhaps they were meant to learn more about the diverse dietary needs of settlers coming from different regions in the colony. The officials were dispatched to investigate diets in southern regions of Sikasso and Koutiala, as well as regions to the east of the Office near San and Ouahigouya—all areas targeted for Office recruitment. Regardless of the reason for their participation, their resulting reports paid special attention to households employing a plow and the production of cash crops such as cotton.

Strikingly, Africans in the colonial service who conducted surveys for the Guernut commission more often included information about consumption, taste, and the seasonality of foods. For example, Bakary Timbo surveyed the Ouahigouya administrative district with a focus on Fulbe and Rimaibe households, and he noted the importance of treats for children between meals such as fruits and beignets.[92] Bouillagui Fadiga surveyed Bamako, and he noted the addition of new luxury food items to the diet including tea, coffee, and the Senegalese ginger drink *jinjiber*. His report also described distinct forms of sociality and eating practices by gender. For regular meals, men sat around one bowl and ate quickly, while women sat around a different bowl taking time to eat and gossip.[93] All of their reports also made special mention of a variety of flavorful fermented drinks and beer in the diet. The picture of daily life in these reports is one animated by the pleasurable consumption of delicacies and specialty beverages, as well as regular socializing around meals.

For the Macina, where part the Office was situated, Aguibou Dembélé's report voiced complaints from men about how administrative labor requisitions had disrupted their work routines. Many men in the region were no doubt conscripted for labor related to the project's construction. Like Timbo and Fadiga, he also noted local tastes, describing a local preference for different types of cola nuts from the Côte d'Ivoire colony to the south. He also amplified the Macina region's reputation for agricultural abundance. In particular, he equated its rice with wealth (he specifically notes that rice is the region's currency). However, Dembélé did not credit the Office or its rice production, which did not have the same reputation. Interestingly, he also included a specific reference to pregnant women: the region's inhabitants acknowledged that the food these women ate was supposed to "flatter their tastes."[94] His observations suggest a relationship between personal food preferences, appetite, and health, as well as further connections between women's fertility, food, and social reproduction. Unfortunately, the attention to regional trade, labor concerns, consumption, and the enjoyment of eating evident in all their reports was not something that Office officials noted in their own observations.

Administrators and Office staff were aware that they ought to pay more attention to food supply concerns at the scheme, but they did not have the same attentiveness to food preference and the pleasures of eating as the African surveyors. Other officials were outright dismissive of any critique. In response to accusations that Office policy aggravated and even caused food shortages, Vincent Bauzil, one of the head engineers and planners, responded that the poor material conditions of settlers were due to their own improvidence. He explained this via the suggestion that families sold their household goods before moving and upon arrival complained of poverty. According to Bauzil, these were fraudulent claims: "In fact, most of the time, at the announcement of their departure for colonization, the indigènes, who otherwise would have lived from their own resources in their home villages, hurry about selling all that they own and declare upon arriving at the irrigated lands to be completely without the means of subsistence."[95] Office planners imagined that settlers would arrive with stores of food and basic possessions. In fact, officials often stated that rations would be distributed only to families in need. However, it is unlikely that men and women forced to move to the new Office villages, often from long distances and without much notice, would arrive with enough food for six months and all their necessary domestic items. Bauzil's accusation of African imprudence was questioned even by his contemporaries, but he was correct that many families may have left behind necessary but not easily portable household items.

The shift to rice from millet as the main staple merits greater scrutiny. A few medical experts at the Office attributed the high number of dietary-related illnesses and general poor health of Office residents to an abrupt dietary transition from millet to rice. Indeed, there was substantial evidence that the shift from millet to rice was responsible for the poor health of many villagers. Settlers preferred eating millet, but the grain was also important on a nutritional level. While some administrators believed rice was healthier than millet, other scientists such as the agronomist Auguste Chevalier noted that millet was more nutritious than rice. In fact, international researchers were learning that eating only white rice was related to several diseases such as beriberi.[96] Chevalier expressed his concerns about farmer diets at the Office in a 1939 report commissioned by the metropolitan government. In particular, he questioned the claim that rice production at the institution would address local nutritional needs: "[Rice cultivation in West Africa] will never satisfy local food needs as they are already constrained [by relatively limited protein consumption]. As I wrote elsewhere the base of Soudanese food is sorghum, millet, and fonio. These grains are richer than rice in nutritional terms." He continued, "Millet has more vitamins than rice. The latter eaten alone causes nutritional deficits for populations who do not supplement it with meat and vegetables. Sorghum on the other hand in the form of couscous, cakes, or cooked dough is a complete food to which the [African] is well adapted." His conclusion was clear, "We must therefore permit the Soudanese to eat sorghum, and the other related grains, as their staple food and only secondarily develop rice cultivation."[97] Chevalier strongly opposed the idea promoted by Office officials that irrigated rice cultivation was a solution to famine. It was creating hunger.

Inspector Carbou had similar concerns, but he also emphasized the problem created by strict controls over rice sales in Kokry. He pointed out that men had no control over their household's harvest. This impeded their ability to provide necessary and diverse foodstuffs for their families. In particular, Carbou reported the complaints of men who wanted to give their wives or daughters small amounts of rice to exchange in the market. He wrote: "With respect to the rice belonging to him [the settler], the Office must leave him free. If he wants to send his wife or his daughter to the market to exchange a calabash of rice for dried fish, seasonings, milk, or something else, it is his business, and he should not be obstructed. If he is tired of eating rice, and if he wants to eat millet[,] he should be free to buy some millet with a little rice. The monitors and cercle guards should not bother him for that."[98] Carbou also noted that the men he met were not opposed to using their stores to constitute food reserves or seed reserves. They simply needed to be able to sell some of their rice.

Importantly, Carbou's observations draw attention to the question of taste, not only with regard to the staple grain but also seasonings and other locally produced sauce ingredients. Despite the strict controls over the harvest sales, farmers routinely sent their wives or daughters-in-law to the market with small amounts of rice to exchange for the provisions they needed for cooking: millet, shea butter, soumbala, fish, and dried leaves for sauces. In Kokry, many women walked to the Macina market to sell small piles of rice, called *dorome, dorome*.[99] Women in Niono also sold dorome piles of cotton for the same reason.[100] These sales helped to supplement meager rations and to add flavor to the diet of Office families.

While most of the settlers grew up in millet-growing regions of the colony, rice was a matter of particular attention because of its value in the colonial export market. French agricultural scientists and merchants appreciated the long history of rice cultivation in the Macina and elsewhere in the colony. To begin with, floodplain rice farmers produced more than one harvest. This fact was extremely interesting to officials who first and foremost emphasized production. The first rice crop was cut while the fields were still flooded, and it was transported by pirogue to the farmers' settlements, where women dried the cut paddy for three to four days before they threshed it. Farmers then harvested a second crop a few months later. In some areas of the flood plains, rice was exclusively cultivated, but in other areas farmers planted rice, millet, and other grains.[101]

French researchers first observed women's dominant role in rice cultivation in the colony's southern districts. In these areas, and in neighboring French Guinée, rice was a woman's crop often grown in marshes and swamps.[102] In the 1920s, an early agricultural official named Jean-François Vuillet observed that rice was sown by elderly women.[103] Another agricultural observer described specific tasks in rice farming that were performed only by women. He wrote, "The rice sprouts with the arrival of the rains. Weeding is done when the plants reach 20 to 30 centimeters, and it is most often done by women. They do this work with care."[104] Notably, southern-based rice cultivators employed irrigation techniques. Women's established expertise with rice cultivation is one reason that Somono fisherwomen were recruited for the Tienfala irrigated rice experiment.

In developing the agricultural export market, colonial officials and merchants wanted to substitute white rice for red rice. This was because white rice was more competitive on the export market. To do so, researchers, administrators, and commercial agents looked to distribute white varieties of rice from Guinée and the Sikasso region across the Soudan. In 1922 the European rice merchant Danel, based in Mopti, wrote to the colonial administration to request white rice seeds from Guinée. He explained in his letter that he wanted to "improve" rice quality in the region by expanding white rice cultivation. His motivation was the expansion of his export business.[105] Another European merchant in Mopti, Mr. Simon, successfully grew a white variety from Guinée in his fields. In 1927 the administration of the French Soudan requested a ton of paddy for seed from Guinée to promote similar agricultural efforts being encouraged by "the possibilities in many Niger Valley regions of replacing red rice varieties, presently grown in the region, by better looking and more profitable white rice varieties."[106] That same year, the agricultural station in the Macina region at Diafarabé conducted trials with a white rice from the southern Sikasso region of the colony.[107] White rice was consistently substituted for red varieties by European farmers and researchers in the colony.

Emile Bélime similarly was convinced that white rice would eventually replace local red rice. He surmised:

> In effect, the abandonment of local rice is inevitable. They are mediocre plants and regular irrigation permits the cultivation of higher value grains. In the Soudan and Guinée, there are good market-quality white rice varieties. The most well-known are Kakoulima and Sikasso rice. . . . Given that Macina rice is first intended for domestic consumption, we should consider whether it is necessary, or simply advantageous, to improve a grain that is perfectly satisfactory locally. However, there are several plants likely to acclimate, that are more productive than local varieties. Thus, they are more profitable.[108]

Bélime was perhaps thinking that the white varieties from Sikasso and Guinée would easily satisfy local consumption needs. He certainly disdained the local preference for a food crop he termed "mediocre." At the same time, Bélime believed that through the introduction of any white rice variety, the colony could compete more successfully with Asian rice imports in Dakar and in other markets. Certainly, for Bélime and the European merchants, the choice of rice variety was a question of market value and superseded local preferences.

For farmers, the choice of which varieties of rice to cultivate had more to do with variety in the diet, adaptation to local conditions, and flavor than simply market value. Studies of farmers who were not under Office constraints found that they selected rice types for their resistance, particular flavors, and cooking and storage properties just as they did with millet. When administrators and merchants initially distributed and promoted white rice seeds along the floodplains, farmers planted the new seeds in a few fields. By harvest time, European officials were disappointed that farmers appeared to devote little attention to these white rice plantings. One survey of rice cultivation in the colony suggested that farmers were reluctant to adopt these new varieties: "The administration has at different times freely distributed large quantities of seeds with the idea of introducing new varieties in the Mopti region. These trials were without result. This is due to a spirit of routine and distrust. Farmers are happy with their diverse red rice varieties, and they have given no care to the cultivation of new types."[109] It seemed that farmers were willing to test the new seeds but not at the expense of compromising the year's food harvest.[110] Much to the dismay of the French observers, these farmers, having experimented with the new varieties, appeared to prefer eating and selling local varieties of rice.

Researchers noted a great variety of rice types, just as they had with millet. In the southern region of the colony a variety called *bintou bale oule* had grains that plumped well with cooking. It sold very well in the southern Siguiri market. Another called *fossa* was known for its pleasant smell, referred to as *simba bô*.[111] In the Macina, two types of quick-growing rice called *timba* and *bougna* were widespread. *Sima baléo* also ripened quickly but produced fewer grains than other varieties. *Tomo* and *sima odéo* were harvested in the second harvest in December.[112] When harvested, these local rice types were a variety of colors: red, brown, black, and white.[113]

In both the southern and northern rice-growing regions, women processed rice paddy,[114] which was a skill that many women at the Office would have to learn. Vuillet described two distinct methods:

> Some of the time, locals are happy to pound paddy in a wooden mortar. The rice is dark red, covered in dust, and contains a great proportion of broken grains. In the second method, before pounding the paddy, it is soaked for two days in cold water, then boiled and dried. The process plumps the grains, separating the parts of the husk. This way, the red outer covering is loosened, making the paddy easier to thresh. Not only are the grains less breakable, but the seed fragments, which when the

harvest is collected too late break inside, are also loosened. If done with care, which is all too often not the case, the process produces an almost white, sort of pointed, rice with very little breakage. It is easy to cook and digest. This is the parboiled rice of the British colonies. While the boiled rice has an improved appearance, it loses flavor and preserves less well than unprocessed rice.

Vuillet was most interested in the process of parboiling and added the following comment to his observations: "The Soudanese know very well that rice loses some of its nutritional value in the process, and this is perhaps one of the reasons that the technique is not more widespread."[115]

Certainly, Vuillet was not impressed by the women's preparation technique that did not involve the preliminary boiling step. He even hinted that he thought women who employed this processing method did not properly clean the rice before cooking it. However, Vuillet did suggest that women had a scientific understanding of nutrition and food value and regarded the first technique as preferable in that regard when compared to the method of parboiling rice. Moreover, parboiling was time consuming for women, whose labor was in high demand. On the whole, Vuillet was not generous to Soudanese women in his assessment of their labors, but it is apparent that the women he observed possessed sophisticated knowledge about the processing of grains.

In the above description, Vuillet glossed over his assertion that parboiled rice was not as palatable as rice prepared by other methods. It is not clear how Vuillet arrived at this conclusion—was he relying on his own palate, or reporting the assessment of local informants? Yet for women this must have been one of the most important considerations when choosing how to process rice paddy. Vuillet also suggested that parboiled rice was easier to cook, which of course does not consider the time it takes to parboil in the first place. One craves a deeper understanding of how the cooking technique emerged here, how it catered to women's needs, and how reduced preservation properties were balanced against the benefits of faster cooking. Vuillet was interested in the parboiling method because it produced a rice that was "almost white." The Soudanese cooks he observed were probably less interested in his aesthetics and the export value of white rice than they were in cooking it to meet the tastes of the people who ate their food.

Vuillet and other researchers observed a great deal about local tastes and the methods of food preparation. However, this knowledge did not translate into a similarly informed understanding of local food supply and agriculture at the policy level. Their research was intended to facilitate the goal of substituting white rice for red rice. In the process they disregarded the information they had collected about the relationship between cultivation practices, taste, nutrition, and cooking preferences. Instead, they looked for ways to promote the cultivation of white rice at the Office. In the cases of both rice and millet (see "Colonial Intervention and Gendered Labor Dynamics" in chapter 1), the research agenda was geared toward the standardization of more marketable grains that might also produce the highest yields. The farmers who tested and then rejected the white varieties promoted by the Agricultural Service demonstrated that they had a choice over what they would eat. Unfortunately for farmers at the Office, their ability to choose what to cultivate and cook was far more constrained.

MISSING WOMEN AND DEMOGRAPHIC CRISIS

A few years into the establishment of the scheme, officials came to realize that there were very few women in Office towns. When administrator Floch visited the project in 1937, he noted the conspicuous absence of women in its settlements. Responding to this observation, a local staff member dismissed Floch's remark and suggested that it was normal for married women to be in their natal villages.[116] Certainly it was not unheard of for women to leave or even abandon their husbands, but they only did so when the conditions of their marriage were untenable.[117] Floch was noting a problem that would lead to a social and political crisis.

The demographic situation of Dar Salam in 1938 was telling. Families in this Office village tended to be smaller than the ideal size for a household in the region. For instance, only four out of the twenty-five families in Dar Salam counted twenty or more members. Another thirteen families had ten to twenty members. At least eight Dar Salam families had fewer than ten members (including children).[118] Ideally, the household was a multigenerational unit bringing together several brothers (and sometimes male cousins), their wives, children (and the wives of male children), grandchildren, and other dependents. Small households were often signs of discord, political trouble, or similar crises. A family with fewer than ten people would have had even fewer members to do the household chores and the work in the fields.

Beyond the question of productive labor, large households had a rich social and affective life. The need for such ties is suggested in comments from Kariba Tangara, who told the administrator conducting the household survey in Sangarébougou that he was "happy to have made some money" but that he "was alone" and "missed his family." As a matter of fact, Tangara was quite alone in the fields and in town, as he was listed by the administrator as a "family" consisting of one person. Moreover, he was also the only man listed from his hometown of Karagoudie.[119]

This sense of social isolation specifically relating to food was echoed by women reflecting on a more recent pattern of household breakup at the Office. Eating groups, according to interviewees for this book, were much larger than the immediate family and comprised both men and women in the household.[120] Women in Kouyan-Kura remembered that (especially for their parents' generation) all members of the household ate from one large bowl. This difference is significant because eating is one way that social hierarchy is expressed. And as hierarchy shifts, rituals shift. Today men and women often eat separately, which is an indication of the dramatic conversion to Islamic practice in daily life following a wave of conversions in the 1950s and 1960s. Women in Kouyan-Kura expressed displeasure with changes in eating practices, but they were more vocal about smaller groups of people eating together than the separation of the sexes. In fact, they blamed the change on younger women who wanted their husbands to focus on the needs of his immediate family (wife or wives and children) rather than those of the larger household, which included their mothers and fathers, brothers, cousins, and sisters-in-law.

Part of the reason for the small size of many families in Dar Salam was that they were not complete households. Many men in Dar Salam indicated to the interviewer conducting the household surveys in 1938 that they left much of their family behind to come to the Office. In response to the question "Did your entire family migrate?" nine men specifically said that their brothers "stayed behind" or "did not come." These statements should be interpreted to mean not only that those households were missing the brothers of these men but also that their brothers' wives, children, and other dependents were absent. Even some men who reported households ranging from ten and twenty members indicated that their households were missing certain people. Some of the same men in Dar Salam also noted that they had left their mothers behind, indicating that Office households were not only numerically incomplete but were also missing several generations of members. Put another way, the recruitment of these men for the Office broke up larger households and disrupted rural social life.[121]

At first glance, the population numbers collected from the 1938 individual household surveys for Dar Salam give the appearance of an equal gender distribution between working-age men and women. Dar Salam had a recorded population of eighty-two men and eighty-four women (without counting the numbers of children and elderly residents). In addition, ten of the twenty-five total households included more women than men; seven other households had equal numbers of men and women, leaving only eight households with more men than women. Demographically speaking, the biggest concern for Dar Salam residents seems to have been the small size of some households and the limited number of workers across the generations. However, further examination of the records reveals that many households clearly lacked sufficient numbers of women. For example, Garantigui Tangara's household consisted of fifteen total people: five men of working age, eight elderly members, and only two women to cook for the entire household. Moussa Traoré's household included four men, one elderly member, one small child, and only one woman to cook for all seven people. This means that one woman (often a young woman or girl) would have had to cook every day and perform every food preparation task until she found other women willing to share cooking duties with her (see figure 2.1). Mama Toure's household had no women at all. When the administrator surveyed this household, he reported only Mama (recorded

FIGURE 2.1. Young women pounding millet in an Office town. Courtesy of the Archives Nationales de la France d'Outre Mer. FR ANOM 8Fi 417/55 Office du Niger Aménagement rurale, 1935 to 1954.

Body Politics, Taste Matters, and the Creation of the Office du Niger ⟿ 97

by the administrator as an elderly resident) and two young children. Toure and his children must have relied on adult women from other families to prepare their food every day.

The small "family" headed by Toure was not uncommon in Dar Salam. Several families had only one adult male and one or two women. For example, the head of family number nineteen, Demba Kamara, was joined by one woman (presumably his wife) and three young children. The couple would have had to do all the fieldwork and the household labor with little help. In Sangarébougou (the first Office town) more than half of the seventy-four families consisted of one to two men and one woman—or sometimes one man and one to two women. Three families in Sangarébougou had no women at all. Families of similar size were common in the eight Office towns but rare in the hometowns of men and women who came to the Office. Indeed, women who grew up at the Office remember their childhood households as being extremely small.[122] By 1944, when Inspector Lenoir visited the Office, he noted with concern the seeming predominance of men across the Office. He drew attention in his report to a study done the preceding year by Dr. Ethès of the health service who similarly reported that the demographic situation was unfavorable for young men seeking to marry.[123] Few girls or young women ever actually migrated to the Office with their parents or for marriage between 1935 and 1945.

Senior family members in the villages of origin overwhelmingly refused to send young girls to the project. Assane Pléah was a young girl when she moved to Kouyan-N'Péguèna in the cotton sector, one of the few girls sent to help relatives. Assane distinctly remembers that she saw immediately that her new household was small when compared to her parents' large household in Macina. At the time, small households were to her a sign of poverty.[124] Given Savineau's description of conditions for women at the Office in the late 1930s, it is not surprising that grandparents hesitated before sending their granddaughters there or that some women would flee. Indeed, Inspector Lenoir ignored the fact that many men also fled Office towns.[125] Office villages, even into midcentury, had few residents. When Fodé Traoré, a worker who participated in building the dam at Markala, traveled to Niono for work in the 1950s he remembered seeing hyenas. He decided to return to Markala because, as he said, "There were no people."[126]

Male farmers who did stay could not help but notice that their wives were leaving. Some even tried to retrieve their fleeing spouses. For example, in 1938, Nambolo Dembélé left his Office town to reclaim his wife, who had returned to her home in San with their child. Nambolo was attacked by two brothers of the local canton chief when he reached San.[127] Nambolo then traveled to Macina town to lodge a complaint with administrator Joseph Rocca-Serra. In Dembélé's recorded testimony, he claimed that he had been attacked because his presence in San might provoke interest in colonization at the Office.[128] It is just as likely that the two men were protecting Dembélé's wife from the prospect of returning to the Office. In fact, few settlers, whether men or women, thought of the Office villages as home. Most viewed them as akin to seasonal farming encampments where a few family members went to work temporarily during the farming season.[129]

When the colonial government released its major inquiry into the Office in 1945 (the Mission Reste), it included a demographic study by G. Lefrou. He reported a massive exodus of women from the town of Dar Salam. He also recorded low numbers of girls in the population at Baguinèda (one of the experimental stations). For Lefrou, the sparse female population across the project was an early indicator of depopulation. It was not only women leaving: there were few young girls still present to ensure the next generation.[130]

Longtime cotton sector resident Djewari Samaké remembered that people did not wish to have their daughters married in Office towns because of the harsh labor conditions. She explained that it was because there was too much work for women.[131] The first generation of Office women who did stay worked extremely hard, especially harvesting cotton. It was a difficult task customarily accomplished through women's and children's labor. When only small amounts of cotton were grown for domestic use, the harvest was manageable. With the introduction of industrial cotton agriculture, women remembered that picking and cleaning cotton became unending tasks.[132] This was in addition to all the work women did to cultivate and prepare food. When women were pregnant or caring for newborns, they still had to work in the fields.[133] Many settlers in the rice sector even refused to cultivate cotton when instructors arrived with new tools to prepare additional fields for cotton.[134] Even without the burden of cotton production, women in the rice-producing areas also worked hard during the harvest. They collected rice cut by men and transported it long distances to the granaries.[135] Across the Office a small number of women shared a large labor burden.

Parents unwilling to send their daughters to Office villages may have feared more than just the heavy labor regime. In the early years, some towns were so isolated that lions roamed the immediate countryside. Tchaka Diallo, a former tractor driver, also remembers lions in the wilds as he and other workers prepared the land for growing crops. Some of the forested lands around Office towns were also off limits because residents believed that the areas were inhabited by dangerous spirits.[136] Moreover, illness was rampant, and those who lived outside the Office villages knew this. The Office was not regenerative or plentiful as imagined by Bélime; it was dangerous and deserted.

Intense colonial surveillance in the fields and in towns by unsupervised male staff members was also a danger to women. In 1944, investigators collected testimony that women were at risk for physical or sexual abuse at the hands of African staff. That year, Mamadou Traoré was one of thousands of disgruntled farmers in Niono who threatened to leave. When he was interrogated about why he wanted to leave, one reason he gave was that a monitor had "abused" his daughter. Another man, Nianzon Coulibaly complained that a monitor "took" his wife.[137] As the sociologist Amidu Magasa noted in the 1970s, such abuse was not without precedent in the Ségou region during colonial rule. In an interview he conducted with an elderly Fatumata Kulubaly, she recounted how the colonial census was conducted. During the census agent's visit, several women would be requisitioned to cook or provide sexual services. Those who performed unsatisfactorily were publicly stripped and beaten.[138]

At the Office there were simply few young girls, young wives, or unmarried women who stayed as settlers. Many families even chose to marry daughters who grew up on project lands to men in outside towns.[139] Customarily marriages connected families and towns, and parents sent young women where they already had family or social relations.[140] Families living great distances from their home regions faced difficulties in maintaining those reciprocal family ties. A small number of girls married into the Office, but at first there were very few other Office towns around and even fewer where families knew one another. In such cases, young women were sometimes married in their own Office town.[141]

Djewari insistently recalled that there were just no women available for marriage to men in the family. It was a "*ko ka gelen*," or a trying affair. As a result, young men were not getting married. Djewari explained to me that even men with beards (signifying they were no longer young men) had not yet married because there were no young women around. She emphasized that this was the case "*yoro be, jamana be*," meaning everywhere in the region. It was not just that there were no women in a few households (*dukono*); the problem was much more widespread.[142] Djewari's recollection that there were no women to marry anywhere is perhaps explained by the increase in female pawnship across the region in the 1930s and early 1940s.[143] Her memories also speak to the extent of the crisis.

Indeed, administrator-ethnographer Henri Ortoli recorded the predominance of women and girls as pawns in the 1930s, as well as the general increase in the practice of pawnship during times of famine. Men in charge of a *du*, or household, could exchange the labor services of a junior member of the household for grain or other needed items, especially in moments of severe crisis.[144] As an administrator for the Ségou region during this same period, Ortoli documented a series of poor harvests, drought, and famine.[145] These were exactly the conditions that contributed to pawnship. For the Sansanding region, immediately neighboring the first Office installations, Ortoli recorded that "the poor cantons in the north of the Ségou Cercle along the left bank of the Niger are peopled by Bambaras and a small number of Markas. After several years of bad harvests, several families (about seventy) have left their villages for good and moved to more favorable lands on the right bank of the river."[146] His observations suggested an atmosphere of general population instability at precisely the moment when administrators began recruiting. This was also when finding brides in the neighboring Office area was a "trying affair."

There is evidence to suggest that many of the first women at the Office were pawns, and in fact these were the most vulnerable girls and young women. When families were selected to send people to the Office it was likely they would send a pawn whenever possible (who would have been an outsider to the family) over a daughter or other family member. Indeed, a 1935 institutional report suggested pawns had been sent to work at Niénébalé.[147] They were often the unlucky ones sent to the Office.[148] During the same period, administrators recorded female pawnship resulting from poor harvests in Tougan, which was one area severely impacted by Office recruitment.[149]

Inspectors and local officials took notice of the demographic crisis because young men were vocal in their complaints about marriage, and their frustrations were potential fuel for political unrest.[150] One proposed solution was to ensure that unmarried women migrated with families. For example, in 1941 recruiters in Ouahigouya, Tougan, and Koutiala were specifically instructed to ensure that greater numbers of young girls migrated: "Office du Niger recruiters working with Cercle Commandants will, in the same interest, also serve as intermediaries between settlers who want to marry from their home region, or want to send for remaining members of their families. With the agreement of Cercle Commandants, they will bring together the settler's family members and the parents of young girls sought in marriage to inform them of conditions at the Office and convince them to leave their villages."[151] Other officials suggested the Office distribute travel passes to young men so they could go home, find wives, and then return to the Office.[152]

In other towns, a similar demographic situation led men who were in charge of an Office family to pursue other strategies to increase the size of their households. In Sangarébougou, Seriba Samake reported a satisfactory harvest but suggested that it had been possible only because family members from his village N'Tama—located not far from the office in the Macina region—came to help him with his work in the fields. Some men specifically told the administrator conducting the 1938 survey that they had asked for a brother to join them. Also in Sangarébougou, Semougou Coulibaly reported that his brother and mother would join him that year. At the time of the survey his household consisted of two men, two women, and two children. Bokari Coulibaly, who was in Sangarébougou with only his wife, also called for a brother to come help him. In all these cases, the men who requested male family members to join them probably also anticipated that the wives of those brothers would also come. In one case, the specific need for women came up during a survey interview. N'Fa Bary, who lived in Dembougou, explained during his interview that he would use the money he earned from selling his crops to the Office to increase his household size by paying for his brothers to marry.[153]

Still, recruiters who had to recruit single men promised that they would find money and food at the Office. This is apparent in the 1938 testimony of Koké Samaké, who told the Ségou administrator Robert Léon how he was recruited: "The propaganda started some time before our departure from Sériwala. It started with Sériwala village where Mr. Blanc came to tell people that they had nothing to eat and that it seemed better to enter colonization. People at first refused." To persuade them Mr. Blanc reportedly added: "Come with me, if you stay you will die from hunger. Me, I have money and food. When you come you will find things to eat. *All those who do not have wives will have them.* If you come, I will give you money. You are dying from misery. With me you will have things."[154] This kind of propaganda misrepresented what settlers could expect (including an abundance of food and the possibility of finding women available for marriage). Even where Office statistics recorded large households, these consisted mostly of males, as many of the women who were initially recorded in censuses later returned home.

Viable households were very much an issue at the Office. Colonial ethnographers frequently noted the predominance of large patriarchal households in Africa. This was true of the well-known ethnographer-administrators Maurice Delafosse, Charles Monteil, and Henri Labouret, all of whom worked in or studied societies in the French Soudan. In fact, supporting patriarchal authority was often a political tactic.[155] In a region where well-being was secured by living in a household with large numbers of people who could work, the households at the Office were only supported by a small number of men and even fewer women. Anthropologists and historians attentive to the family and household have long theorized that value was expressed through social relationships and reproduction in Africa, a dynamic frequently referred to as "wealth-in-people." Jane Guyer and Samuel Belinga updated this theoretical model, pointing out that those who brought a range of specific skills and knowledge were particularly valuable—"labor" was not generic.[156] Across the French Soudan, women's knowledge of food production and preparation—specifically the cooking labor of young wives—was highly valuable. This was particularly the case at the Office, where the conditions for food production and preparation were materially challenged.

Upon closer examination of the 1938 survey records from Office towns, it becomes clear that several men in charge of an Office household tried to expand and extend their social and work networks. In Dar Salam, Amadou Koita was the head of family number eighteen in Dar Salam. He came to the Office with two women (probably his wives) and five young children. His household, constituted as such, was relatively small, with only three adults and five children to feed. Rather than move to the Office with only this small family unit, Koita called for another man and his small family to join them. Mamady Togoba, who was in charge of family number twenty-four, told the administrator that he came to the Office because Koita had "decided the question for him." The administrator did not write down the specific relationship between the two men (friendship, patronage, debt, etc.), but it is likely that the two "families" ate together and helped in one another's Office fields. Togoba brought two women and two children with him to the Office. Together the Koita and Togoba families would have constituted a viable household of thirteen people (two men, four women, and seven children).[157] In Dembougou two other larger household units were likely formed between two groups of brothers. Fassoum Tangara, who headed family number seven, was the brother of Ba Tangara, the head of family number sixteen. They each brought one wife to the Office and several children. In the same town, Nene Coulibaly (family two) was the brother of N'Tio Coulibaly (family eight). In all, the Coulibalys were a household of twenty-three members (three men, six women, and fourteen children).[158] These households were the exception, as most men (and women) struggled to form viable households, with few family members willing to live at the Office and even fewer women willing to migrate for a marriage at the Office.

By the end of the first decade of the scheme's operation, the absence of women had prompted a demographic crisis with political dimensions. In 1944 persistent poverty at Niono gave rise to the threat of a mass exodus, resulting in a series of negotiations and public meetings. At least one thousand male farmers asked for property rights and monetary compensation, a demand that administrators proposed to address through the distribution of rice or cooking oil.[159] This response showed some recognition of common complaints from residents about food shortages. It was also an acknowledgment that farmers could not grow enough food to feed their families nor could they earn enough to purchase it. That year, the Office and local administration shipped food assistance and increased the prices paid for cotton.[160] It was a long overdue response. The crisis in Niono had been building for years. In fact, the previous year, 960 people had already left Niono. It was a loss of significantly more people than the 692 new settlers who arrived that year.[161]

All the protestors complained of food shortages, but Investigator Pruvost, who was looking into the 1944 disturbance, noted a generational divide in the demands. Married men complained that they saw little profit and that the work regime for growing cotton was too harsh.[162] Young men were more concerned that living at the Office made it hard for them to marry. The young men who voiced frustrations over not finding wives also demanded cloth.[163] This request was ignored because local reports cited an abundance of cloth available in the area.[164] Officials no doubt failed to recognize that this demand was related to young men's complaints about marriage. Men gave gifts of cloth to new brides, their mothers-in-law, and other new female in-laws.[165] Cloth was a symbol of a man's material circumstances. In 1944 young men at the Office protested that they had no cloth to give. In describing the conditions leading up to the demographic crisis, Inspector Pruvost wrote: "This sector [Niono] does not have a good reputation among the natives; it is far from the river, located in harsh woods, and farmers concentrate essentially on cotton cultivation. This requires a lot of work with little remuneration unlike rice cultivation. It is subject to frequent flight, almost a quarter of those who entered since its origin."[166] Another investigating official noted that only a few families had food stores in Niono that year and that often even those stores were insufficient.[167]

Djewari remembered the 1944 crisis as being exclusively about women. In her description of the event she explained that the young men got together and vowed to leave if they could not marry.[168] The young men's complaints were also a concern for senior men who needed cash to help their juniors marry. Clearly, the problem was not just the lack of food stores: it was the absence of women to prepare even the small amounts of food available, despite the gifts of cooking oil from the administration. For Djewari, the entire crisis centered on the absence of women. They were the ones who sold small bits of cotton in the market for sauce ingredients. They pounded millet for the main dish, toh, and carried the meal to the fields. Vernacular demographic knowledge associated women with food production and rural prosperity. Without women there was nothing to eat, and without wives (who were recipients of gifts of cloth and potential future mothers) young men would have little to indicate their material status. Office settlers protested during moments of food crisis, and the absence of women only exacerbated prolonged crises. It also brought to light how little the planners understood about the relationships between labor, agricultural wealth, rural social life, and women. For the settlers, however, the absence of women signaled more than a food supply crisis. They worried over the size of households, the potential for future generations, and daily well-being. Women, in particular, animated rural life with their food labors to provide daily meals and support harvest celebrations. In the subsequent decades it was this critical work that occupied women in the Office villages. For without these women's efforts, the Office simply lacked the flavor of life.

3 ↫ "We Farmed Money"
Reshaping the Office and Reclaiming Taste

SOMETIME IN the mid-1940s, when Mamu Coulibaly learned that she would marry a farmer at the Office du Niger, she had already been told that "wari be sene"—or that people farmed money there. She knew that farmers grew rice and cotton, but most importantly they acquired cash. Women also farmed money at the Office. Because of this, Mamu heard that "things could be found," including many new metal household goods and foodstuffs. Mamu's husband had been forced to go to the Office along with his parents. Mamu knew this family history of displacement, but she was happy to marry there. When as a new bride she rode to her new home on a cow, she took it as a symbol of the life that money could offer at the Office. Mamu was among the second generation of women to come to the Office and joined a cohort of women who were also trying to "farm" money for themselves. After arriving in her new home in Kankan, like many other women Mamu brewed millet beer and sold it in town. She had some free time to brew in part because she lived in a town where farmers grew rice, which at the time required less female labor than cotton.[1] Brewing beer was Mamu's way of farming money. Although colonial planners of the project emphasized agricultural production and the disciplined organization of labor, women's everyday lives suggested that the history of colonial agriculture was also shaped by consumption and celebration.

The first generation of women at the Office met with unfamiliar surroundings and a lack of nearby rural resources such as plantations of specific food-bearing trees and limited access to garden plots. In time these women, their daughters, and succeeding female migrants set about turning their new environment into a usable resource. Wherever possible they transformed their new surroundings by planting gardens and essential trees. They also integrated the most dramatic feature of the Office landscape—the canal running along the edge of their new towns—into their foodscape. Now many Office women like Mamu also needed to earn money in order to do marketing on a regular basis for food production. It so happened that her income-earning strategy also supported nighttime leisure. Women quickly learned where to purchase the foodstuffs they simply could no longer produce themselves. Making use of a variety of resources had always been a part of how women managed their food-related labors. Increasingly, provisioning a household required detailed knowledge of nearby social networks and new colonial markets. In addition, rice would slowly become a staple of the rural populations who came to work at the project. Women learned to cook the crop that their families cultivated; however, when millet was preferred, they purchased it when possible. The foodscape was shifting, but women shaped it as much as possible to fit their own needs.

THE OFFICE AFTER 1944: RESHAPING THE FOODSCAPE

By the mid-1940s the Office had greatly expanded its geographical boundaries. Many of the oldest Office towns were now surrounded by men's cash-crop fields and other Office towns. Women still did not have their own irrigated fields, but they had to grow or buy food somehow. The new conditions women faced called for a reordering of the agricultural world that included new rural social ties and the re-creation of harvest celebrations. For some women like Mamu, farming money was new and intriguing, but it was also the only way for some women to feed their households, especially if they wanted the meal to taste good. Before 1944, the Office was not a place where women (or men) wanted to live. During the Second World War, especially during the period of Vichy control in French West Africa (1940–44), men and women were still subject to strict labor controls. Travel outside of Office towns or fields—even to their own millet or cornfields, or to forested areas—required staff permission. In this environment women had difficulty cultivating food because it required significant labor outside the designated Office territory. The rain-fed fields belonging to settlers were a long way from project towns, as were the useful trees and bushes found in the increasingly distant bush.[2] Moreover, men who were in charge of a family's Office fields often did not earn enough from the sale of cotton or rice to buy what was needed to eat or cover other household expenses. During the war, farmers increasingly protested policies regarding labor constraints, travel, pricing, fees, and access to land. Then at the end of Vichy rule in 1944, the Office began to ease travel restrictions and to compromise on other demands. Between the mid-1940s and late 1950s, many men decided to stay and farm at the Office, and more women settled as wives.

CLAIMING THE OFFICE

Even though only men sought land ownership and were responsible for the fees and debt payments to the project, these and other issues affected entire households. One of the most salient concerns was the water fee, which together with the charges for other agricultural services sometimes cost farmers as much as 50 percent of the harvest.[3] Men and women opposed these fees as well as irregular weighing practices, unjust evictions, and the wrongful arrest of farmers.[4] Men sometimes recorded grievances with the government official in Macina or with their local agricultural instructor.[5] Women were also active protesters. For example, in 1955 three women in Mamu's new Office village, Kankan, were arrested for insulting the town's chief. Alimata Dembélé, Fatoumata Barry, and Téné Diarra were all angered by the chief's failure to address their families' grievances with the Office. Subsequently, the three women and their families were evicted.[6] After the war men and women wanted improved conditions but did not want to leave their Office homes. By this point many women and their families had lived at the Office for over a decade. Eviction would have deprived them of Office land they had worked on and maintained for years. Women in particular had planted gardens and trees and established market relationships. Through their labor, men and women were beginning to claim the Office as their own.

Farmer protests were supported by a growing number of African labor activists and political leaders in the colony.[7] Political newspapers under African management published reports of evictions, arrests, and other complaints. In particular, they accused the Office of poorly managing land allotments. For example, in the 1950s many farmers were relocated to new towns only to find that their new fields were extremely unproductive. They then attempted to return to their original Office towns. Distressingly, the original farmers would find that new Mossi settlers from Upper Volta had taken their places.[8] Each group had a stake in land that they had worked on and improved, making for uneasy relations between different generations of settlers. Major African labor groups and political activists petitioned the Office administration to respond to settler needs for good land, redress past wrongs, and change its policies on behalf of farmers.[9] In response to repeated criticisms, the Office did reverse several eviction orders in 1955.[10]

Markala, the site of the central dam, was a center for labor and political activism. It had become an industrial workers' town and home to clerks and other African professionals working for the Office.[11] A great deal of labor action in this period addressed the pay and labor concerns of these workers. Worker protests began in 1945 and continued until 1947, when Office employees went on strike.[12] As the protests gained strength, one of the leading local labor activists, Jacques Doumbia, was arrested. Prior to his arrest, a large crowd of workers and their wives blocked the local guards from reaching Doumbia. He was arrested only after the administration sent troops from Ségou to Markala. Then the crowd of men and women walked to Ségou in protest.[13] Following the strike in 1948, the Office did agree to allow for African representation on the Administrative Council to the Office du Niger.[14] Despite the increasing visibility of labor organizers among Office wage workers, the Office refused to recognize a similar labor union for farmers.[15] As such, they had a much harder time petitioning the Office administration or colonial government. Settlers, especially women, had to stake their claims by other means.

Farming at the Office was far from the ideal of the industrialized agriculture once imagined by its planners. By the end of the war, the heart of the Office—the irrigation system—was neither efficient nor economical. In particular, farmers vigorously protested the high costs of water. Meanwhile, officials countered that what they collected in fees did not even cover the cost of providing water or other agricultural services like field preparation.[16] Moreover, the pace of settlement and production over the first decade did not match the administration's hopes. Officials blamed these setbacks on wartime reductions in staff and financing.[17] Yet during the war, rice from the Office helped provision military troops in West Africa and in the West African capital of Dakar.[18] In fact, many Office employees who came to the institution after 1944 believed that the war had brought the Office into being.[19] For women, the Office became a viable place to live only after they had transformed it into a resource for local needs rather than for provisioning a distant war.

GENDER AND OFFICE INFRASTRUCTURE

The establishment of the Office du Niger brought considerable change to the rural landscape. The large dam fed large and small irrigation canals cutting into the landscape, and new roads linked major project towns and political centers. The Office dominated the agricultural landscape. In particular, its carefully measured and ordered fields were an attempt to erase older forms of land distribution and customary claims to natural resources. To use the language of James Scott, Office administrators sought to control agricultural production by making rural life "legible" to its agricultural staff.[20] From this perspective, the network of canals and roads created by the Office helped to consolidate the scheme's territory. The irrigation works not only watered the crops but also demarcated Office fields and lands. Every rectangular field was nominally the same size and treeless.[21] In regional agricultural practices dating at least to the eighteenth century, trees served to mark land claims and boundaries. They also functioned to integrate production from the fields and forested areas. Under Office management, an administrator assigned and recorded the allotment of orderly hectares to only one male head of household. Additionally, new Office roads connected project towns and centers but made travel to important non-Office towns more challenging.

The project design created other kinds of boundaries by delineating two distinct cultivation areas. The cotton sector was centered around the administrative center, Niono, and the rice sector near the Kokry and Kolongotomo centers (see map 2.1). Though Emile Bélime believed each sector would be profitable, prior to 1945 the primary difference between the cotton- and rice-growing areas was overwhelmingly economic. Farmers who grew cotton earned less money even though it was a more labor-intensive crop than rice. Women living in the cotton sector also had less time for gardening or other cash-earning activities. This separation was reinforced by the fact that there was substantial physical distance between the cotton and rice sectors. The two main canals, the Canal du Macina and the Canal du Sahel, fed each sector, and a large area of non-Office territory separated the two. Niono and Kokry-Kolongotomo were only connected by Office roads that passed numerous non-Office towns. Nevertheless, the production landscape in both areas was unlike that found in surrounding regions.

The Office was not just in the fields. The technological aspects of the project were accompanied by a considerable administrative bureaucracy that influenced how men and women interacted with this infrastructure. The new towns also reflected a changing political order. The entire Office became an increasingly dominating element of daily life and work. Even so, women created their own infrastructures for food production. Women also cultivated ties with fellow settler communities, and together they re-created yearly agricultural celebrations marking the transformation of the Office from a colonial project into a local resource.

WATER FOR THE HOUSEHOLD

The concepts of canal building and irrigation were not new to farmers in the French Soudan. Rice was long cultivated by irrigation to the south of the Middle Niger region, and the floodplain cultivation technique prac-ticed in the Macina was a water management system that maximized river water use. Oral traditions from the Segu Empire also record a major canal-building project a few miles from the site of the Markala dam. Sometime in the mid-1800s a female bard named Musokura Jabate pres-sured the Bamana prince Nci Jara to bring the river to Banbugu where he held his seat of power.[22] Musokura shamed Nci for not providing his followers at Banbugu with water. She claimed that there was not even enough water for people to bathe. After her interview with Nci, his father sent workers to dig a canal from the Niger River to Banbugu, passing through Tio and Diamarabougou.[23] For three years workers used axes and hoes to dig the canal. About a hundred years later, when the French oversaw canal construction for the Office, they similarly relied on large amounts of manual labor.[24] Neither canals nor the technique for their construction was especially revolutionary to women just moving to the Office.

One important aspect of the Bamana canal to consider is the role women customarily played in water management. Bamana oral accounts emphasize that water was a woman's concern. Women could not ensure the well-being of people in the empire without water. Indeed, an abun-dance of water was one marker of status. The bard Musokura claimed that in the empire's capital Segu (which was located by the river) women bathed three times a day. In other words, access to water afforded a com-fortable life for women in Segu. At the same time, Musokura exclaimed, "Thirst is about to kill Banbugu Nci's slaves and the people he adminis-ters!"[25] In short, water was also essential for a ruler to effectively rule—here it is also associated with enslaved labor. Strikingly, Musokura associated water with women and daily life. Women with easy access to water in Segu had the luxury of many baths. The female bard also reinforced that it was women's duty to ensure the domestic water supply. When Nci failed to provide water in Banbugu, it fell to a woman to protest such condi-tions and suggest the solution. The oral historical narrative anticipates the likelihood that when women came to the Office, water management would be one of their first concerns.

The Musokura tradition serves to highlight the fact that those who designed the Office canals did *not* center women's labor needs or concerns for the availability and quality of water. In fact, the Bamana project extended the river to Banbugu specifically as a means to aid women's work. Yet, it was not meant to alter *how* women did that work. By contrast, the French system of canals was heralded by its French promoters as a manmade achievement that would alter agricultural practices. Moreover, it was assumed that male farmers would be the primary users of the water. And unlike the Bamana canal, the water supply to the Office canals was shut off in the dry season, when the fields were fallow.[26] They were not at all intended to address household water needs. Women, of course, understood that water was a year-round domestic necessity.

To women, the canals were most obviously a source of water to be used for household purposes including cooking, drinking, and washing. Cooking millet dishes or rice (as well as the sauce) required quantities of water in addition to the amount needed for drinking. Because the canals were located near Office towns, women did not have to walk far to reach them. It made water collection from them relatively easy. For example, when Sitan Mallé was a child in Kouyan-N'Péguèna her mother and other women used canal water for all their cooking and washing, especially of household items and clothes.[27] Women in other towns similarly used the canals for all their water needs when it was available.[28] They often treated the canals as they would have a river. Women piled big rocks from the bed of the canal up to its edge to form something like a staircase. This provided women better access to the water, and the rocks scrubbed against the clothes during washing. This practice employed by women living near the Niger River was replicated by Office women in the canals.[29] The rock staircase also created a meeting point for women to work and talk together. In each town women had access to a water resource like this because the new towns were all lined along one of the major canals or a significant offshoot. The major difference was that river water flowed constantly (even when the waters were low), while canal water could be stagnant.

Canal water was cut at the end of the agricultural campaign—ironically it was highly seasonal. For several months during the year, however, water was abundant. Women had access to canals and wells; for women living in the region from Kolongotomo to Kokry, there was also the river.[30] In some cases women distinguished between the quality of water from each source. In Kouyan-Kura women collected well water for cooking and drinking but washed with water from the canal.[31] Women in Molodo-Bamana similarly used canal water only for washing and relied on wells for their other needs.[32] Canal water was frequently unhealthy to drink, as many early families discovered. Nevertheless, it was suitable for washing cooking pots and other tools, serving dishes, and clothes. The canals eased the labor of water collection for these tasks. When the canal water was drinkable, women's water collection labors were all the more lightened.

Women who did use canal water for drinking and cooking often stopped using it in the period just before the water to their towns was cut off. They had to judge when it was safe for consumption and learned the seasonality of the water irrigation system. Ordering labor around a cycle like this was not new, as women were accustomed to dealing with the seasonal nature of other resources. Indeed, swamps and marshes were seasonal water bodies common in many areas of the French Soudan. This new seasonal water resource could easily be adapted to a new work calendar. It made sense to use the canals when the water was available and clean enough for either consumption or household use. During the canal's dry period until the waters returned for the next agricultural season, women near the river in towns like Koutiala-Kura walked to the river for their water.[33] Elsewhere, as in San-Kura, women switched to well water.[34]

Despite the irregularity of the canals, compared to the surrounding regions, water was readily available at the Office. Across the Office (even in some areas as yet unprepared by the Office for irrigated cultivation) the available groundwater supply increased. For example, in Molodo-Bamana, wells had always been very deep. People had to use animal labor—donkeys or camels—to help with the digging of a new well. To draw water from the wells also required a lot of force. An adult had to pull about twenty-two arm lengths (both right and left) for one measure of water. After the Office was established it became much easier to dig a well and draw water (see figure 3.1).[35] In towns further outside the Office like Monimpébougou, digging a well and drawing water from it required a great deal of labor. Women there still pulled fifteen double arm-lengths or more for one measure of water well into midcentury.[36] Farmers in towns unaffiliated with the Office but situated close to the canals—such as Macina—could use water from the Office for their fields because canals frequently overflowed and flooded nearby fields.[37] In those cases the canals became a shared resource.

FIGURE 3.1. Women working at a well in Niono. Courtesy of the Archives Nationales de la France d'Outre Mer. FR ANOM 8Fi 417/137 Office du Niger Aménagement rurale, 1935 to 1954.

Because of the availability of water at the Office, women had choices over how to manage it for the household. In Nara, women chose to use well water for all their needs. For example, they built an in-ground basin to use for household washing.[38] The women just had to draw enough water from the well and pour it into the surrounding basin. It took less time and energy to fill the basin than to carry water from the canal. It also meant good quality water for all their work and family consumption. Finally, it allowed women to stay closer to home. Where women used canal water they did so because it relieved their labor. In general, Office women had greater choices over how to manage their water resources and associated labor than women in many non-Office towns.[39]

Moreover, the canals brought valuable food resources: fish and rice. Men and women could literally catch fish within a few steps from where the women cooked. Office residents could also catch fish in flooded rice fields.[40] According to Kono Dieunta, "Jege tun be yalla," or "The fish were out wandering." What she meant was that fish for the sauce was readily available. Women remembered always cooking fish. While other ingredients may have been hard to come by, for much of the year they were assured of the availability of fish. More broadly, Dieunta was conveying that women were able to create good quality food if they used the resources nearest to them. Kono, who comes from a fishing family, credited the water for that abundance. In Sokolo the northernmost edge of the Office, residents remember the canals "brought" water, fish, and rice. All these things were previously scarce in Sokolo, a region that was historically rich in millet and cattle. With the arrival of the Office the new water wealth was especially remarkable.[41] Fish and rice were also increasingly in abundant supply after the Office moved into the region. This newfound wealth of water was not without disadvantages, however: waterborne diseases plagued men, women, and children. Like the memory of finding jéba in a time of famine (see chapter 1), nostalgia for abundant water and fish should be set against a background not only of increasing illness but of the poverty of other necessary resources.

THE LOSS OF WOMEN'S GARDENS AND WILD FOODS

With the loss of space for gardens and the diminishing bush in many areas of the Office, women began marketing to earn cash and purchase foodstuffs they could no longer produce themselves. To do so, women spent more of their labor time traveling Office roads to get to markets. Women traveled these routes by foot, for the most part. Near Kokry and Kolongotomo the Niger River continued to be important for all transportation, especially when the main road that ran alongside the river was flooded.[42] In this case, women paid for passage along the river. In the same region, women could also pay for a quick passage across the canal that ran parallel to the main road. Otherwise they walked a great distance on the path from their town to the Office bridge that connected with the main road.[43] Once on the road, even women new to the region could follow them to Office-sponsored markets or other regional ones.

The official purpose of the roads that were maintained by the Office was threefold: to transport crops, facilitate the maintenance of the irrigation infrastructure, and supervise agriculture. The roads were made from dirt, and the major ones were wide enough to support the traffic of four-wheeled vehicles, digging machines, and other equipment.[44] From the 1930s, the colonial government of French West Africa concentrated several road and other transportation improvement efforts around the newly planned scheme.[45] Planners laid out the roads as they added new towns, and they paid more attention to commercial transport needs and access for local staff than daily use by residents.[46] Only small numbers of vehicles were available to women for transport when they arrived for the first time with their families or as a new wife. Other vehicles and even carts were not widely available for travel until after the war.[47]

First and foremost, women needed to get to markets where they could purchase foodstuffs. These market activities were more successful as they learned to navigate the shifting geography of the Office and the surrounding areas. In the early years, women were restricted to travel within the Office. Knowing how to get to the Office-operated markets was critical. In Molodo, women quickly learned the new route between their town and Niono when they were conscripted to bring provisions to workers building the road. This was even before their town had been fully integrated into the Office.[48] For most women, the first journey to the town in the Office where they moved or married took them along some of these new roads. This was their introduction to the routes around their new town. Learning to navigate these roads was especially important for women who arrived in relatively isolated towns. For example, as a child Kaliffa Dembélé traveled by foot and car with his mother and father to San-Kura. Their home region San was a considerable distance from the Office. Along with a large group of new settlers from the San region, they were led to San-Kura by Office agents.[49] Their whole journey was laid out according to how Office planners and agents perceived the geography. The roads, markets, and region were all new, meaning that women did not readily know where the popular non-Office markets were located.

Even after women had greater freedom to travel, those new to the Office continued to use this first trip to orient themselves. When Kadja Coulibaly married a wage worker at M5 in the new mechanized rice sector near Molodo-Centre, she traveled from her hometown outside the Office to Niono by car. The Office driver sent to pick her up quickly joined the first administrative road and then followed project routes. He dropped her off in Niono, the major Office market town in the region. From there to M5 she rode on the back of a bicycle, which was a slower trip but one easier to follow.[50] Even before arriving in M5 she knew where she would go to market, and she had an idea of the long distances she would have to walk.

RENEWING RURAL TIES

Once at the Office, women expanded their geographic knowledge of the Office and region. First- and second-generation women attended yearly Office parties where they met other women. Women and men tended to look for other people from the same region to exchange news from home with. In so doing, they mapped where they each now lived in the Office.[51]

Women traveled to different Office centers every year because the location for the parties rotated regularly. At the same time, women gathered information about towns and sites outside the Office where they could obtain millet or other desirable foodstuffs. For example, longtime residents of the region, or women who had arrived earlier than others, exchanged market tips and details with newly arrived women about where specific items were available. Women also learned about older routes that connected important regional centers such as Monimpébougou and Sandsanding to other towns (Monimpébougou is marked on map 3.1).

First- and second-generation women created new social and market maps as they lived and moved around the Office. This mental (and social) labor helped them to manage their engagement with the changing economic and agricultural layout of the region. Many second-generation women grew up in one Office town and married in another.[52] Women also moved when the male head of their household requested land in a more fertile region under Office control.[53] Wives of workers also became wives of farmers when former Office employees requested Office land.[54] As women moved they built and reinforced social ties across Office towns. The generation of young women who grew up in the Office villages also brought specific knowledge about their hometown's surrounding markets and resources to the town they were married in. As noted above, they also learned from other women already residing in their new households about the particularities of that area. This type of exchange was further integrated into the practice of senior women mentoring new wives.

This process was not seamless. The Office administration frequently (and often by force) relocated some households to found new Office towns, or in some cases entire towns were moved.[55] For example, Fouabougou—a town that had been relocated once because of flooding along the Gruber Canal—was divided after the Second World War. Some families stayed, while other residents moved to the nearby town of Sagnona. Many young men from Fouabougou chose to move because they would receive their own land allotment and have control over their own Office harvest. These young men either brought their wives or began looking for one to move with them to Sagnona.[56] Women did not always have the same choices. As in the case of the first Fouabougou move, relocations were sometimes driven by hardship.

New Office towns and markets were plotted over existing socioeconomic geographies. For example, young men in Nara sought wives from the Sibila region located between the two major Office agricultural zones.

Most of the first families in Nara came from a small town near there and maintained special ties with the remaining families.[57] Families in Nyamina also maintained a similar social relationship with Tièmadeli (a town also outside the Office). Many women from Tièmadeli married into families in Nyamina, and young women from Nyamina often married men in Tièmadeli.[58] These types of social exchanges were common across the French Soudan.[59] At the Office, they helped foster economic relationships separate from the Office structure. Relations living outside the project benefited in turn through the exchange of millet or sauce ingredients for cotton or rice from the Office. These types of mutually beneficial arrangements were not common to all Office towns or residents, but where they existed they were quite important.

In the postwar years, the Office became a dynamic socioeconomic space for workers who had access to cash and for young residents of farming towns. Young men working for wages at the Office helped create new economic niches. After the Office opened a new town at km 39 (also known as Madina), it became a well-known center for after-hours leisure. Machine operators and other workers traveled to km 39 at night to purchase drinks and host dance parties. To foster a festive atmosphere, the workers paid musicians to play and encouraged local young women to dance. Women who sold food or drinks also made money during these parties. It was an atmosphere not unlike rural work parties. In the same region of the Office, workers frequented the new market at the town called B6 (also known as Bolibana). As wage workers, these young men had cash year-round to purchase goods (many sold by women) in the market.[60]

For tractor driver Tchaka Diallo the Office was very "cosmopolitan." In his words, "You could always find people from your home at the Office." He further explained that this was because there were people from everywhere (French Soudan and the wider region) living and working at the project. This was his definition of cosmopolitan. Tchaka traveled to every sector of the Office for his work and always looked for people from his home region of Koutiala. He was a Christian and went to church in Koutiala-Kura (New Koutiala) where many other Christians formerly connected to the Protestant mission in Koutiala lived. Many of his fellow Christians came to the Office as forced settlers, whereas he was a voluntary wage laborer. His association to the Office differed from that of the settlers, but they connected because they came from "the same place" and shared the same religion. When Tchaka began looking for a wife, he approached a man who was several years his senior from Koutiala. The man was then working in Macina in an administrative post and had befriended the younger man. Tchaka never had to go home to look for a wife because he married this man's daughter.[61]

By the same token, daughters of farmers sometimes married wage workers. Adam Bah was a young woman whose father settled at the Office to farm after working on the dam in Markala. She won the local (Office-sponsored) beauty pageant and was sought after as a potential bride. Adam eventually married an African agricultural staff member after being courted by many other workers.[62] The point here is not that the Office became an idyllic "îlot de prosperité" (island of prosperity) as Bélime and other officials alleged, but that men and women worked to make the Office an economically and socially viable place to live.

THE IMPACT OF AN EXPANDING OFFICE

In the postwar years, the irrigated areas of the Office and the number of its towns expanded. Plans dating from the 1930s for the Office included a vast amount of territory that by the late 1940s was not yet open for irrigated farming. Each year more of this farmland was cleared, and new towns were established with the idea of reaching those initial estimations. Following the opening in the 1950s of the new sectors, based at Molodo-Centre and Kouroumary, the French official Inspector Mazodier approved the institution's plans for even more expansion in 1957.[63] This pattern continued into the 1960s with the establishment of the N'Debougou and Dogofiry sectors to the north of Niono.[64]

During this process, existing towns were integrated into the irrigation infrastructure, and more new towns were created. For example, Nyamina was created near Molodo in 1945. A young Sékou Coulibaly and his parents were among the first group of settlers there. As Sékou grew up he also witnessed the concomitant growth of the Office. When Sékou was still young he watched the construction of neighboring Sokorani. Office workers dug the canal until it extended to Sokorani, just down the road from Nyamina; this is also what happened with Bo, a town that predated the Office but in these years began farming cash crops by irrigation. What Sékou saw was that once the water reached a town, it would then become part of the Office.[65] In many ways, Office geography followed the water. Baba Djiguiba, a driver for the Office, similarly described the order of events in the construction of new towns. First, the canal was dug, then the town was built. People were then brought in to cut the trees, followed by the machines that created the fields.[66]

A continually expanding Office disrupted and reordered the regional landscape. More and more nearby towns were compelled to formally associate with the Office or lose their access to land. For example, in the mid-1950s the towns of Tomi and Niaro were pressured to join the Office for fear of forced relocation. Farmers in both towns had used excess water from overflowing canals in their own fields. In response, the Office administration threatened to evict the entire populations of both towns if they did not pay for the water and join the Office.[67] Towns at the edges of the Office that refused incorporation frequently lost their customary rights to land. Boky-Were, an old town near Kolongotomo and Kokry, resisted the incursion of the Office and ended up losing some of their land holdings as new towns were created (Boky-Were and Tomi are designated on map 3.1).[68] Other towns that refused to farm for the Office simply lost their claim to settlement under a colonial government supportive of Office expansion; these populations were relocated outside the expanding Office territory. This expansion impacted Office residents whose towns were also moved without the consent of farmers. As late as 1955, labor and political activists were requesting information from the administration about forced relocations of Office towns.[69] Despite the best efforts of many Office residents to transform the space of the Office, they still faced the constraints and uncertainties of living at directly under colonial rule.

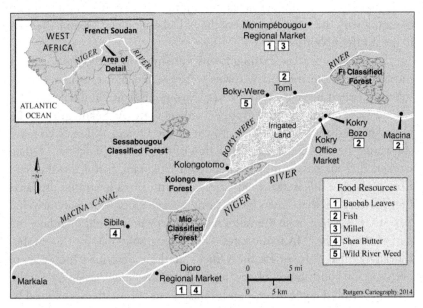

MAP 3.1. Locations of regional markets, wild food resources, and sources of wood fuel near the Kolongotomo region of the Office.

Even though the Office claimed a large territory, access to land was a problem for women and men. Office allotments per household were not always big enough for large households to earn enough profit and feed their members.[70] Women had little access to land to cultivate sauce ingredients or common land to collect wood for fuel or to harvest fruits and nuts. Residents were even prohibited from entering some nearby lands that had been classified as protected forests.[71] For example, the government classified territory from Mio (near Sansanding) to N'Zirakoro as a protected forest. This move outlawed cutting firewood from the forest (see map 3.1).[72] As the Office expanded, women encountered more irrigated fields and canals than wooded areas.

Office administrators were aware to some extent that settlers wanted access to more trees. Agricultural experts in particular promoted the planting of fruit trees to supplement the diets of settlers. Indeed, some new Office towns benefited from mango and other fruit trees that were planted by the Office.[73] These trees were appreciated, but they did not solve the problem of specific tree loss as it related to women's tasks. They especially needed wood for cooking fuel and access to products from baobab, shea, and nere trees. In fact, when new Office fields were prepared, baobab trees were routinely poisoned to make it easier to clear the land.[74] This created a serious void in the available resources for women's food production.

Few women who moved to new Office towns could easily find all the trees and plants they needed for the manufacture of sauce ingredients or other food. Office towns were sometimes located near a plantation of *one* of the kinds of trees that women needed to make foodstuffs. Other women came to towns without any nearby tree resources. Whenever possible, women planted the trees they needed. In Nara, women purchased *namugu* (made from baobab leaves) for sauces until they started planting their own baobab trees in town (see map 3.2).[75] Their choice of location was restricted by the way the town and fields were already laid out, but they planted where they could. Most women also lost access to rainy season foods like wild rice. As late as 1957, women living in the former experimental town of Niénébalé collected wild rice along the banks of the Niger River. After the Office formally integrated Niénébalé into its structure, the administration subsequently granted several farming concessions to European owners on nearby lands. In 1957 a European wife living on one of the concessions tried to chase women from Niénébalé away from an area along the river. The women were there to collect wild rice and even had to compete for access to the uncultivated land at the edge of the river.[76]

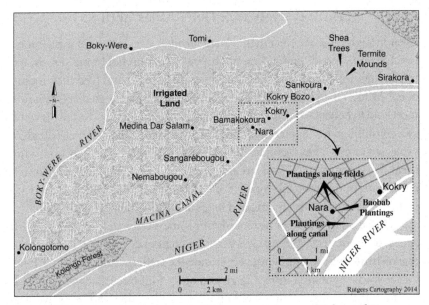

MAP 3.2. Women's gardens, trees, and wild resource claims near Kokry and Nara.

Ever since the initial founding of the Office, women worked to re-make the irrigated farmland and surrounding areas for their own purposes. One of the first things that many farm families did upon arrival was to clear land beyond the Office for additional rain-fed fields where they grew millet, corn, and other food crops. The first settlers arriving quickly claimed land for this purpose. This meant that farmers in towns established in later years had less access to land for additional farming. However, farmers in towns at the edges of the Office were able to claim fields just beyond the reach of the scheme's claims of land ownership. Even though farmers depended on rain for these additional harvests they, like some neighboring towns, benefited from excess water in the irrigation system. Whenever possible they even channeled the water to their neighboring food fields.[77] Early labor restrictions on women and men meant that farmers had little time to devote to their rain-fed fields. Much of this work had to be accomplished at night, after the highly supervised work in their official Office fields was completed. The same was true for women's cultivation in gardens or in the available bush.[78] Mamu Coulibaly remembered only working in the millet fields and that at first women in Sangarébougou had no gardens.[79] This was perhaps one reason that women increasingly purchased ingredients that required extensive labor time to produce.

Despite these restrictions, women did eventually cultivate small gardens.[80] To do so, they had to carve out space wherever it was available. Some women grew food plants in small plots they prepared next to the canals. In Nara, women grew onions and okra in the space available between the rice fields and the canals (see map 3.2).[81] Even though the land was not always optimal,[82] it enabled women who were expected to be working in the fields to cultivate some sauce ingredients. Many of the women cultivating garden crops in the early years were older women (perhaps in their forties or fifties) with daughters-in-law. They were less frequently monitored by guards and not expected to work as hard in the Office fields. However, they had to work if the household was going to eat.[83] In Mossi towns, women also planted cotton, okra, corn, sweet potatoes, and *ngoyo* (small bitter eggplant) in small open areas next to their houses. This had been common practice in their home region of Ouahigouya.[84] Replicating this practice at the Office made good use of what little land was available. In San-Kura women planted cotton in small nutrient-rich dirt mounds called *tungo* built by termites and other insects. They lived in the rice sector but still needed cotton to make clothes or earn cash. Tungo dotted the land in and around the rice fields, and women in the same region also grew calabashes, *datu*, and okra in the mounds (see map 3.2).[85]

At the end of the Second World War, the Mission Reste (1945) investigations resulted in a series of reports that recommended, among other changes, that the institution henceforth officially provide women and elderly residents with garden plots. This particular recommendation came on the heels of widespread food shortages at the Office, and the investigating officials could not ignore the devastating impacts of these shortages.[86] Over the following decade women's gardening activities increased significantly, and the shift was still fresh in women's memories in 2010. For example, Fatoumata Coulibaly came to Sabula in the late 1950s from another Office town called Tongoloba, where her parents had been brought by force. Growing up in Tongoloba, few women had plots to grow sauce ingredients. By the time she arrived in Sabula, women were planting gardens with onions, peppers, garlic, and tomatoes. To water the gardens Fatoumata remembered that women at first filled calabashes with water and splashed it over their plants. After she arrived in Sabula, women also began to build small channels in the plots to facilitate small-scale irrigation.[87] One possible reason for the latter change might be that gardening had become established to the extent that women could mark ownership over their plots. This was all the more necessary following a 1955 order for bulldozers to flatten the termite mounds in rice fields where women had previously gardened.[88]

During this period, men gradually allotted women garden plots in their Office fields, implicitly acknowledging women's gardening activities as integral to a functioning household. When Hawa Diarra arrived in Nara in the late 1950s, her husband gave her a parcel of land to grow peppers, garlic, and tobacco. She grew the tobacco for sale in the Macina, Kokry, and Kuna markets.[89] In Kokry, the wives of workers also maintained small plots behind the row of workers' houses. These plots were granted by local Office officials.[90] By 1960, it was common for men with Office land to also provide their wives with a small garden. Even a woman whose husband did not grant her access to land could rent space in another man's fields. In Djewari Samaké's estimation, half of women's gardening produce was for home consumption, but the remaining half was sold.[91]

The assignment of garden space to women was by no means inevitable, and some men retained sole control over the garden plots on their Office lands. For example, men in San-Kura grew peppers, onions, and tobacco for sale; they also gave some of these foodstuffs to their wives.[92] As labor constraints in the fields eased after the war, many men increased the amount of time they spent cultivating grains and vegetables for home consumption and sale. A 1954 manual for European instructors even directed staff to offer men small parcels of land for food cultivation.[93] Gardening soon provided income to supplement what men earned from cotton or rice. In Kouia gardening was so extensive that the town reputedly produced almost all the sauce ingredients sold at the large Niono market.[94] By 1959 Office staff members in the Kouroumary sector were even urging men to cultivate gardens as an extra source of income.[95] Another change following the war and the Mission Reste was that more of the rice grown in Office fields was retained for consumption or for sale in local markets.[96]

When the French agricultural expert René Dumont toured the Office in 1950, he noted that gardening families fared especially well. He described a household in Lafiala:

> A large family does a better job farming and is better off than the average family. Therefore, visitors are often taken to see the Bandiené Tangara household counting 13 men, 14 women, in total 44 members [including children]. They cultivate 28 hectares and are still in need of land. The gardens here are fertilized and in the rainy season are planted in corn, millet, dah,

and cotton. Thanks to early watering in the dry season peppers, calabashes, tomatoes, potatoes, and assorted sauce ingredients are grown. At the edge of swampy or stagnant waters it is tobacco, garlic, and onion. . . . With well cleaned earth from the best garden land, they manufacture bricks for their houses.[97]

As observed by Dumont, farmers used whatever land was available to grow as many grain, vegetable, and other crops as possible. Canal water was also used for dry season gardens before it was cut off. In effect, the farmers extended the irrigation season to benefit their dry season activities. It is also worth noting here that the model Office family was large and included as many (or more) adult women as adult men. In many ways, Office men and women drew on a pre-Office agricultural world that they remembered as producing food security and diversity in the diet, as well as economic and social wealth.

During the period observed by Dumont, many men like Tangara did see their incomes from cash crops briefly rise, especially in the cotton sector. However, profits quickly fell as prices declined. In response, many male farmers in the Niono area increased their gardening activities. In fact, 28 percent of cultivated land in the Niono sector was taken up by garden plots during the 1958 to 1959 agricultural season.[98] Large-scale cash-crop farming was simply not enough to pay taxes, maintain social ties, and support a family's every need. In those years, some men began producing some of the crops that women needed for cooking, but women's efforts to provide sauce ingredients continued to be essential.

BUYING THE SAUCE

Women's food production was greatly altered by the labor and geographic constraints of the Office, and marketing was beginning to occupy more of women's overall labor time. In some areas women still grew much of what they needed to cook and maintain a household; those same women might also have surplus to sell. When women did not have access to fields, or the specific trees and plants necessary for making a meal, they purchased the raw or processed ingredients. As a result, they could devote more energy to cash-earning work. Women who did well with their market purchases prepared tasty meals, while women who were less successful at preparing their own foodstuffs or earning cash had trouble making quality meals.

Women in towns that predated the Office and were later integrated into its farming structure continued to produce many of the ingredients necessary for cooking and household maintenance. For example, women in Molodo-Bamana were fortunate to have personal fields because the town had rights to land predating the establishment of the Office. When the town leaders agreed to join the Office, they retained rights over common lands and fields beyond the irrigated areas.[99] As a result, women maintained much of their previous labor regime. They grew millet, peanuts, beans, and other crops beyond the limits of the irrigated Office fields. They also tended the surrounding trees and bushes for sauce ingredients.[100] Women's harvests in Molodo enabled them to easily continue cooking for local tastes.

When Molodo joined the Office, women maintained control over many food resources, but they also did more work in the millet fields. This was a recognizable change in their work rhythm. Under the Office scheme, farm labor was now divided between the millet fields and the irrigated Office rice fields where men were largely occupied with the latter. Men were responsible for following the Office labor calendar, paying the institution's fees, and delivering the harvest for sale. The varieties of cotton and rice cultivated for the Office required properly timed fertilizer and pesticide treatments. This work followed a strict schedule that was new to most farmers.[101] Though men still helped prepare the millet fields and bring in the harvest, women were largely responsible for the crop. Women also weeded the Office fields and helped during the harvest. Field labor was intensified for everyone in the household.

The nearby Office center Niono offered an attractive marketing option for women. After Niono's founding in 1937, women began to travel the new road to its growing market. There they purchased fish from Bozo sellers (representing long-standing fishing communities).[102] Many women from other Office areas fished, whereas women from Molodo chose to purchase fish. They were also busy farming, cultivating, and processing wild foods. The trip to Niono on foot added labor time, but it was quicker than traveling to the older regional markets.[103] Fresh and dried varieties of fish were now readily available in a nearby market and added to the flavor and nutrition of the sauces prepared to go with the main dish. Molodo women also bought salt, which had long been available from Saharan traders but was then also offered in Niono. To afford these market purchases, women sold small quantities of rice from their husbands' fields. Women with their own fields also sold millet and peanuts from their own harvests. Many Molodo women had a cash income, but unlike many other women at the Office, they were less reliant on it for food production.

In other towns where women had less access to common lands or personal fields, they purchased more of what they needed to cook. Women who did garden had only a small plot. Many new towns also lacked the cultivated plantations of trees like shea or nere that existed in long-established towns like Molodo. Women from other towns near Niono purchased basic necessities like millet, cooking oil, shea butter, soumbala, *tegedege* (peanut sauce paste), okra powder, and soap. Such extensive purchasing was unnecessary for most women in Molodo but common in towns created by the Office.

For women who did need to buy more cooking ingredients, the Office marketplace offered a diversity of foods for familiar dishes and new sauces. Women in the cotton sector especially remarked upon the appearance of tomatoes and onions in the Niono market. Women similarly found a variety of goods at the Office market in Kokry. Sellers from non-Office towns near Kokry, like Bolodi, for example, offered peppers, onions, tomatoes, garlic, and potatoes, some varieties of which had been introduced by colonial botanists looking to expand garden production.[104] In the late 1950s women could purchase a large sack of sweet potatoes for 400 to 500 FR, which was a good price for women.[105] Across the Office lands, women also cultivated guava and mango fruits from trees planted in new towns by Office staff.[106] Women's lack of access to fields and the absence of many products from cleared forests constrained food production and preparation, but nearby markets offered a somewhat more extensive selection of foods for women who were successful at earning cash. Seasonal variety from the fields and commons had long been important in local diets, and the new types of foods helped women to maintain diversity in what they prepared. For example, Bintu Dembélé, who grew up in the Office town B1 (also called Niobougou) eating mostly the sauce made from datu, started purchasing ingredients for peanut, onion, and okra sauces after she married in Sokorani.[107] The changes to her sauce ingredients were noticeable (but not necessarily undesirable) in the meals she prepared for her husband and their household.

Women usually met their cooking and household needs through a combination of market purchases, home manufacture, and farming. For example, many households cultivated datu. Families growing rice for the Office also increasingly ate rice, while millet was now a cash purchase. Most women and men also fished in the canals, meaning that the sauce frequently included ingredients made from fresh fish. Just south of Niono, shea trees were scarce, but baobab trees were plentiful. Here, women bought shea oil and butter, but dried and pounded baobab leaves for use in sauces. Elsewhere, women purchased this baobab powder, called namugu, but made their own cooking butters and oils. Women in San-Kura, not far from Kokry, used the nuts from shea trees near their town to make butter. They also used shea butter to make soap but had to purchase *sege* to do so. In the same town women purchased nere grains but made their own soumbala from the grains.[108] Women from both the irrigated cotton and rice areas produced cotton thread, but those in the rice zone purchased cloth when they went to the market for peanuts, soap, and other goods.[109] Similar to the practice of women from Molodo-Bamana, most women earned cash for purchasing foodstuffs by selling small piles of rice or cotton from their family's Office fields.[110]

Eventually a specialized regional trade developed between Office towns and non-Office centers. For example, the non-Office town Sibila (near Sansanding) traded in shea butter. The Somono fishing town Tomi produced smoked fish. Sellers from near San supplied soumbala. Farmers in millet-growing areas like Monimpébougou, Dioro, and Bako often sold millet, as well as sege, tegedege (in the rainy season), okra, shea butter, and soumbala in Office towns or markets.[111] Mossi women also purchased *boombo*, a tree product from their home region in Upper Volta. Boombo was necessary for cooking a specific sauce and grew south of Ségou in Konobougou. From there, sellers brought it to Office markets catering to Mossi towns.[112] Women from Office towns also brought goods to exchange in the regional trade. Women near Niono sold cotton and cotton thread to traders and weavers.[113] In Kokry, women sold calabashes and rice.[114] Assane Coulibaly, from Sirakoro in the rice sector, exchanged rice directly for namugu with sellers from Monimpébougou, Boky-Were, and Kuna.[115] Women were more successful in these kinds of trades if they were familiar with the regional market geography and the Office roadways. It also helped if women were able to manage their time to allow for earning cash and marketing along with their farming and cooking work.

Many sellers from the non-Office regions traveled directly to Office towns to sell their goods. Aramata Diarra grew up in Siribala just south of Niono and as a young girl traveled as far as the market in Sansanding. It was at least a day-long trip from her hometown to Sansanding. After she married in Fouabougou, she no longer needed to travel to purchase goods. She bought everything she needed from traveling sellers.[116] Bozo fishermen and women from Kokry-Bozo supplied fresh and smoked fish to the Kokry Office market. They also traveled to nearby Office towns selling basic items like salt, shea oil, soumbala, and namugu.[117] Cattle-herding men and women also frequented many Office towns selling milk and meat. Mamu Coulibaly remembers paying only 10 FR in Kankan for a packet of meat from traveling herders (which may have been a sign of hardship on the part of the herders).[118] Most of the goods sold in Office towns were affordable. Purchasing from traveling sellers meant women saved time, especially in the first decade when travel was more difficult and free time was rare. Moreover, Office staff sometimes confiscated farmers' guns, meaning men had fewer opportunities to supplement the household meat supply through hunting.[119] It was often easier for women to purchase basic ingredients like meat, fish, salt, or tree products from sellers who came directly to their town. However, women exerted less control over prices in these situations.

Other women chose to buy and sell in regional markets outside the Office. Women persisted in traveling to these markets despite early constraints on their movements. In the first decade of the project, guards working for the colonial administration often stopped women precisely because they were selling rice and cotton in other markets.[120] Travel restrictions were lifted after 1945, and it eventually became common practice for women to sell cotton and rice in markets across the region for cash used to purchase necessary foodstuffs.[121] Women from Nara near Kokry often traveled to the large regional market in Dioro where they purchased namugu, soumbala, unprocessed nere seeds, salt, and shea butter. The Dioro market offered a wide variety of ingredients women needed for cooking. Women could also sell rice to local merchants and consumers for a better price than what the Office offered their husbands.[122] Rokia Diarra, who grew up in Nara, also traveled as a young woman to Macina to buy large dried fish.[123] Both the Dioro and Macina markets were at least a day's travel from Nara, whereas the Kokry market was less than an hour's distance by foot (see map 3.1). Women weighed the time for travel against greater selection and often chose the market with more to offer.

Simply put, the Office markets only addressed some of women's household needs. Certainly, better prices for cotton and rice in outside markets were one good reason to travel to a distant market. In fact, the regional market at Siengo gained local prominence due in part to the sale of cotton and rice by Office farmers (see map 2.1).[124] However, some markets' operations also fit better into women's labor schedule. For example, all Office markets were open for business on Sunday. It was the day all European and African staff were off work.[125] Women who cooked throughout the week often needed to go to market on other days too. In many cases, women also had family relations near a preferred regional market. Thus regional markets were sometimes more attractive economically and socially.

BUDGETING CASH EARNINGS AND LABOR TIME

Producing a meal at the Office required the management of cash resources as well as knowledge of the markets, the changing bush, and agriculture. A woman's food budget included contributions in kind from the male household head who was in charge of the Office cotton or rice fields. Women still labored in those fields but no longer received land for their own fields in exchange. Each woman who cooked received small quantities of the Office harvest to sell. Along with the cash gained from these sales, women also added income earned from their own enterprises.

When René Dumont toured the Office in 1950 he took particular note of women at the scheme. In his report to the Office administration, Dumont highlighted women's responsibility for the household budget: "Women who are essential to the prosperity of small farms not only because of their labor but also because they control the farm's budget, participate only minimally in agricultural activity in Africa."[126] While Dumont certainly overlooked women's farm labor, he was right about the importance of women's budgeting. Indeed, the ability to earn cash and plan for essential food purchases became increasingly important for a family's well-being. The more cash available to a woman to purchase ingredients, the better her sauce would be. Even if she could make some of what she needed to cook, she still had to buy many basic ingredients. Women with more cash, either from the sale of Office crops or from other work, inevitably made better food. Most everyone in Bamana villages, for example, were eating toh, but the better-off farm households had more appetizing sauces.[127] In other cases, women in some towns were better situated to benefit from tree and other wild resources than others or had better access to markets. These economic distinctions among women transcended the larger divide between cotton growers and rice growers.

To make the best use of divergent resources, women's work routines had to allow time for going to the market, fieldwork, production of ingredients, and of course cooking. Women's work calendars varied across the Office. Around Niono, the calendar for cotton cultivation influenced much of women's time, especially during the harvest. In areas where Office farmers grew rice, women devoted less time to field labor, and time spent in the field was more directly related to food production. Additionally, women incorporated new considerations into their calendars such as knowing the days for certain markets or the schedules of traveling merchants. Work varied according to how much purchasing women needed to do each week and which tree and other raw resources were immediately available. For example, young women like Fatoumata Coulibaly, who grew up manufacturing soap in the Office town of Tongoloba, started purchasing soap when she married into a household in Sabula. The same ingredients were no longer available to her, so she altered her manufacturing and purchasing schedule.[128] Fatoumata also grew up with a sense of the natural geography around Tongoloba. She knew where to find certain wild nuts and other plants. After she married, her sense of the Office resource geography expanded to include where to buy some of those same products.

"WE FARMED MONEY"

Mamu Coulibaly (whose story opened this chapter) and other women in Office towns found many ways to earn cash. Mamu was a beer brewer and seller: women in the region had long brewed beer for income. When women arrived at the Office many continued this work, selling to men who had ready access to cash from the Office. Farmers earned cash from their crop sales, and workers like Tchaka Diallo earned significant wages. Young men who earned wages at the Office were ready to spend some of it on leisure activities like beer drinking. However, many of the women's clients were farmers from their Office hometown. For example, in San-Kura near Kokry, women sold beer every Sunday. The sellers opened small beer houses where they could reportedly make the sizeable sum of 10,000 FR in one day. They sold mostly to men, but other women and brewers also drank on Sunday, which was a lively day at the Office.[129] Men were free from field labor, and a convivial atmosphere was steeped in the familiar taste and smell of millet beer.

Successful brewers often had enough cash from their earnings to pay for their household needs and for purchasing new clothes.[130] Prior to the Second World War most married women at the Office wore only a long skirt. Older women who were interviewed for this project attributed this "nakedness" (as perceived today) to a lack of money.[131] When women began to earn more cash, they often purchased cloth to make long skirts and *duloki* (blouses) in addition to making clothes for their children. By the 1950s and 1960s, most women at the Office wore duloki as a testament to women's earnings and savings. Now older women laugh at the memory of the days when, in general, not as many clothes were worn.[132]

Later in this period some men stopped drinking beer for religious reasons,[133] but brewing remained a major cash-earning activity into the first decade of independence (1960s). In fact, administrative reports from the 1950s suggest beer drinking was the predominant leisure activity in many towns.[134] During local harvest festivals women also began to prepare a lemon and ginger drink called *lemburuji* for men who prayed. For these men, women also made teas from citronella and other herbs such as *shukolan*. Clearly women worked hard to reproduce the rituals of abundance even at the Office where much about daily life had changed dramatically. By the 1960s local harvest parties featuring beer drinking waned, in some measure due to reduced millet beer consumption.[135] None of the non-fermented drinks offered the same kind of cash-earning opportunity for women as beer brewing, but their preparation demonstrated the variety of beverage-making skills that women incorporated into their culinary repertoires. Diversity was always important to local diets and something women continued to provide during festivals and other special occasions.

Women engaged in several other activities to earn money. As previously mentioned, some women had their own fields because of their location in older towns. They sold crops from these fields as well as small amounts of produce from the Office fields. In fact, an administrative review of household incomes at Kokry and Kolongotomo for 1948–49 listed the sale of paddy by women as an important element in family budgeting. By this period, many women in the Kokry and Kolongotomo sectors also gardened along the edges of Office fields or in separate plots.[136] Women with training in regional crafts practiced trades like pottery making, hair dressing, or praise singing for special events. These long-standing women's trades were particularly lucrative.[137] Women without these special skills had fewer options for earning money and were more dependent on their agricultural production, access to a garden plot, and the ability to travel to markets.

Most women earned money through agricultural and household tasks performed outside of their own households. During harvest time, men in need of extra field laborers would pay women to help bring in the cotton or rice harvest. Every three days or so, during the harvest, one or two men called for groups of women to assist them in the fields. Around Kokry, field owners gave women one calabash of rice for their work.[138] During the same period, young boys were called to help cut crops, and young girls could also earn rice or cash by bringing water to the boys working in the fields.[139] In the cotton sector, women's major cash-earning activity was spinning cotton thread for sale in Niono and other markets.[140] In Kouyan-N'Péguèna and Fouabougou, first-generation women also cut wood for sale and used what they earned to purchase ingredients in the nearby Niono market. As the trees available for fuel were located farther from town, women pursued other means of earning money.[141] Some women hand-threshed rice or pounded millet for households with few women. They also washed clothes for unmarried workers who lived nearby.[142]

Young unmarried women were interested in earning cash for other purposes. In the mid-1950s, a young Adam Bah spent her free hours preparing for a future marriage. Her family farmed cotton and rice on Office land near Niono. When she was not working in the fields she joined small groups of girls who collected rice from the pig pens of the local French Office staff. Adam remembers that the guard in charge let them go into the pens to pick up any uneaten rice left by the pigs. In one week the girls collected enough rice to fill a sack that could be sold for 1,500 FR in the market. Once the girls filled a sack, they loaded it on a neighbor's cart headed for the market. When their seller returned, the girls split the profits. Adam Bah had earned enough by the date of her marriage to buy her own serving bowls. All the other girls also purchased goods for their weddings from their rice profits. Like other women in the French Soudan, Adam's mother prepared and purchased most of the kitchen and other household goods needed for her daughter's married life. However, Adam, like other young second-generation women at the Office, used her own cash to supplement what her mother could provide. Such cash earnings increased the quantity and quality of goods a young woman brought with her to a new home.[143] Indeed, Adam's mother also gave her daughter twenty *tafew* (measure of cloth for a long skirt).[144] She even had gold left over from her own saved inheritance to give to her daughter.[145]

The cash that Adam and the other girls earned was possible because of a unique set of circumstances at the Office. Nearby markets were readily buying rice, and the girls had access to this market. They grew up learning how to make money using anything that was available. In many ways the second-generation women at the Office grew up with an awareness of the particular resources of the project such as the possibility of rice collection in the pig pens of French staff. Pigs were not widely kept by farmers. Yet, girls and young women knew leftover rice could be gathered from their pens. As the young girls worked and prepared for marriage they were also preparing for life as wives at the Office. Knowing how to earn cash had become a prerequisite for provisioning the household with basic necessities, but this was not necessarily a sign of increased wealth.

Even men had greater needs for cash. Men in charge of Office households needed money to finance marriages for their sons. Office staff remarked in the 1950s that farmers sometimes sold the cattle and the plow they had received on credit from the Office for this purpose.[146] Cash income was a necessity, but having cash on hand did not guarantee more than subsistence. The head of the household could easily become indebted to the Office. Indeed, many women were not always able to buy all the foodstuffs that they needed, let alone cloth or other value goods.

While many households struggled with debt or with simply getting enough to eat, some women earned enough cash to not only feed their families but to invest their earnings.[147] René Dumont observed that a few women in Kokry even purchased gold and other jewelry. The women able to save such stores of wealth were most often wives of workers with easy access to garden plots in town and reserved employment winnowing rice for the Office.[148] Based on observations from a tour of the Office, Dumont believed that areas with high rates of beer consumption were also the wealthiest.[149] Certainly his observation would attest not only to the presence of men with enough cash to purchase beer but also of women who made a profit from brewing. In those same towns the beer drinking must also have reinforced older associations between beer, abundance, women's labor, and agricultural wealth. In Molodo-Centre, wives of workers who earned money in town were less able to invest in garden activities because of the scarcity of plots. They often saved their earnings to purchase cattle. Women in Kolony and Koue-Bamana also aspired to cattle ownership. If women had just a little cash set aside, they tended to purchase cloth, which could be gifted, resold, or used to make items of clothing.[150] Mothers of young women also saved their extra cash to purchase household goods for a soon-to-be-married daughter.

While all women may have "farmed money" (as articulated by Mamu Coulibaly) at the Office du Niger, the goal of farming money was not always accumulation. Of course, a few women were able to save enough money to purchase cloth or even cattle. However, the phrase "farming money" captured the fact that women's agricultural labor was bound up in cash earning and the market to a greater degree than in the earlier part of the century. It had also become an essential element of daily life at the Office. Women needed money to provide food for their families, especially if they wanted to cook tasty food like millet toh or prepare a sauce with namugu or shea butter. The value of agricultural production was still connected as much to taste and consumption as it was to the amount of money that women earned.

As women integrated aspects of the Office infrastructure into their labor routines, they also began to switch from cooking millet to cooking rice. This transformation in the diet of Office settlers first occurred in the predominantly rice-producing sectors Kolongotomo and Kokry. The cotton and rice grown intensively at the Office created new labor conditions and constraints on food production, but the burdensome labor time spent on the cotton harvest determined to some extent the amount of time left for food production. In the rice sectors, farmers grew a crop that could be eaten. Still, the white rice variety promoted by Office staff members was different from many rice types already cultivated in the region. It may have been less flavorful than other kinds of rice, but women put a great deal of effort into making it a more familiar food.

For the most part, the switch to cooking Office rice was most noticeable for women accustomed to processing and cooking millet. In some towns women exchanged rice for millet because of this preference. For example, women in Nara exchanged rice from their Office fields for millet from a town near Monimpébougou called Fy.[151] Women in Togolo-Kura similarly acquired millet locally for cooking and beer preparation. Kono Dieunta, who grew up in a fishing family, was familiar with rice dishes, but she remarked that women in the predominantly Bamana town preferred to cook millet because they believed it was more filling than rice.[152] This preference for the way food made a person feel had long been an important consideration for the region's cooks. Slowly, the eating habits of people accustomed to eating millet changed as they began to appreciate the taste of Office rice.[153]

Gradually, rice consumption increased across the Office villages.[154] The shift occurred not just because farmers were growing rice but because women who previously cooked millet learned to make use of the rice they cultivated for the Office. Women who had previous experience preparing local varieties of rice quickly adapted their knowledge to the new Office variety. For example, Assane Coulibaly had learned rice preparation techniques from a Fulbe woman in Ségou prior to moving to the Office. After she moved to Sirakora, she also learned a technique for making rice toh, which was a Bamana dish ordinarily made from millet.[155] In this way, cooks approximated the texture of millet toh that had been so pleasing to local tastes. By the late 1950s, when Hawa Diarra arrived in Nara from a millet-producing town, she found that people in her new town only ate rice. She had to learn how to cook it. To this end a senior woman in the household was assigned to teach her to prepare the new staple grain.[156]

In time, women began to use rice to make most of their staple millet dishes. Rice was pounded like millet and cooked into the stiff porridge texture that characterized toh. It was also transformed into a warm, sweet porridge called seri and the colder porridge called dege. For these dishes women handled rice as they would have handled millet. They pounded the grain to transform its texture and then cooked it.[157] Rice was also prepared simply, as the cooked grain with a sauce poured over it. In Kokry farmers lived at the edge of the Macina region where many Fulbe and Bozo women were already making specialty rice dishes. Bamana farmers in the Macina region were also accustomed to eating rice cultivated in the floodplains.[158] In these cases, the grain variety changed but women prepared familiar sauces to go with the new rice.[159] For example, Kono Dieunta continued to make Bozo sauces, especially *chouchou*—a red sauce made with copious amounts of fish and topped with heated oil just before serving.[160] In Sirakora older women made a specialty smoked rice, or *malokene*, that they sold in local markets like Dioro.[161] These techniques for making satisfying meals with the new and widely available staple were shared with new women when they married into Office households.[162] Second-generation women especially grew up learning to prepare rice in a variety of ways.[163] In this way, women adapted the cash-crop rice to their cooking, generation by generation.

The switch to rice came about a decade later in the cotton sector where farmers continued to grow millet in fields outside the Office. In the project field designated as the *baloforo* (roughly translated as the "food field") farmers also cultivated mostly millet and maize. Many residents recall that people living in the cotton sector did not even have wild rice to eat before 1945.[164] However, a few farmers grew rice in the baloforo.[165] Because cotton was so labor intensive, some towns that were integrated into the cotton sector actually refused to any cultivate cotton and negotiated with the Office to grow rice.[166] After the major protests from young men in 1944 the Office changed its policy and placed less emphasis on cotton production. Beginning with the 1945–46 agricultural campaign, rice was integrated into the cotton sector's crop rotation schedule. Farmers continued to sell the cotton they grew but kept the rice for home consumption. This move alleviated to some extent the ongoing food crisis in the sector.[167] The improved food situation was also due to increased gardening.[168] For example, when farmers began planting rice, women started planting okra by the side of the rice fields.[169]

After 1945 settler households increasingly planted rice in the Niono sector, but the switch to eating rice was gradual. Initially, some of the Office fields were set aside for rice, but crop rotations in cotton fields still allowed for some millet cultivation in off years. A decade later some farmers were still growing millet in their Office fields. Then in 1954 a new policy called for the substitution of all millet and sorghum rotations with rice in Office fields. The policy shift did not necessarily reflect a taste preference for rice among farmers. Official documents indicate that the move was meant to eliminate any remaining labor overlap between the millet and cotton harvests.[170] However, as the historian Monica van Beusekom demonstrated in her study of the Office, this shift was due in large part to ongoing dissatisfaction with cotton as a cash crop and farmers' efforts to improve their economic prospects.[171]

Even though the substitution of rice for millet as the primary grain in the diet was a gradual process, women remember it as a major event. For one, the switch radically changed and eased women's labor burdens in the fields because rice cultivation was less labor intensive than cotton (which required intensive labor in addition to women's work in the millet fields).[172] Hawa Coulibaly explained that with cotton cultivation women could only spin cotton at night to make money for foodstuffs, soap, and clothes. When her husband began cultivating rice in Fouabougou, she had time available during the day to make money in other ways.[173] Women who arrived in Fouabougou after the switch to rice also had a much lighter labor load than had experienced by their predecessors. The switch became a major generational marker among women.[174]

Rice cultivation also helped women to feed their families. Aramata Diarra remembered that it was much easier to get foodstuffs for the family once they started growing rice.[175] Rice was not only a food but a resource that provided women with the means to acquire grains they preferred or good sauce ingredients. By the end of the 1950s women who arrived in the cotton sector found that in towns like Kouyan-N'Pequena people now ate mostly rice.[176] In this same postwar period, other farmers in the region came to associate Office food production with rice. Some early French scientists had (incorrectly) believed that rice was a better food than millet. This was a notion picked up and retained by Office and other colonial officials. For them, a shift from millet to rice consumption was a step forward for the overall economic and social development of the region. Other colonial scientists like August Chevalier had earlier tried to point out the high nutritional value of millet and the dietary risks of eating mostly rice (as had been the case for many early Office settlers).[177]

When a noticeable number of farmers at the Office actually started to eat more rice than millet, Office administrators claimed that it was due to their policies. In 1955 the Office Director Georges Peter gave a speech promoting the Office, emphasizing the role of the scheme in providing rice to other colonies. He triumphantly proclaimed, "Finally, Africans are turning away from millet and now prefer rice."[178] That same year the controller of the Niono sector reported to the administration in Ségou that the Office had helped prevent famine in the colony in 1954.[179] Ironically, these same officials emphasized the production of broken rice favored by Senegalese consumers over the local preference for unbroken rice.[180] By 1960 Office officials were pitching the scheme as a development program specifically aimed at providing food security. The renewed institutional mission was laid out as follows: "Create in the heart of Africa a real breadbasket. It will be free from dependence on climate and ready to alleviate local food shortages still frequent to this day. It will also provide food for the urban centers."[181]

Eventually, the Office gained a reputation as a center of rice production. Sékou Coulibaly and other farmers who grew up at the Office also came to see rice cultivation as a means of assuring *"securité alimentaire"* (food security) for the immediate region. The political phrase securité alimentaire had become a part of the culture of interaction between the farmers and Office instructors.[182] Sékou explained securité alimentaire in this way: "Farmers from the rain-fed regions grow the millet. Because of *ja* (drought) in the region, it was also important to have rice. The rain may stop before the millet ripens but with the water from the canals Office farmers can still give water to the rice."[183] In his framing, rice did not replace millet but was more akin to foods for the lean season or even a famine food eaten in times of extreme shortage, especially due to drought.

In a way, Office rice became like the grains that women previously found in the bush. When surrounding towns suffered from poor harvests, Office farmers often sold rice to their neighbors.[184] This they did clandestinely, even though Official rhetoric had long associated the Office with ensuring the food supply. While the shift to rice was due in some measure to the constraints of Office cash-crop production, project farmers actually turned the Office into a regional food source. A significant outcome was the change in consumption habits (though not necessarily taste preferences) among project residents, the only rural population in the colony to predominantly eat rice outside of the urban areas.

REPLACING THE BUSH AND RENEWING THE HARVEST FESTIVAL

Food consumption at the Office changed in other unforeseen ways. For many years foodstuffs distributed by the Office substituted for some of the products women previously produced. First-year settlers in Nyamina and other towns received somewhat regular rations of millet, oil, canned meat, and canned fish. The canned items required extra labor to prepare the meat and fish for cooking, and women had to adapt recipes that called for fresh meat and fish, or dried fish.[185] Assane Coulibaly remembered that rations in Sirakoro continued well after her first year at the Office. She received what she thought was good-quality peanut and cotton oil, locally caught fish, other fish varieties, and salt from the local African monitor. For her, the rations were indispensable to preparing meals and stocking up on necessary items.[186] Her memory of these items offers a striking contrast to the first years of colonization when the rations were of poor quality and insufficient in quantity. As new towns popped up around the Office from the late 1940s into the 1950s, first-year settlers in those towns probably received rations similar to what Assane remembered receiving. It is also possible that her household qualified for the rations because of poor harvests or an exceptional relationship between her husband and Office staff members.

For Assane, the institution of the Office itself became a food resource akin to the bush. For example, the Office made much of the oil it distributed. An oil factory in Niono processed oil from cotton seeds and manufactured cooking oil and soap for distribution and sale. During the 1950–51 agricultural year, the Office sold a liter of oil for 80 FR, which was lower than the advertised Ségou market price of 120 FR per liter. The Office also processed peanut oil from peanuts grown in Riziam (S-8).[187] Undoubtedly the oils had a different taste than the oil women produced from shea nuts or even from their own peanuts. Assane certainly found the oils to be good for cooking and thought she was able to make delicious meals with her oil rations. Ready-made soap and oil also saved her and other women time otherwise spent manufacturing these goods or traveling to market for the same ingredients. However, these distributions were not consistent in every town and not reliable resources for all women. Many complaints from male farmers during this period, in fact, were about favoritism shown toward some farmers because of their political affiliation or social relationship with Office monitors or workers.

Wives of wage workers received greater rations than farmers' wives. Kadja Coulibaly, who was married to a thresher operator in the late 1950s, received a meat ration of mostly fresh mutton every three days. At the end of the month she also received milk, salt, soumbala, oil, powdered milk, and sugar.[188] Kono Dieunta, who was married to a worker in Kokry, remembered receiving oil and European soap, although she also made her own soap from shea butter and sege. Sometimes the monthly ration also included millet, which was then a luxury for people living in the rice sector.[189] The rations in Kokry were distributed from a large warehouse that housed rice processing equipment and bags of processed rice. When it came time for the distribution, the worker's wife in charge of distribution rang a bell. All the wives expecting to receive a ration walked to the warehouse. Then the woman in charge called out each worker's name, at which point his wife stepped forward to receive the ration.[190] It was a regimented system that workers' wives incorporated into their food planning.

Men, women, and children also supplemented their personal store of basic goods from distributions at yearly parties. Office residents attended New Year's Day and Bastille Day events hosted by the Office and the colonial administration where they competed in friendly games for gifts of cash, soap, cloth, and foodstuffs.[191] All the items were basic necessities. In addition, the Office organized yearly harvest parties that offered food to Office families and sometimes to people from towns outside the Office. For parties in Niono, people came from as far as Monimpébougou (non-Office) and Diabali to the north (incorporated in the early 1960s). The Office even offered transportation. Aïssata Mallé remembers that when a group arrived "someone immediately would lead the whole party from our town to a place to eat saying: 'Kouyan-N'Pequena! Here is your food!'"[192] People went in part because of the food but also because they associated the parties with harvest celebrations (see figure 3.2). The parties recognized the hard labor of settlers in the fields and were full of music and conviviality. While celebrating, settlers recalled another world that celebrated the physical labor of farming and rural generosity.

Certainly, such celebrations were also the occasion for public ritual demonstrating the Office providing for its farmers. Indeed in 1939, the Office announced the end of harvest party at Baguinèda in *Les Cahiers Coloniaux*; the publicity served to garner sympathy for the Office at a time when it was sharply criticized in France.[193] For rural residents, the parties were also part of the constellation of ways to obtain and enjoy food. The festivities were perhaps necessary rituals because many women struggled during the rest of the year to produce or purchase soap, let alone cook regular meals amidst still problematic food shortages. Not all women were able to seamlessly integrate their marketing, gardening, farm work, and cooking. At the same time, the parties were evidence of a reshaped Office environment and infrastructure, as well as the cultivation of new tastes and the creation of new rural social ties. Office residents farmed and cultivated multiple harvests worth celebrating. Indeed, their renewed celebrations recalled the older ideals of generosity and a rural society that strove to produce food security no matter the material nature of the harvest.

FIGURE 3.2. Celebration with drummers in an Office town. Courtesy of the Archives Nationales de la France d'Outre Mer. FR ANOM 8Fi 417/16c-19 Office du Niger Aménagement rurale, 1935 to 1954.

4 ⤳ Reengineering the Office
Cooking with Metal Pots and Threshing Machines

IN THE late 1940s people living in agricultural towns run by the Office began to hear the new sound of young women preparing meals in metal cooking pots. As the large wooden spoon, the *toh pasa*, passed through the dough, it struck against the metal pot and generated a noise loud enough to be heard from outside the courtyard. The sound signaled that the meal would soon be ready. The noise also sounded the call of a subtle revolution not only in labor-saving household technology and the aesthetics of cooking but also in the daily experience of eating. This transformation unfolded slowly, the *whip* and *thump* heard in one compound, and then another, each owner developing her own stirring style. Within a few decades, cooking looked and sounded different in the new pots. *Negeda*, as the new pots were called, cooked food faster and required less wood on the fire than the clay pots women had previously used to prepare meals. Shiny new metal household goods from the market were also durable and needed to be replaced less often than clay pots or calabashes. From the late 1940s into the mid-1960s, young brides bringing metal pots and buckets into households slowly changed women's cooking routines and the way people ate.

This modest technological revolution had a significant impact on the material experience of daily life but was less dramatic than the technology directly associated with the agricultural project at which the women lived. In the late 1940s the Office established an all-mechanized sector employing only wage workers and several semimechanized towns where new settlers employed industrial farm machines such as tractors and threshers. More than ever, Office planners put their faith in large-scale technologies to improve agricultural output in the colony.[1] Yet the canals sometimes failed to deliver water, the new tractors did not guarantee high yields, and the threshing machines often arrived in the fields several months after the harvest. To manage these unpredictable elements of Office infrastructure and their impact on daily life, women adopted new household goods such as metal pots, which were well suited to the unique resources and constraints of the Office. Women had always managed their labor with modest technologies. Their work routines at the increasingly industrial Office were no exception. In the postwar years, the Office was indeed a technological landscape. But for women it was a landscape where large- and small-scale technologies intersected. As the new sights and sounds of cooking became part of the texture of daily life, metal pots and buckets slowly became quotidian elements in the making of a new agricultural landscape where rural and industrial overlapped.[2] The new pots and buckets were also daily reminders of the significance of women's food production labors. The Office was a new material world but one that women still managed and would reengineer with their expanding repertoire of embodied labor and knowledge.

DEFORESTATION AND WOOD FUEL SHORTAGES

The arrival of metal pots in particular signaled a range of changes taking place in the rural French Soudan at midcentury. These changes were connected to what people ate, their daily welfare, the agricultural economy, and the environment. Before the founding of the Office, the Niger Delta region was sparsely populated and had far fewer fields compared to the postwar years. At the time, the region had been rich in wood for fuel, not to mention trees and bushes that provided nutritious fruits, leaves, and nuts.[3] In the thirty years following the creation of the Office, thousands of hectares were cleared for new towns and fields (see figure 4.1). More wood was also cleared to fuel the first tractors, river steamers, and railroad engines: the latter two transporting building materials and crops. This intensified activity resulted in a rapid clearing of the forested areas, which had an immediate impact on people living in the region.

Women were quick to understand the problem of deforestation because it made the collection of wood for fuel and wild food resources more challenging. At first, some women collected wood for cooking fuel where

FIGURE 4.1. Large-scale clearing of fields for irrigated agriculture. Photo by George Rodger. Courtesy of the National Archives. NARA RG 286 Photographs of Marshall Plan Activities in Europe and Africa, ca. 1948 to 1955. Photo no. 541632.

tractors and bulldozers had uprooted trees.[4] Over time good fuel wood became more and more geographically inaccessible from Office towns. Women spent four to five months in the dry season collecting wood for the year. In towns founded by the Office, especially in the cotton production zone, new settlers often had to cut down trees for millet fields, inadvertently contributing to the loss of other necessary foodstuffs.[5] While the Office planted fruit and other trees in new towns, the trees planted by the French were not the varieties that residents used most frequently for food, fuel, or medicine. The need for very specific trees by rural residents was lost on Office planners. Even the Office botanist Guy Roberty, who expressed great interest in local flora and fauna, was puzzled as to why local farmers in the Ségou region generally left some species of trees uncut in their fields.[6]

The immediate effects of large-scale technology and heavy machinery were felt by women in their increased workload. They were also felt by everyone living in Office towns due to the shortage of appetizing food products. When I interviewed Djenebu Coulibaly, she remembered pounding and processing leaves for different sauces as a child with her mother. When as a young bride she came to Kolony (km26), the first Office town in the cotton sector, she could no longer find the trees that provided those leaves. She recalled that Europeans planted mangoes, now a common market fruit sold during the dry season, but that the French cut down the trees that provided sauce ingredients. Even today, although there are many trees visible in Office towns, she is adamant that there are "no longer any trees." The ones Djenebu remembered from her childhood and in her hometown are no longer growing: the important trees were no longer there.[7]

The loss of foodstuffs from the bush was profound across the region, and the dwindling supply of these resources was a process that many regional residents witnessed. The historical fear of such losses is contained in the local account of how the town Molodo-Bamana joined the Office in the 1940s. It is the oldest existing town in the Molodo sector of the Office, and according to town elders was founded at least two hundred years before the French conquest. The leaders of Molodo-Bamana initially refused official requests and later orders to submit to Office control. However, once new project towns began to surround Molodo, and the surrounding wooded areas began to be deforested, townspeople began to worry. Harouna Bouaré, a town counselor, explained that it was then that their notables brokered an agreement with French administrator Joseph Rocca-Serra to join the Office. Importantly, Rocca-Serra agreed not to cut down specific trees important to the town's food supply. Because of this promise, Molodo-Bamana joined the Office. The protected trees included baobabs, the leaves of which are pounded for various sauces; tamarind trees, which provided fruits; and a small tree with bitter fruits called *béré*, which was consumed only in the event of famine.[8] Ultimately, these elders remember, the town had no choice but to join the Office and try to maintain the landscape that they relied upon for their sauces and year-round food supply.

The people of Molodo had good reason for concern. The botanist Guy Roberty noted the problem of deforestation on lands occupied by the institution as early as 1937. Even if he had been puzzled as to the local practice of leaving some trees in fields, he understood the dangers of too much tree loss for the long-term supply of wood.[9] Elsewhere in the colony, the administrative demand for wood fuel by the military and transportation services led to forced woodcutting for administrative use and bans on cutting by African residents in state preserves along the railroad tracks and near major rivers. These measures to control wood use dated to at least 1915. During and after the Second World War, the administration's need for wood fuel only increased.[10]

The increasingly denuded landscape was a major concern for women, who not only collected fuel for cooking from the forest areas but also gathered ingredients for the daily sauce. This environmental shift was connected to technological change. At the same time that young women began cooking in metal pots, they started using the time saved on wood collection and cooking to earn cash to buy the sauce ingredients from women in neighboring towns outside of the Office.[11] It was not surprising that when a woman had more cash to purchase good ingredients, her sauce was often better. Most likely, the women in those households also had metal pots.

METAL POTS: GENDERED LABOR, HOUSEHOLD GOODS, AND VALUE

Young brides were in the vanguard of a new technological rural world at the Office. When a new wife moved into her husband's household she brought with her a wealth of pots, buckets, cloth, stools, and calabashes, many of these items purchased for her by her mother.[12] During the rainy season, which was also the wedding season, the major Office market in Niono was flooded with mothers buying household goods for their daughters.[13] When it came to domestic tools and technologies, women were savvy producers, consumers, and users. Some items, such as a cotton mosquito net, a woman could make for a new bride herself. This item was especially necessary because the Office canals attracted mosquitoes carrying malaria and other diseases even during the region's dry season.[14] Many of the items necessary for food preparation such as the mortar and pestle were ordered in advance from artisans and carefully examined for quality before purchase. Local female potters produced the necessary clay goods such as water jars and steamers.[15] Calabashes and locally produced cloth were purchased in the market and increasingly so were metal pots and buckets. In the postwar years mothers still assembled the basic items their daughters needed for marriage, but their daughters were beginning to purchase additional items for themselves.

In the first half of the century, a new bride's husband would order a *bogoda* (clay pot) for his wife. This pot might vary in size with the anticipated number of people in the household for whom she would be cooking in her new home.[16] Beginning in the late 1940s, mothers began purchasing for their daughters large metal pots that were big enough to cook for households of thirty people.[17] The first imported metal pots were so heavy that when filled with water or toh they had to be carried by two women.[18] Although the new pots were clearly cumbersome, they provided significant labor savings in the time and effort women expended collecting wood and cooking. The metal pots took less time to heat, cooked at least twice as fast as clay ones, and as a result used much less wood to fuel the fire.[19] All of these benefits made it worthwhile for a mother to invest in a metal pot for her daughter, even though the groom was expected to purchase the bogoda. While a few men purchased metal pots for their wives, these purchases were generally made by middle-aged and older women preparing for a daughter's wedding.[20]

The pots, calabashes, cloth, stools, and other items a bride brought to her new household were displayed for the wedding. This display was socially important for young women. As historian Barbara Cooper has demonstrated for twentieth-century Niger, these gifts were more than useful household items. The quality and quantity of the goods were a demonstration of her social network and value, as many of the items were gifts from her mother's friends.[21] This message was reinforced by the mother's additional expense in buying a cooking pot and the visual aesthetic of the new metal good.[22] Once the bride headed for her new home, she was accompanied by a host of female relatives and friends who carried the items on their heads to make a show of the young bride's arrival. This journey was often made by foot, but during this period some Office brides arrived atop a cow, by horse, and in the later years by car.[23] New wives of African staff such as tractor or thresher drivers might even undertake a portion of the journey in one of the Office vehicles.[24] When women made the journey by foot, it was tiring, as the journey was often a day's travel from the bride's hometown.[25] Nevertheless, the bridal parade was a festive event.

Significantly, the bride's arrival was always a spectacle. First, it was a fun event for a nervous young woman being transported by novel means and for her friends who were carrying the bride's new collection of household goods. Spectators knew right away that the parade of travelers was escorting a bride. The metal pots shined among the collection of goods that the young woman's friends carried on their heads. In this way, new domestic technologies such as metal pots and buckets entered family life through a celebration expressing hopes for the future. The event was also a carefully crafted presentation of a technologically savvy bride. Moreover, the new pots still linked women's cooking to reproduction and (in the case of young women) sexual allure. A cow might symbolize the success of a new husband's household, or a car might be a sign of his position at the Office. The metal pots and buckets were all about the bride.

Once the young woman was settled in her new home, her metal goods played a further role in helping her to establish a comfortable place in the new household. A new wife was often considered an outsider until she demonstrated a willingness to work and support her husband and his extended family. This meant working in the family's fields, drawing water for the household, collecting or purchasing ingredients for cooking, preparing meals, and bearing children.[26] Women who came to the household equipped to do this work with pots, calabashes, buckets, and other household tools earned a measure of respect for their willingness to prepare for and take on their new role.[27] When women brought metal house goods with them into marriage, they also signaled to their new family members that they were going to shape the conditions under which they carried out their labor obligations.

From this perspective, the shiny new metal pots and buckets marked a new generation of young women, and the metal items made individual women highly visible to their female in-laws. This group included a mother-in-law and sisters-in-law, all of whom a young woman would want to impress with her domestic and culinary skills. Fatoumata Coulibaly remembered her desire to be perceived as adept at using her metal pot not long after she arrived as a new bride in the Office town of Sabula. One day while cooking she accidentally burned her leg on the hot metal. The pain was intense and, in fact, the burn so severe that Fatoumata still has a scar. Yet, that day she held herself back from crying out so as not to let the more senior women in the household learn of her clumsy mistake.[28] In Fatoumata's thinking, a woman who knew how to use the new metal pots was a successful wife and valuable contributor to the household. She wanted to maintain this image.

New wives often apprenticed with an older female in-law to learn how to cook rice or specific sauces,[29] but a young woman entering with new metal cooking pots would have had the opportunity to showcase her innovative use of these items. One way for a young woman to show off her new metal pots (and cooking technique) was through the sound her cooking made in the compound. Fatoumata quickly learned that this was not always so easy. Even if a young wife knew how to make the toh pasa sound in her new metal pot, she could also make a mistake and burn herself on the hot metal. In that case, she knew she could not scream because it would be the sound of inexperience—or poor cooking.

COOKING WITH A METAL POT: A TECHNIQUE OF THE BODY

The newness of the pots and the special qualities of the women who possessed them were both displayed through embodied action. Arriving in visible possession of one or more metal pots was just one step in the process. By showing off their cooking skills using a metal pot, women reminded their household (and the whole town) of their social value and economic potential. For example, a woman with metal pots often had more time to devote to income earning. At the same time, she dutifully fulfilled her cooking duties. This was the case even though these younger women spent less time cooking. Moreover, when a woman with a metal pot was cooking, everyone within earshot knew that she was making a meal. It was an audible reminder of her arrival in town, the wealth (of household goods) she brought, and the network of women who supported her in her new married life.

Residents at the Office were told over many years of living at the scheme that they were engaged in bringing "modern" farming and "civilization" to the French Soudan. Women may not have used the term modern to describe their metal pots, but they were cognizant of the changes brought about in their daily life because of them. Metal pots became a pervasive audible and visible testimony to change and displaying the values and meanings bound up in the metal pot was a sensory and bodily experience for women. Toh preparation in particular became very much a performance as more women cooked with metal pots. Cooking faster meant stirring faster. Assane Pléah remembered when her younger sister began to cook with the metal pot: "It was as if she were dancing."[30] As toh cooked, it became thick in the pot (clay or metal), and to stir it quickly required a great deal of force. Women used all of their upper body strength to do the stirring: pulling and pushing the toh pasa through the mix and then passing the spoon around the edges of the pot. This is why Assane's sister looked to her like she was dancing. She was moving her whole body to cook. It was a scene visible across the Office, as women vigorously stirred: the toh pasa struck the side of the pot rhythmically and women "danced" while cooking.

Marcel Mauss theorized that people educate their bodies to perform certain daily tasks from the most basic to those requiring special skill. In this way, the body itself is an instrument for human action. He further argued that bodily habits for a given task are subject to change. Mauss used the example of swimming, but the analysis is apt here. The first generation of women learned to slowly stir toh in clay pots but adjusted their bodily movements to do the same task in a metal pot. The learned body movements of cooking changed and were passed on to younger generations of women. It so happened that cooking in a metal pot required bigger and more dramatic movements, making cooking a more spectacular endeavor even as it remained a mundane daily activity.

Assane laughed heartily when she described her sister's technique to me, but then she added that her father complained about the taste of the toh. She explained that if you went too fast and forgot to add water to the top of the dough after mixing, the food lost some of its flavor.[31] For some young women, the fast stirring and cooking went hand in hand with skipping steps during meal preparation: these cooks perhaps paid more attention to the whipping and dancing than watching their toh. In any case, even older men in the household noted the arrival of the new pots: the whole domestic atmosphere changed, and men could not help but notice. In fact, toh tasted different when cooked in a metal pot, even if the cook remembered all the steps.[32] Women certainly noted the change in the taste of their food, but mothers and daughters continued to purchase and use metal pots despite complaints from some men in the household.

Young wives were adamant about their cooking expertise despite the different taste of food cooked in metal pots as opposed to that made in clay pots. This was especially true for women who arrived in rice-growing towns and learned to make rice toh. Their food tasted doubly different. Yet they insisted that they always made the same food. These cooks made millet or rice toh, with an emphasis on toh.[33] In the women's recollections, it did not matter whether they cooked millet or rice. Neither did it matter that they cooked in metal or clay pots. For them, the important fact was that with their skill they continued to prepare familiar and appetizing meals. As an object of study, food preparation and consumption has been perceived to be an area of social stability even conservatism.[34] On a superficial level, the insistence by women in this generation that they prepared the same toh seems to support this notion of social continuity through food. Yet women readily adopted rice as the staple grain for making toh and added new ingredients to their sauces. Women who owned metal pots were made aware that their meals tasted new or different. In the face of such perceived criticisms, they perhaps insisted that the food was the same.[35]

At the same time, women were concerned with satisfying the tastes of their fathers and later husbands and in-laws. It was part and parcel of fitting into the new household as a young wife. Nevertheless, the younger generations of women at the Office also had a role in reshaping their family's food tastes. For example, cooks safeguarded the texture of toh while adding new ingredients to its preparation or to the accompanying sauce. They also prepared the meals in new cooking pots that altered the resulting flavors of the dish. By the time the second generation of women arrived at the Office, women at the project had developed their own definition for what constituted toh. Good toh was an elastic concept, and the new definition was in some ways a response to changing technological resources, the environment, and women's labor interests.

When I asked women to contrast their own work during this period with the work of their mother's generation, many responded that the only work their mothers did was make toh. When pressed, the same women remembered the only other work that their mothers did was to carry toh to the fields to feed family members working there. The women's mothers also helped in the fields, but in their daughter's minds cooking took a predominant role in their mother's work lives. Certainly for their mothers it was a more time-consuming task to cook in a clay pot. For both generations it was also a technically complex portion of the meal. A good cook knew when to begin stirring and for how long. She also knew when to add water to the top of the mix to give the toh the right consistency: it had to be easy for family members to take a handful from the bowl, the textures had to be just right, and it had to taste and feel good in the mouth. Where women saw change between the generations was in the range of activities they could accomplish that their mothers could not. They cooked *and* earned cash.[36] Indeed, women's successful use of a metal pot recalls the remarkable work of the fast cook of the early twentieth-century folktale.

The cooking pot may be a modest technology, but as Judith McGaw observed, the history of gender and technology demonstrates that simple technologies (sometimes called "feminine technologies") require complex knowledge.[37] Indeed, owning a metal pot entailed more than simply knowing how to cook with it. A woman had to remember that the metal pot near her legs could become extremely hot. Knowing how to use a metal pot also meant knowing how to keep it clean. Women who had these new metal items maintained them carefully. A new clay pot had a red-brown color but was quickly charred by the cooking fire. With the introduction of metal pots, young women added an additional step to the way they cooked and cleaned so that the metal would not blacken like earthenware.[38] Women worked to keep their pots looking like new.

Before placing the pot on the fire, the woman cooking coated it with a few handfuls of earth from near the hearth and mixed it with a small amount of water. She covered the bottom and sides of the pot almost up to the opening. When the pot was on the fire, the mud was burned, not the pot. After the meal was cooked and she was cleaning, she scrubbed the mud off. The pot retained its metallic shininess, a visible testament to her other domestic skills (see figure 4.2). When women described this process, the word they used for the result, *jè*, could be translated to mean "clean" or "bright."[39] Here, I draw another insight from McGaw, who pointed out that by examining feminine technologies we see that "technology" is often as decorative as it is functional.[40] The look and materiality of the pot had a lot to do with its success. The new pots may have saved labor time, but they were also heavy and unwieldy, especially the biggest ones. One other consideration was that they were attractive. A metal pot brought a pleasing new visual aesthetic to cooking to accompany its new sounds. This aspect of the metal pot also made women's work reengineering food production highly visible.

FIGURE 4.2. Metal pots in varying sizes showing the sides scrubbed bright and the bottoms blackened by the cooking fire, 2010. Photo by author.

THE POSTWAR METAL HOUSE GOODS MARKET

After the Second World War the metal house goods market expanded in French West Africa. As historian Emily Osborn documents, growth in this sector was rooted in the war itself. During the conflict, the colonial government operated two foundries in Dakar for producing iron goods to support the war effort. Much of this metal production involved a new technique called "sand-casting." Men working in the factories (many of whom had previous metalwork training as blacksmiths) discovered that the technique worked well with scrap aluminum, and scrap metals were increasingly available in Dakar at the time. When these metalworkers left the foundries after the war, many of them opened their own blacksmith workshops across French West Africa and employed the new sand-casting technique with scrap aluminum. They also trained apprentices from across the region. This technological know-how among men was thus transmitted across most of French West Africa.[41]

Interestingly, the first goods that these blacksmiths made from aluminum were cooking pots and other household items. Prior to the war metal cooking pots and buckets were introduced in colonial markets by French merchants, but the goods made up only a small proportion of French imports. These merchants imported more European manufactured cloth, soap, and cooking oil.[42] After the war, the newly trained blacksmiths slowly increased sales of their metal wares in local markets.[43] In the first few years, it is likely that some of these items produced in Senegal were sold to traveling merchants who frequented markets across the region. Aluminum is a light metal, and when compared to imported iron pots the new aluminum ones were much lighter. The older iron pots had been affordable because as some Office women put it, "The tubabs [Europeans] wanted women to buy [the imported iron pots]."[44] The aluminum pots were even cheaper. At the same time, the male blacksmiths had to compete with women potters who produced clay pots.[45] One reason the blacksmiths might have focused on goods made specifically for women was that there was a ready market for domestic goods and cooking pots.

Gradually, the local production of metal cooking pots increased and spread to the French Soudan and the Office region. Osborn's work helps to further situate this history. First, more scrap aluminum was available following the war. Then in the 1950s aluminum sand-casting technology was transferred to artisans and workshops in Bamako. By the 1960s, the number of these workshops increased as the availability of aluminum spread more widely because of the emergence of scrap-metal dealers across the interior markets.[46] Markala, with its workshops for the repair of industrial machines and for the manufacture of agricultural equipment, was a site where these new artisans could easily acquire scrap metal.[47] The Office also offered several expanding colonial markets where blacksmiths could sell their wares. From the late 1940s to the 1960s, women at the Office increasingly purchased these aluminum pots and other metal goods (serving spoons, buckets, etc.) for their daughters. This technological shift was made possible by new metalworking techniques transferred from France and Senegal during the Second World War. It was also made necessary by accelerated deforestation associated with the rapid expansion of farmland under the project. The modest technological innovations in the French Soudan were tied both to global industrial transformations and to women's food preparation needs.

METAL HOUSE GOODS AND THE OFFICE ECONOMY

By the end of the 1950s, Office administrators regarded metal cooking pots as items of primary necessity and paid close attention to the prices for pots and other basic goods in the markets. Their particular concern was that project farmers ought to be paid enough for their crops to purchase such items. In 1959 François Wibaux, the director of the Office, wrote to the colony's Ministry of Commerce, Industry, and Transport requesting an increase in prices the government paid to Office farmers for rice.[48] He explained that the daily cost of living had increased over the past decade, while the price for rice paddy had dropped.[49] Clearly, this was a problem for an Office administration that had long tried to make good on its claims that the scheme would improve the standard of living for its farmers. In the letter, Wibaux cited the costs for basic items like sugar, salt, cloth, and metal cooking pots. In a little over a decade, metal house goods had become as essential to an Office household as salt or cloth, both of which were widely recognized by the colonial administration as basic necessities.

In the decades following the Second World War, colonial officials sought to improve living conditions at the Office more directly. This had been one of the main objectives of the investigation carried out by the Mission Reste in 1945 and succeeding policy changes that directly addressed farmer complaints. More broadly, the colonial government was concerned with the demands for greater political, economic, and social freedom across the colony. It is, therefore, not surprising that Wibaux expressed concern for the cost of living at the project. It was a matter of politics. His aforementioned 1959 letter clearly demonstrated this change in attitude toward farmers.

However, the 1959 letter also revealed Director Wibaux's assumption that it was male farmers who brought new technologies into the household. From the point of view of Office administrators, the economy of the Office revolved around men's incomes. This was represented more specifically by the income earned by male household heads rather than the incomes of junior men. Wibaux seemed to have been unaware that cooking pots and many other basic necessities were purchased by women. For the region's blacksmiths and merchants, the women's visible presence in the marketplace was unmistakable.

In general, Office women earned less than men at the project, but their incomes and savings were enough to support the expense of a metal cooking pot. In the same 1959 letter, Wibaux listed the price of an average metal pot as 305 francs CFA in 1959, up from 255 francs CFA in 1953. In 1959, a sack of twenty-five kilograms of salt cost 405 francs CFA, or slightly more than the cost of a metal cooking pot.[50] While the cost of the new pot was an investment, it was well within the price range of common bulk foodstuff purchases. More to the point, the purchase of a metal pot was within the realm of possibility for many mothers. Even with inflation, the pots remained accessible technologies. In fact, the purchase of a metal pot for a daughter became even more necessary as wood for fuel in areas surrounding the Office became scarce. By the 1960s, middle-aged mothers were reliable rainy season customers for a new market in metal household goods.

By contrast, senior men spent a larger proportion of their incomes to support the infrastructure and associated agricultural technologies of the Office. The plow, which was the farming technology most consistently promoted by the Office, was a major expense. Most men only obtained a plow on credit from the Office, or they borrowed one from a fellow farmer. In fact, the purchase of a plow was out of reach for many farmers already indebted to the Office.[51] In the 1950s, as industrial technologies and mechanized farming increased at the Office, it became even more difficult for a farmer to afford the plow. From 1953 to 1959 the price for a plow rose from 8,000 francs CFA to 10,000 francs CFA.[52] When Wibaux appealed to the colonial government in 1959 to raise prices paid to farmers, he also raised the issue of the high cost of irrigated water and mechanical services. In 1959 these fees were out of proportion with men's earning power. The director was voicing a common complaint of farmers. Men's decreased earning power, as he observed, did affect the welfare of families. Given the hard economic times, women's access to new domestic technologies helped alleviate the burden of household expenses, especially those associated with food preparation.

PLOWS AND CARTS:
GENDER, TECHNOLOGY, OWNERSHIP, AND LABOR

At the Office the plow was promoted as a means of increasing cash-crop yields. It was not necessarily intended to address the needs of families concerned with daily survival and the quality of their food. For agronomists and planners the plow was a male technology. In later decades, the promotion of the plow was expected to reduce the demand for women's labor in the fields. Even development specialist Ester Boserup (who, in 1970, directed the attention of scholars to the predominance of women in agricultural activities in Africa) emphasized men's use of the plow. She argued that increased plow use would lead to a reduction in women's labor in the field and an increase in their nonfarming income-earning activities.[53] By the time Boserup was writing, women at the Office had been farming *and plowing* for decades.

Office farmers, both men and women, also associated the plow with men, even though women regularly worked with it. In a telling example, from the Office town Sokorani, Fanta Sogoba and her sister plowed their family fields because the family had no sons. In fact, the Sogoba girls were known for preparing their fields earlier than most other families who did have sons.[54] It was widely recognized that the Sogoba girls were skilled at preparing fields with their family's plow. Wives also plowed when a husband or brother-in-law was ill, or when there were no young men in the household to help with field labor.[55] Both men and women at the Office were fully aware that many women plowed. Yet the masculine associations of the plow persisted among settlers. In fact, improved plows produced by Markala blacksmiths and metalworkers were given distinctly masculine names such as the *wuni*, or "something rapid." The same plow was also known as the *Kamelenbani*, a hyperbolic term referencing the perceived superhuman stamina of young men who do not break a sweat while working in the field.[56] Even women who plowed regularly thought of it as men's work.

Scholars in feminist technology studies over the past twenty years have argued that technology and gender are coproduced, meaning that each is shaped by the other. Some technologies become expressly associated with a masculine or feminine identity.[57] Thus, gender ideals shaped the plow as a technology even as men's and women's roles were defined (and redefined) by plow use. Fanta, who was famed for her plowing skill as a young woman, explained many decades later that plows were associated with men because to work the plow you had to be quite muscular.[58] Yet work associated with women also required great strength: carrying heavy loads of wood, drawing and carrying water, and moving heavy metal pots to and from the fire. In addition, the use of the plow for farming was as novel for men as it was for women. Clearly there were other reasons for the plow to be associated with men.

On one level of analysis, the Office distributed plows only to men. This fact influenced to some extent how men and women perceived the plows. They were also owned by men—just as metal pots came to be a woman's possession. Men's ownership of plows was an important unstated factor in how the technology was locally perceived. On another level of analysis, it becomes clear that whoever worked with the plow (a man or a woman) made a statement about their family's fortunes. A man plowing spoke to the good standing or health of a household head who was able to purchase or access a plow. A woman plowing was more often an indication of illness in the family or other misfortune than a testament to uncommon female strength. In this way, families in possession of a plow had an interest in reinforcing its masculine association, even if it was sometimes operated by women from the household.

Ultimately, the introduction of the plow at the Office impacted how women participated in farming and agricultural production. Despite the fact that some women could be skilled at using a plow, a technological differentiation was introduced between men and women. Men most often used the plow to prepare fields while women continued to use a small hoe for subsequent tasks like weeding. When there were no sons or other young men in the household, women performed all this labor. The introduction of the plow by European agricultural staff brought a set of European gender ideas to agriculture. Men farmed, plowed, and controlled land. This ideology was reinforced by the way the Office distributed land permits, seeds, and farming technologies. While the successes of the Office's early techno-agricultural interventions were inconsistent, globally they ensured that women had little access to land to farm or garden for home consumption or sale. Owning a plow had become a claim to land, and women possessed neither.

Women's technologies were also highly gendered. They purchased most of a household's domestic technologies. Those same tools tended to assist food preparation rather than cultivation. A new metal pot enabled women to cook more quickly, and the new cooking techniques associated with the metal pots produced a finished meal more quickly. In fact, older women around the Office often noted that young women would not be able to cook in the old earthenware pots because if they stirred too vigorously clay pots would break.[59] Moreover, women owned their metal pots (as opposed to men purchasing the clay ones). This gendered shift in the ownership of a specific domestic technology translated into more control for women over their own labor time.

The audible and visible everyday use of metal pots made plain the changing gender-labor dynamic. Indeed, the shift from clay to metal pots sheds light on the history of a technical and sensory process that at first glance did not appear to change much about women's labor, and yet such subtle shifts were motivators and signals of broader social transformations. The most obvious transformation was that the woman's workday was gradually modified with the introduction of metal house goods. Women continued to labor long days, but the distribution of their labor was new. The alteration in women's temporal regimes was something that men and women noted in their daily lives. It entailed changes in the activities of younger wives, who had more time for cash-earning activities such as cotton spinning, rice winnowing, etc. Therefore, women who obtained small plots started market gardening, and women generally spent less time around the cooking fire or collecting wood. New sauce ingredients like onions and other purchased foodstuffs like tegedege (used to make peanut sauce) were cropping up. Men and women at the Office could literally see and taste the innovations in the younger women's cooking.

By the 1950s, another significant labor shift was connected to the introduction of carts at the Office. For several months after the harvest and before the next rains, women carefully calculated the amount of wood fuel they needed for the year. In an unrelated development, administrators promoted the use of carts by men to assist in field labor. Carts pulled by donkeys slowly appeared in Office towns, and they were largely purchased by male household heads. Like the first large metal pots, the first carts (dating to the 1930s) were slightly unwieldy. They had heavy metal wheels and a similarly cumbersome wooden frame.[60] Only large oxen, a breed foreign to West Africa, could pull the first models. Eventually, these early carts were replaced by lighter ones that donkeys (a cheaper animal) could pull.[61] A cart was first introduced to facilitate farm work. Tools, seeds, and grains were more easily transported from town to field and field to granary by cart. However, by the 1960s, the increased purchase and use of carts at the Office had the additional effect of reducing women's workloads.

The gendered shift occurred when some men began collecting large quantities of wood fuel that they then transported using their carts. This change coincided with the time when wood was found at distances greater than men's wives could walk. The men who collected wood did so seemingly out of a need for additional income. After a man returned with a load of wood, he often gave his wife (or wives) some of the wood, and then he sold the excess in the big Office markets. A few women explained this change by saying that the men who collected wood for their wives loved them.[62] This characterization alludes to the fact that these men were doing women's work (and even transforming it into a male income-earning activity). By these women's assessments, the men who collected wood were helping women with their household labor duties. For the men to do so, the elderly women who recalled this change surmised, they must have acted out of love, just as a woman's cooking might demonstrate love for her husband. By the end of the 1960s, this was a major unforeseen shift in the division of labor across Office households, and it was a change that was not necessarily embraced in other rural areas of the French Soudan.[63]

Office planners (including the project's main architect, Emile Bélime) had promoted oxen-drawn plows and carts from early in the project's history. They anticipated that the cart would promote cattle raising alongside farming because it required the use of oxen. For many years, the well-established local distinction between herding and agricultural livelihoods troubled administrators. Worried about political unrest among mobile herding populations, the promotion of farming accompanied by cattle raising was also designed to settle nomads and seminomads. One aspect of the rhetoric describing the benefits of such a cultural and economic shift included the suggestion that it would improve settler diets by bringing steady supplies of meat and milk into farm households.[64] Indeed, Mamu Coulibaly, the beer brewer who married into a family in the first Office town, Sangarébougou, remembered an early emphasis at the project on cow ownership: plenty of cows and milk. Mamu described the abundance of milk by explaining that her son drank straight from the cow's udders.[65] It was for her a novelty to have daily access to milk. However, Mamu's experience is not easily generalized for the Office, since Sangarébougou was the model Office town. For many years official visitors were brought to her village to observe successful African farmers. Elsewhere, families struggled with the scarcity and expense of livestock, plows, and carts. As with other technologies, the cart impacted the local labor regime and food supply but not in the ways that planners expected.

Even with help from men, the difficulty of acquiring wood fuel continued to be worrisome for women. For example, the loss of many trees translated into increased prices for a type of wooden serving bowl called a *kuna*. Just as many Office residents associated toh with good and filling food, this type of wooden bowl was associated with mealtime and proper eating. By the late 1960s, the wooden bowls were hard to find in the market.[66] Many women recalled how the taste of toh was different when cooked in metal pots, but they overwhelmingly appreciated the technological change. However, men and women were consistently nostalgic for the wooden bowls. When metal bowls first came to the market (1950s–60s) many older men simply refused to eat from the new vessels.[67] Many decades later, residents at the Office recalled having regrets about the substitution of the older wooden bowls for new ones made from other materials.[68] It was not a change prompted by the benefits of labor savings or the market cost of such goods. The much-desired wooden bowls simply appeared in fewer quantities in the markets, and women were obligated to purchase other serving bowls.

Visually, metal bowls looked different. A kuna was dark brown, the wood blackened over a flame to preserve the material. The largest of these wooden bowls allowed big family groups to eat around the same dish.[69] New metallic bowls came in smaller sizes meaning fewer people ate from one bowl, changing the social habits of eating.[70] Perhaps toh made in a metal pot had a displeasing taste to some at first, but the memory of the technological shift and the associated economic and labor improvements overshadowed this change. The metal material of the pots also provided tangible cooking benefits. Metal serving bowls did not offer the same conjunction of technological, environmental, and aesthetic benefits.

CANALS AND BUCKETS: INTEGRATING INDUSTRIAL AND DOMESTIC TECHNOLOGIES

Water technologies were integral to the Office as an industrial agricultural project, but the canals also affected domestic life, especially with regard to water collection. The canals also unexpectedly impacted daily food production. Sékou Coulibaly, whose parents were among the first families at the Office town Nyamina, remembers that men and women fished in the canals and in the fields when they were flooded. Everyone engaged in fishing, but as with most activities, fishing had its own gender dynamic. Men most often fished with nets while women used a round wooden cage or basket.[71] The biggest catch of the year, he recalled, was when the water supply was cut off, leaving an abundance of fish near the drains.[72] Villages lost an important water source during this part of the year, but residents were able to catch and prepare a store of an important food resource. The fish caught and dried during this part of the year supplied not only a much-needed sauce ingredient but also essential proteins. Many older residents who grew up at the Office were nostalgic for the food from their younger years, and such memories most often centered on the abundance of fish in the sauce.[73] For families coming to the Office from the river region, fish would have long been a staple of their diet. Women in these families helped to ensure the familiar taste of their food when they cooked with fish. For new families who arrived at the Office from other regions, the abundance of fresh fish was a tasty and perhaps luxurious addition to the sauce.

The availability of food and other resources necessary to produce food was an ongoing concern for Office women. They could spend an entire season calculating wood needs, but they measured water requirements daily. Water was essential for preparing the breakfast porridge, the toh and its sauce, not to mention for drinking, washing the cooking pots, doing laundry, and bathing. To supply the water needed to feed and care for everyone in the household, women made numerous trips over the course of a day to draw water. Among the new metal goods in the market, buckets were another modest labor-saving technology. With the new metal buckets a woman made fewer trips to the canal or to the well because she could now carry one bucket (or a large baarakolo) on her head and carry a bucket full of water by hand because of its handle.[74] Water collection now took less time.

Like the first metal pots, the buckets were weighty, but they proved useful. For example, a metal bucket full of water on top of a woman's head was less likely to break when she moved the bucket from her head to the ground. Baarakolo and calabashes tended to break with the same motion. With some of the calabashes women also had to cover the top with a smaller calabash to prevent spilling; this was an old labor practice. Buckets eliminated this extra labor step.[75] Cooking was a physical task, as it required collecting water, threshing, and pounding grain. For women who counted the number of arm lengths needed to draw water, fewer arm lengths expended at the well meant more energy for other important tasks.

Drawing water from the canal (when it was possible) even eliminated some of this labor. To get water from the canal women simply filled their buckets (or calabash) by dipping it into the irrigated water source (sometimes with the aid of a rock staircase). These quotidian changes in women's domestic labors were unforeseen by the men who designed the dam and its canals, but they proved important for daily life. Women were best able to take advantage of these man-made resources with the introduction of buckets. A bucket full of water may have been slightly heavier than a calabash full of water, but women were more assured of not losing much water, thereby eliminating unnecessary trips. In essence, the buckets served to mediate women's engagement with the water and the canals, both of which were hallmarks of industrial agricultural technology at the Office.

Buckets, like the metal pots, were another new item gifted to women at the time of their marriage by mothers willing to spend a little extra for more efficient and sturdier household goods. Buckets were a similarly substantial investment for mothers, but one for which the positive benefits were felt daily. The adoption of both domestic technologies (metal buckets and the metal pots) was unlike the dramatic entry of industrial agricultural technologies (e.g., tractors) at the Office. The latter anticipated radical changes to work processes in the fields and required a great deal of expert maintenance. Buckets and pots fit well into women's already established work routines and required little upkeep.[76] In addition, the new metal items ensured that young women would not need to spend money to replace their household items as often. Calabashes, spoons cut from small gourds, and clay pots all broke fairly easily.[77] Over time, as with the metal pots, the buckets were understood to be a necessity. In effect, mothers who bought new metal buckets and pots for their daughters assisted the second generation of women in managing their access to two necessary natural resources for cooking: wood and water. They also, perhaps inadvertently, helped satisfy the need for appetizing food in the face of a radically changing environment.

TRACTORS AND THRESHERS: DEBT AND EMPTY BELLIES

In the postwar years, the subtle changes in women's purchases and daily use of new domestic goods contrasted sharply with their engagements with the often intrusive, large-scale technologies that planners assumed would modernize farming. From the outset, the technological centerpiece of the Office, its irrigation system, was prone to flooding. At other times, administrators cut off water to particular towns when farmers there failed to pay project fees.[78] Despite such inconsistencies in the irrigation infrastructure, promoters of the Office continued to herald irrigation and later mechanized farming at the project as major achievements. In the late 1940s, Office administrators pushed for massive mechanization, including the founding of an entirely new mechanized rice sector in Molodo-Centre (near Molodo-Bamana). The project was funded in part by the American Marshall Plan, which was intended to promote global economic recovery after the war. About a decade later in 1955, the director of the Office, Georges Peter, was promoting the Office as a model for colonial development based, in particular, on its technological apparatus. He claimed that the Office had in fact turned a desert into profitable farmland and raised the standards of living and level of civilization for its African farmers.[79] These were ambitious and highly value-laden claims. On the ground, the impact of project technologies was not so easily measured.

The midcentury push to increase mechanization at the project was part of a larger effort by the administration to improve the institution's production figures and profit.[80] Following a pause after the war, the fully mechanized rice sector called Centre du Riz Mécanisé (CRM) was opened in 1945 at Molodo-Centre.[81] At the CRM, agricultural machines were employed for field preparation, planting, and harvesting. There were no farmers with individual Office fields. Instead, wage workers helped to operate the machines and also to bag the harvest. In the following year, the director of the Office (at the time Pierre Viguier) even investigated the purchase of equipment for mechanized tobacco cultivation.[82] Tobacco was a secondary crop that many Office farmers planted because it was a profitable garden crop, which explained Viguier's interest in the possibilities for its industrial cultivation. Large-scale tobacco cultivation was not pursued further, but the example speaks to the larger trend toward mechanization.

Throughout the 1950s mechanization increased, and the large technical infrastructure was expanded and reinforced. Additional American Marshall Plan funds supported the 1952 opening of the semimechanized cotton sector called Kouroumary, where farmers tended to individual plots with the aid of new agricultural machines. This new sector was located just north of Niono and Molodo and drew water provided by the Canal du Sahel. In Kouroumary all the field preparations were done by tractor. Office instructors did not encourage plow use as had long been the case elsewhere at the project. Additionally, farmers in Kouroumary paid for the use of the harvesting and threshing machines. In this semimechanized sector, men and women still planted, weeded, and picked some of the cotton harvest. They also grew rice.[83] Agricultural equipment across the Office now included not only cattle-drawn plows but bulldozers, tractors, and the double-task harvesting-threshing machines.[84] Finally, the French colonial development fund called FIDES supported repairs for the irrigation infrastructure and its expansion during the 1949–50 agricultural year.[85] The Office had become an even more industrialized space.

In 1953, a little less than a year after farming began at Kouroumary, Tony Revillon, a high-ranking metropolitan official who served on the advisory board of the Office, reported to Paris that the mechanized sectors of the institution were profitable.[86] Revillon's optimistic financial assessment proved hasty: only a few years later in 1957, an unofficial report of the Office's financial standing signaled that the institution was still running at a loss. After the 1956–57 agricultural campaign, the total recorded losses were listed at five hundred million francs CFA. Moreover, the auditors who authored the report cited the extensive costs of mechanical agriculture as one of the major reasons for the institution's financial difficulties. In fact, the Office had not even covered its costs as far back as 1953, when Revillon was touting the successes of mechanization.[87] The Office had a top-heavy bureaucracy and was deeply invested in the promises of a costly infrastructure. These sobering facts were glaringly apparent by the late 1950s.

Agricultural interventions elsewhere in the colony during the war years had similarly emphasized mechanical production methods. Metropolitan researchers and agricultural officers encouraged mechanization across the French Soudan in an effort to promote greater production to support the war. For example, in 1942, the Institut Colonial de Marseille funded research for expanded peanut farming using mechanization.[88] Industrial oil production in the colonies was a particular concern during these years. Local officials even looked into the possibility of industrial shea butter production (locally a female-produced foodstuff). Most European observers still believed that female producers would not produce large enough quantities of shea butter for the market. They also assumed that the butter manufactured by African women was of inferior quality. As at the Office, mechanized production implied a process by which European men supervised the wage work of African men using imported machines. Indeed, the proposed scheme for the mechanization of shea butter production anticipated excluding women from the market altogether.[89] In the end, the trial scheme for shea butter manufacture failed to come to fruition, which signaled a trend: mechanization in the French Soudan was reserved for cultivation rather than manufacture. Industrial mechanization did not target women. Nevertheless, women clearly engaged with the process of industrializing agricultural production at the Office.

"Modern" agriculture at the Office in these years referred specifically to mechanical or motorized farming. This stated goal for the Office anticipated more machines for field preparation, planting, and harvesting, which was believed to be more feasible after the Second World War. In the 1950s and 1960s, agricultural staff continually looked for new ways to revolutionize farmers' methods of field preparation, planting, pesticide and disease control, harvesting, processing, or crop storage, even when it was unnecessary or inefficient. The ongoing assumption was that there must be some technological input that would finally realize the Office's ultimate goal of increasing production and profit. This was one reason that the Office organized some factory processing of the cotton and rice harvests before the raw goods were shipped. In addition in 1946, agricultural staff members investigated the possibility of installing new metal silos in project towns in the hope of reducing storage losses. These commercial silos were large and made of metal rather than mud and thatch. However, the new silos did not take off with farmers. One reason might have been that the company selling the metal silos charged 55,600 francs CFA for a five-ton capacity silo and an additional 9,000 francs CFA for transport.[90] It should have been clear to the Office staff that a farmer already indebted to the project for cattle, plows, water, and mechanical labor fees was not likely to also purchase a metal silo on credit, even if the purchase was made through the local farmers' cooperative. Moreover, granaries made from local materials stored grains and other crops well.

Internally, Office staff worried about improving the institution, but externally they promoted the project as a tremendous success. When in 1955 Georges Peter publicly promoted the Office as an example of successful French colonial development, he did so on the basis of investments made in the technological infrastructure. More specifically, he claimed that the scheme had created an infrastructure that benefited residents. That year he wrote: "The Office is a pioneer in public service for the Middle Niger region as a provider of water and electricity and by building and maintaining roads and schools."[91] Certainly the Office provided water for irrigation, which incidentally made daily water collection easier. Yet, the irrigation system itself was still inefficient and prone to malfunction. Moreover, the dam only provided limited electricity to European staff members' homes and some worksites.[92] The roads he referred to only covered Office territory and bypassed older local routes and regional economic centers. Finally, the Office had actually been criticized for the lack of education available for the children of its farmers. Peter's claim that the Office was a model of colonial "assistance" spoke less to the reality of 1955 than to an imagined project.

In truth, the Office infrastructure was highly unpredictable. Men and women appreciated when the water flowed in the canals. They struggled with the consequences of flooding or water shortages at other times. The new agricultural machines such as tractors and threshers, as it turned out, were similarly unpredictable. The heavy government investment in technology did not ensure men's ability to pay for even basic elements of the Office infrastructure such as the irrigation network, let alone road maintenance and schools. As with the metal silos, some proposed Office technologies were mere window dressing: something new that made the Office appear more modern.

By 1961, the move to intensify mechanization was recognized as a financial failure, and the CRM was reconverted to family farms. In those same years, the Office ceased using tractors for field labor.[93] Privately, the upper levels of the administration admitted that even where mechanical labor and harvesting produced more crops, individual farmers did not see any increase in income because the technology was so costly. As a consequence, numerous farmers refused to pay for these services. In some cases entire towns refused to pay, as was the case for Segou-Koura and San-Kura in 1958 and 1959.[94] Other towns sent formal delegates to request assistance from political officials in lowering the fees.[95] The Office consistently faced problems collecting the water and fees from farmers, and the project operated in the red for decades.[96] It was in this context that some officials even questioned the ability of Office farmers to adapt to rural industrialization (rather than blame management).[97] Large-scale technology at the Office had become a spectacular failure in the same years that women were enthusiastically adopting new but far more modest technologies.

In many ways, the Office du Niger was similar to other large-scale agricultural development schemes focused on controlling agricultural production over the immediate welfare of participants. Office fields were planned according to the rational aesthetics of what looked like modern agriculture to the French colonial eye. This meant straight lines and uniform plots. Office staff members recorded field allotments, projected yields and population numbers, and advised farmers on the use of proper inputs. Yet, what enabled the Office stay in operation over the years was the labor of farmers and their families. In the vocabulary of James Scott, much of this activity was illegible, or out of the range of what the Office administrators chose to see.[98] The vast majority of families cultivated fields outside the prescribed and orderly Office zone. They also regularly avoided collection centers and sold their cotton and rice in domestic markets. At times the institutional technology of the Office facilitated additional food production such as fishing. Men and women also diverted water from the canals to outside food crop fields. However, the supply of water and fish in the canals was unpredictable. Most of the overtly mechanical elements of the scheme were unreliable from day to day.

As early as the 1950s, many residents were unsurprisingly ambivalent about the possibilities that large-scale technology offered and about the continual push for increased mechanization. For example, farmers were suspicious of the new all-mechanized sector. In a 1957 article in the African newspaper *L'Essor*, the writer accused an Office staff member of refusing farmers the right to expand their cultivation activities in favor of machines. The implication was that the growing number of machines threatened farmers' livelihoods and the overall well-being of Office residents.[99] The newspaper consistently adopted pro-labor views and often printed exposés about conditions for farmers at the Office. This article posed a question: What did it matter for farmers if the Office increased production if the people the project claimed to be helping struggled to make a living? This very tension had been foreshadowed by Bélime's contradictory emphasis on bringing large numbers of farmers to the project and at the same time planning for future motorized cultivation that would reduce the number of fieldworkers. From the perspective of farmers, the Office promoted production at any cost, whether by man or machine. Men living at the Office wanted to increase their own farm production by planting more fields, but their understanding of a good harvest did not necessarily match the official emphasis on cash-crop yields.

Moreover, the increasing presence of agricultural machines challenged the meaning of farm work for men at the Office. What of the social value of physical labor in the fields? The importance of men's farm work had long been associated with the ardor of their labor in the fields. It was a symbolic relationship renewed through annual agricultural rites and dances. The public performances were also meant to transmit agricultural and social knowledge to younger men. Those who excelled in the fields were celebrated as champion farmers.[100] By the late 1940s, when the anthropologist Dominique Zahan was conducting fieldwork at the Office and in neighboring regions, he noted that these agricultural rites were declining and surmised it was due in part to the new technologies. First, the introduction of the plow took emphasis in farming away from the embodied knowledge tied to the relationship between a man's physicality and the hoe.[101] Mechanization at the scheme only amplified this challenge to the meaning of men's farming labor. At the same time, the workers who learned to operate the new plows, tractors, and threshers gained new masculine prestige from their new technological mastery.[102] Increasingly, the machines were celebrated over men's embodied skill and labor.

Tractors, harvest machines, and industrial rice threshers epitomized modern agriculture for the colonial administration (see figure 4.3). They were also embraced by many wage workers. However, these same machines, rather than improving food security through increased production, often served to reduce what farmers and their families had available to eat. Even where people grew rice, families often paid so much in fees that after the sale of the harvest families went hungry. It was for this reason that women's agricultural production and cash earnings for food purchases were so important in the same period. For example, in September 1956, a group of farmers from Kossouka, in the rice zone, requested millet and seeds for the next planting season from their local administrator. The administrator had called the meeting to discuss water fees, but the farmers were more concerned with what they would eat in the season after the sale of the harvest.[103] Once farmers had sold their crop to the Office, they had only the cash or grain they had put aside to last the next six months until the first harvest of the next season. The Kossouka farmers insisted that project staff ought to address their pressing food needs first.

FIGURE 4.3. Mechanical threshing in an Office field. Courtesy of the Archives Nationales de la France d'Outre Mer. FR ANOM 8Fi 417/171 Office du Niger Aménagement rurale, 1935 to 1954.

The material impact of mechanization on the diet was to prolong periods of shortage or uncertainty. Male household heads usually did not receive enough money through sales to the Office to purchase extra grain when the need arose. This generally occurred during the hungry season when grain prices were at their peak. During the rainy season, many families like those of the farmers in Kossouka were forced to eat their seed store to get by. When the income of the household head suffered due to the high price of all the technological services, families had less to eat, unless the woman cooking that day was able to cover the costs of the grain and sauce. Moroever, it was because of the Office tractors and bulldozers that wild food products were noticeably scarce in this season. The strategy employed by most farmers, especially in the area where they grew cotton instead of rice, was to spend more time on millet fields planted outside of the project land area.

The rice harvesters and threshers, in particular, had the added inconvenience of slowing the harvest season. These machines were large and unwieldy, and when they were most needed in the fields, they often got stuck en route because the roads and paths were still muddy from the rainy season.[104] The roads to Office towns were in constant need of repair, and travel to the Kokry area was often undertaken by motorized boats on the Niger River. Boats also transported paddy, but the large unwieldy machines of so-called modern agriculture meant the threshing lasted well into the months of the hot season—when local markets were full of product months earlier. Most farmers were strapped for cash until they sold their harvest. Waiting to sell crops to the Office prolonged the end of the hungry season. The fuel requirements for these machines, and the large number of workers required to thresh the thousands of kilograms of paddy per day also added to costs. The accumulated water, tractor, harvest, and threshing costs that farmers were expected to pay meant that these technologies had an ongoing cost, yet farmers saw little in return.[105] This contrasted with the one-time investment women made in household goods from which they saw immediate benefit in easing their burden of cooking on a daily basis.

THRESHING MACHINES AND THE GENDER OF THE HARVEST

Even as the Office became increasingly industrialized, women relied on their physical labor to make the machines produce food. Women by and large processed rice for home consumption by hand. Whereas machines such as the industrial rice threshers offered potential labor savings for women, they were not generally employed to process grains for home consumption. When the agronomist Réné Dumont visited the Office in 1950, he suggested women would have more labor time in the fields (thereby increasing production) if the Office employed the threshers for home use. The reply from one staff member to Dumont's suggestion was dismissive. His internal response to Dumont's suggestion about giving women access to the project's much-lauded technology was twofold: (1) women do not farm and (2) men at the project were not in favor of their wives using mechanized threshing or other modern technologies.[106] Dumont did not directly propose the expanded use of threshing machines as a means of empowering women. Rather he pointed out that allowing women access to Office technology would also mean an increase in crop production.

From the perspective of the Office official, women did not factor into agricultural production. For that matter, he seemed to think Office technology had nothing whatsoever to do with women. He reinforced this viewpoint with a colonial official's fear about disrupting gender and labor relations in the home and thus undermining African men's authority. This Official depicted African men (and not the European-administered Office) as holding their wives back from accessing technology. He missed one reason men might have opposed greater mechanical threshing. The more rice processed by the Office, the more the household head was charged in fees. The costs reduced the household head's profit, which adversely affected the well-being of his family.

This internally documented exchange between two colonial experts highlighted the institutional shaping of access to technology along gender lines. Initially, men were urged to purchase plows to increase cash-crop production. Men were also expected to pay for a host of mechanical services. While planners anticipated that plows, technical mastery of irrigation canals, and heavy machinery would transform the men into modern farmers, no official expected threshing machines to make modern women. The Office staff member who objected to the mechanical threshing of rice for domestic use reinforced this masculine bias. In this way, the institution determined who was supposed to use and benefit from French agricultural equipment. Moreover, French officials had long expected that the technology they brought to the French Soudan would be used by African men exclusively to increase their participation in the colonial market. Agriculture for colonial export was already a male affair, even when women were ready to adopt new French technologies for domestic use. Nevertheless, women were intimately engaged in the processes of industrialization and agricultural change. Women at the Office chose to draw water from the canals with a metal bucket or prepare food in a metal pot. In addition, women made use of project's industrial machines.

While the Office threshers did not process much rice for home use in the 1950s, women did make use of the machines to support household needs. Women were widely employed to clean the grains processed by these machines. During the harvest, male workers operated the thresher while women winnowed by hand alongside them. They normally did this work in exchange for rice to sell or, less commonly, for cash. The cash that women eventually made from this labor did not lessen the cooking load at home, but it did support the purchase of foodstuffs. Women were not obligated to do this work; rather they chose it as a way to earn extra cash. Male farmers, by contrast, were obligated to use the industrial machines in their fields.

In the cotton-growing region, women also entered the industrial production process by assisting the male household head at harvest time. Picking cotton was labor-intensive work, and the surplus of domestic cotton that women processed for their families did not benefit from any industrial processing, as was the case with export cotton. As with other infrastructural elements of the project, women adapted the industrial farming machines to their labor regime. The introduction of machines for processing the harvest eased some of the added burden on women that the Office had created. For example, up until the late 1950s women manually cleaned all the cotton delivered to the Office. However, the scale of cotton production greatly increased the amount of cotton that women had to clean. By 1960, cotton sold to the Office was cleaned by machine in a factory in Niono (see figure 4.4).[107] Male workers and the machines, in effect, took over some of the women's labor. The increased use of threshing machines similarly translated into a lighter overall load of rice threshing for women in other sectors of the scheme. Machines eventually processed more of the harvest intended for sale, leaving women to process the rice kept for home consumption.[108] Ironically, in the mechanized world of the Office, the value of women's physical labor was higher than ever.

WOMEN AND MACHINES IN MECHANIZED MOLODO

In the 1950s Bintu Traoré moved to Molodo-Centre (the site of the all-mechanized CRM). Bintu remembered that when she arrived all the women in town (who were mostly wives of CRM wage workers) were buying their sauce ingredients in the market. Bintu came from a small town

FIGURE 4.4. Workers processing cotton by machine in Niono. Photo by George Rodger. Courtesy of the National Archives. NARA RG 286 Photographs of Marshall Plan Activities in Europe and Africa, ca. 1948 to 1955. Photo no. 541637.

near Boky-Were where women harvested and processed what they needed to cook. Faced with a new cash-oriented sauce market, Bintu joined other women in town who winnowed alongside the Office threshing machines.[109] In fact, most women at Molodo-Centre had to purchase their sauce ingredients because at the time all the fields within walking distance were set aside for mechanized rice farming. Molodo was meant to be an industrial farming center. As such, there were no individual farm families, only workers.

Thus, women did not even have access to gardening space bordering the rice fields. They also had little access to surrounding common wooded areas to collect ingredients from because many trees had been cleared for fields (or these areas were controlled by women in Molodo-Bamana).

Given these constraints to women's production, cash was especially important in town. Bintu used cash from her husband's wages and her own earnings to buy cabbage, peanut oil, and potatoes in the market, although like most families of workers her family also ate rice produced at the center.[110] Her family's diet reflected the prevalence of new sauce ingredients not common in older recipes for sauces made with *datu* or baobab leaves. She used cabbage and other ingredients purchased in the market to make new sauces to accompany the rice. Another change to the diet was the substitution of rice for millet. Bintu's work for the Office epitomized how women adapted to the particularities of Office life. One such aspect was the prevalence of machines for agricultural production and wage labor.

Increased mechanization provided some women (but not all) with new opportunities to earn cash. Like Bintu, women were paid to winnow or clean rice grains after machines threshed the rice paddy. These machines threshed large amounts of paddy, shooting out piles of rice to be bagged and shipped for sale. It was the women's job to clean the rice before it was bagged, and sometimes they also bagged the rice. In Molodo, where all agricultural tasks were mechanized, wives of workers simply winnowed the threshed rice in large hangars. In Kokry, the oldest center for rice collection, wives of wage workers also did this grain-processing work. Women there were paid every three days with one sack of rice. Kadja Coulibaly was one of the women selected to winnow in Kokry and remembers that she worked with fourteen other women who were also wives of workers.[111] Fewer women than men worked directly for the Office, but the numbers of women who worked alongside these machines were significant. Women in farm households winnowed when the machines came to their fields. Generally, the local agricultural instructor maintained a harvest schedule. When a town's fields were ready to be harvested, he ordered several machines to facilitate the harvest. Usually men carried the paddy to the machines. When the paddy from one household's fields had been threshed, a group of women from that household winnowed all the rice tossed out by the machine. At the end of the day, they each received one calabash of rice. This was the standard labor organization and rate of compensation for women in farm families all around the Office.[112]

The advent of rice threshing machines altered the gender and temporal labor regimes during the harvest. Women were accustomed to transporting cut crops from the fields for storage. While women still transported some of the harvested crops for storage, the paddy that was transported to the machines was carried by male wage workers or by men in the household associated with the field.[113] Vehicles and carts also transported crops for sale from the fields.[114] This traditionally female task in Office fields was now carried out by either men or machines. Some of the men working for wages at the mechanized rice center specifically carried out this labor for pay. In some cases, young men traveled from Molodo-Bamana to the CRM by foot or bicycle to work during the day and return at the end of the workday. They collected the threshed and processed rice to fill the sacks and then carried the sacks filled with rice to project store houses.[115] Even though the nature of the harvesting work was gendered as female, working with the machines was associated with men's labor.[116] Men who worked for the CRM were paid cash wages, and they were listed on employment registers. Women were less formally compensated, receiving measures of rice at the end of the day. Moreover, their day-to-day employment was not officially recorded with the project administration.

On the one hand, the arrival of some of these machines saved women from certain labor obligations. Men provided some of women's customary harvesting labor for the portions of the harvest sold to the Office. On the other hand, winnowing large quantities of rice during the harvest added other labor to the work women generally already did during the harvest. Ordinarily, women processed grains for cooking on an as-needed basis. They did not thresh or winnow great amounts of grain all at once. When harvesting and threshing machines arrived in the fields, women spent all day winnowing so that the bagged rice sold by the Office was clean. This was a practice instituted by Office inspectors in an effort to improve the market quality of the project's rice. In previous decades, officials from the French Soudan received word from administrators in Dakar that some grains furnished in the course of wartime provisioning were substandard. They found fault with the fact that the grains were not cleaned prior to shipment.[117] Employing women to winnow the rice was one way to address this problem. The Office transformed this particular household task into paid labor, even though women were not considered workers for the Office in the same way as men. The women's efforts could also be categorized as household labor because they used the rice they received for their winnowing services to cook or purchase other foodstuffs.[118]

Still, women regularly threshed rice paddy by hand for home consumption. Children and even men also threshed on occasion.[119] Mechanical threshing services were reserved for the portion of the harvest that was delivered to the Office, and farmers were charged considerable fees for such work. Therefore, when women threshed by hand they saved on any further payment to the Office. This resulted in savings for the household head (who controlled the Office fields and all associated fees). While women did not see an increase in their own earnings, their extra work did allow men to save on the fees they paid to the Office, thus leaving more money for the family food budget.

Home threshing also occupied a significant portion of women's time during the millet grain harvest. In Niono this period coincided with the cotton harvest. When farmers in the cotton sector also began to cultivate rice, administrators for the Office hoped to capture more of women's labor time. Mechanical threshing became a solution to the need for more women in the fields (unlike the earlier resistance to women's use of rice threshers). In 1954 Office administrators instructed staff to encourage mechanical threshing for home consumption. The guide to instructors stated the following:

> Threshing among farmers is done by women who collect paddy from the fields little by little according to family needs. Strictly speaking threshing takes place for 2 or 3 months after the harvest. It would be opportune during the cotton harvest, which requires a lot of manual labor to avoid overlap between the two activities. It is advisable for you to persuade farmers to ask for mechanical threshing during the harvest. This way the paddy would be quickly collected and stored away from all sorts of parasites, limiting losses to time and product.[120]

Instead of reducing women's labor time, mechanization in this context only shifted it to the Office fields. Mechanization would most benefit the project—not women laborers. By contrast, women usually chose to manage their labor time in ways that were most efficient for the production of food for the household. They spent valuable time winnowing in order to obtain rice and ultimately cash for going to market. They also hand threshed rice and millet for home consumption. However, they were happy to allow machines to clean the cotton harvest. Mechanization served women less as time-saving technology than as a means to acquire cash.

The late colonial Office was certainly not a triumph of industrial technology or modern agriculture as it was promoted by administration officials, even though some residents of the project embraced aspects of the scheme. From the late 1940s into the first decade of independence in the 1960s, the heavy expenses entailed in large-scale irrigation, field preparation, and mechanical harvesting financially crippled both farmers and the Office itself. The industrialization of the project greatly impacted the availability of food resources, but women made the shifting mechanized landscape productive through their labor and by using technologies that were more modest in design than the great industrial farm envisioned by Emile Bélime. Technologies coded as feminine such as buckets, cooking pots, and other metal items like wash basins increasingly made their way into Office households. These modest objects entered the Office through the regular social rhythm of rainy-season weddings, whereas large-scale technology had the trappings of the colonial (and later postcolonial) state. It bears repeating that households integrated the aluminum pots and metal buckets into their labors more easily than adapting tractors and threshers to daily farming conditions. Being relatively cheap technologies, the pots and buckets had a sweeping impact on women's labor, their economic capacity, and the way the community experienced food and labor time on a daily basis. They also acquired meaning as gendered objects for women's work and in daily use displayed women's ongoing value to a changing and even industrializing rural world.

5 ↪ Rice Babies and Food Aid
Reengineering Women's Labor and Taste during the Great Sahel Drought

FATOUMATA COULIBALY vividly remembers the Great Sahel Drought. On one day, like many others, she rose early to go into the irrigated rice fields that she, her husband, and their household tended for the Office. She called for her sister-in-law to be on the lookout for the guards while she prepared to leave. Fatoumata dressed in a big tunic over her wrap and grabbed some extra cloth. The household food stores were nearly empty, and she needed to get some more rice from the fields. Beginning in the early 1970s the recurrent crises that would be collectively referred to as the "Great Sahel Drought" had already caused famine across the Sahel region. After 1974, the cash-strapped Malian government moved to tightly control rice production at the project. At the time farmers and their families were only allowed a limited ration from the fields they cultivated, and they had to have a paper permission slip to take any rice home. Otherwise, one of the guards stationed between the fields and project towns would confiscate all grains being transported from the field. If a woman did not have the paper providing authorization, she did as Fatoumata did: she either found a way to sneak more rice home or went hungry until the next ration day.

In these years, many farm households at the Office did not earn enough from their official rice sales to the state grain board to pay for any other food. Women like Fatoumata cooked the rice they took from the fields or sold it for fish or salt. If women got their hands on some rice from the fields it was always for eating or was sold to pay for something else to put in the sauce. This particular morning Fatoumata and her companion left for the field, not following the road but winding through other people's fields until they reached their own household's rice crop. Fatoumata had set aside some rice the day before when she was working in the field. Now she wrapped it with her extra cloth into a small round package that she tied around her stomach and under her big overshirt. She was "pregnant" with a *rice baby*. She would later use the same ruse to carry the rice to the market. In an interview, Fatoumata, with a loud laugh, said that when she got to the market "it was like you gave birth!"[1] Certainly it was a happy occasion to have made it all the way to the market. In recounting the story, she playfully alluded to the strong association between women's fertility and the harvest. However, Fatoumata's association of hiding rice—as many women did during this time—with childbirth also speaks to the difficult bodily experience of hunger from the era. As her story makes plain, food politics continued to have a lot to do with women's bodies.

THE GREAT SAHEL DROUGHT AND
FOOD POLITICS IN POSTCOLONIAL MALI

Mali gained its independence in 1960, and not more than a decade later food security was a critical matter for national sovereignty. The Office had maintained its position as a major agricultural project, and state policies continued to emphasize the technical infrastructure of the scheme now as a means to improve national agricultural production. In reality, the dam and canals only assured a bare minimum of food security for its residents and the surrounding region, repeating old patterns of the colonial era. Eating every day required a great deal of effort in the midst of a large-scale agricultural enterprise that was unique in increasing its overall grain production during the 1970s. Out of necessity women reengineered food production again, incorporating rice babies and new food aid products into daily meals. However, women lamented the poor quality of these resources. Like Fatoumata, women more often relied on their own physical labor and their ability to transform their bodies in order to ensure survival.

The Great Sahel Drought was a period of wrenching hunger and extreme fatigue for many Malians. From 1969 to 1973 much of West Africa experienced dramatically low rainfalls, resulting in a succession of poor harvests and widespread food shortages. The extended drought devastated communities across the region. Famine threatened the worst affected regions, especially northern nomadic and herding communities, thousands of whom sought refuge at the Office and further south. In a 1974 meeting of Sahelian states, Mali's new military leader, Moussa Traoré, recounted the horrifying impacts of the ecological crisis still unfolding: "Food crops dried up in the fields before the plants even had the chance to mature on the stalk. . . . Farmers witnessed the loss of long months of labor and the destruction of a year's worth of resources with death in their souls. Formerly majestic rivers carried only unrecognizable and pitiable streams of water. Entire villages sought refuge and a precarious survival in the south. Herders brought skeletal animals to lost water sources and rumored graze lands. Cattle paths were strewn with the dried carcasses of lost animals."[2] Even as rain returned to the region, the ecological damage was severe, and another series of droughts in the early 1980s compounded the damage to the environment. Indeed, the drought was one of the worst natural disasters of the twentieth century and is ever present in local memory.[3]

Malians recall their experiences of living through the Great Sahel Drought in starkly visceral terms. Nana Dembélé, who was a young woman during those difficult years, remembered eating red millet, a widely distributed form of food aid. She, like many others, recalls with intensity its poor quality and disagreeable taste. It was so unpleasant to eat that "you would cry if you ate it."[4] Nana's description of eating red millet highlights the frustration felt by many Malians—and for that matter of other Sahelian residents—with the inadequate aid they received in the midst of severe hunger.[5] Women learned to make meals with red millet. But to those who cooked with it or ate it, the food aid millet signified extreme hunger, disappointment, even shame. Red millet epitomized the hardship of those years to such an extent that many Malians refer to the period as the "Famine of the Red Millet," and the generation who grew up during the drought was called *nyoblé si* (the red millet generation).[6] Tellingly, this terminology highlights famine and hunger, but especially the poor quality of food, over the ecological crisis of drought. Indeed, the experience of the Great Sahel Drought was a distinctly embodied one.

By the early 1970s the international press began to publicize the grim realities of the ongoing crisis. These same journalists also criticized inefficient international aid operations and the slow responses of the Sahelian countries. Reporters covering the drought regularly cited as many as a hundred thousand deaths resulting from the famine.[7] It was not until 1972 that national and international aid was marshaled on a large scale.[8]

At that point, international donors began shipping significant quantities of food aid to the region, and this would continue well into the late 1980s. Malnutrition persisted despite this assistance, and hunger in the region was only exacerbated when another drought hit the region between 1982 and 1985. For the government of Mali, "food sovereignty" became an important political goal.

Yet for many people suffering from food shortages, the first source of assistance was local. As hundreds and later thousands of people left their homes, they expected and received both food and lodging from their neighbors, including farmers with irrigated fields at the Office. The Malian scholar Téréba Togola suggested that the region's well-known hospitality is rooted in the history of past environmental disasters. He writes that the "Mande tell of times of economic stress and interethnic food aid during periods of crisis." Togola further argues that this sharing practice in times of famine has persisted and is "embedded in both cultural and social values."[9] At the same time, such assistance incurs long-standing social debts that are similarly recorded in myth and everyday social interactions.[10] From this perspective, hospitality derives in part from broad social obligations and even indebtedness. It was a broadly shared understanding for responding to regional crises and shaped the way many Malians understood and received food aid.

The national government framed food security as a specifically national goal or, more specifically, one of food *sovereignty*.[11] A nationalized approach to agricultural production was not necessarily a new strategy. Following independence in 1960 the first president of Mali, Modibo Keita, pursued a national economic development policy that stressed increasing agricultural exports through collective production. However, food production was not prioritized, even though the food supply had been a popular concern before the emergence of the Great Sahel Drought. Indeed, despite the fact that Mali had long produced its own grain supply, the country was increasingly reliant on imports by the late 1960s. A proponent of African socialism, Keita pursued agricultural collectivization and nationalized food distribution. City and town residents queued to purchase rice and other basic foodstuffs from the national food board (OPAM).[12] This shift to nationalization did little to ensure access to food for ordinary urban Malians or even for the farmers producing the food. Before 1968, when Moussa Traoré overthrew Keita, popular perception was that OPAM only fed the elites. Adam Bah, who as a young woman collected rice from the floor of an Office pig pen to make money (see chapter 3), was married to a man who worked as an agricultural engineer under the Traoré administration. They lived in Bamako during the Keita years. She remembers waiting in line for rice (with a six-kilo limit per person). At the same time, she observed that many others had no access to rice at all. Remembering those years, she noted that "only the rich ate, and the poor had nothing."[13]

Imported red millet also made its first appearance in the Malian food supply during the Keita years. In Adam Bah's recollection, Mali imported the millet from China because people were hungry. In short the agricultural policies of the Keita administration resulted in nationwide food scarcity. Adam Bah described this millet as particularly unpalatable, not unlike the detested food aid red millet from the Great Sahel Drought. In fact, she believed the red millet was a type of food intended for animals (but being sold in Mali to people).[14] It was a charge suggesting that the elites of the Keita administration disrespected the Malian people despite the socialist rhetoric calling on Malians to work together. In further assessing the food policies during the Modibo years, Adam Bah recalled that these policies "tired everyone."[15] Her language speaks volumes about the embodied experiences of Malians that extended into the drought years. They were tired of waiting in line but also tired of working in collective fields and dealing with chronic hunger.

COLLECTIVE PRODUCTION AND THE
FOOD SUPPLY AT THE OFFICE IN THE 1960S

Between 1960 and 1968 the first postcolonial administration experimented with national collectivization.[16] At the Office this meant the creation of new collective villages. Men and women were organized into work units for field labor, and they turned over the vast majority of their harvest to the state.[17] It was during this era that the state also created the Economic Police. This national military unit confiscated surplus agricultural products (like rice, cotton, or millet). Officers from this force frequently patrolled the markets and took what people were trying to sell. What constituted a surplus was generally decided by the police on the spot.[18] At the Office, additional guards working for the institution worked with the Economic Police. Both groups tightly guarded the state harvest and harassed women. The Economic Police and the Office guards of the 1960s set a precedent for militarized agricultural control at the Office, which would continue for roughly fifteen years.

Broadly speaking, economic development under both Modibo Keita and Moussa Traoré centered on agriculture and called for intensive physical labor. The state-run Office presented the opportunity to control that labor to a greater extent than was possible in many other rural areas. Not long after independence, the Office issued a call for new settlers. A plethora of new towns and irrigated fields followed this recruitment drive. Men such as Kono Dieunta's father were attracted by the promise of productive land. Kono's father moved his entire family, including his daughter Kono and her husband, to Tongolo-Koura. There she remembers that people came from "everywhere" to make use of irrigation for farming.[19] Dominant political rhetoric depicted the arrival and work of the new farmers as overtly patriotic. As an example, in 1962, the state-supported newspaper *L'Essor* published an article extolling the Office as a "national worksite." Photos of extensive cotton fields and piles of processed cotton accompanied the text, suggesting an abundance of wealth both for the nation and for farmers (who were still assumed to be male).[20] The postcolonial Office, like its colonial predecessor, chose which crops new settlers such as Kono and her family would grow. In the 1960s, cotton (rather than food crops like rice, or even millet), as depicted in the article, symbolized the fruits of this newly patriotic labor.

The Office retained much of the colonial-era orientation and focus on cash-crop farming.[21] Agricultural policy at the Office continued to emphasize production for the urban and export markets to the detriment of local food needs. Indeed, the new Malian administrators oversaw cotton production with renewed energy as all households, even in the former rice sectors, were required to grow some cotton. Farmers who refused to grow cotton risked eviction.[22] As in previous years, cotton grown on a large scale was labor intensive, and the commercial varieties promoted by the national Office still demanded a great deal of work that overlapped with the grain harvest. Picking cotton remained women's work, and their food cultivation activities again suffered. For these reasons, farmers who had been at the project for decades under colonial rule still preferred rice as a cash crop.[23] Yet, the Office of the 1960s vigorously pursued *increased* cotton cultivation.

While cotton was the poster crop, the Office also pursued intensified rice cultivation. To some extent, then, the new administration sought to shift Mali's position from an overtly colonial production center to a national one. They now strongly associated rice production with provisioning Mali's urban markets, even though some Malian rice was still sold in the regional export market (predominantly to the Ivory Coast).[24] In an attempt to nationalize the rice delivered to the market, the Keita government declared a state monopoly over the Office harvest. This policy forced farmers to sell rice (and cotton) to the Office at fixed prices, which were generally lower than the prices offered in local markets. Farmers could only keep enough rice for home consumption.[25]

Throughout the 1960s, the poor financial state of the Office inherited from the colonial government persisted and worsened. One result was the degradation of much of the technical infrastructure. Canal maintenance suffered as did the upkeep of tractors, threshers, and the like. The fiscal troubles of the institution were matched by the persistently poor economic state of its farmers. Farmers still sold their harvests for little profit. Yet they paid high prices for basic foodstuffs like peanuts, fresh meat, milk, fresh fish, shea butter, and onions. Prices in Office areas were much higher than the prices for the same products in other regions. For example, in 1961, shea butter was 54.9 CFA[26] per kilogram outside the Office and 65.8 CFA per kilogram in Office regions.[27] In some cases, prices in Office markets were more than double the average rural price; this price disparity was consistent into the 1970s and 1980s.[28]

In the years immediately following the 1968 coup, the region was hit by drought and the threat of famine (roughly 1969–73). In response, policymakers under the new military leader, Moussa Traoré, elaborated a national strategy for food production and agricultural sovereignty. The Office was central to this effort because it made use of irrigation from the Niger River. While Traoré had ended the policy of collective production, great pressure was still brought to bear on farmers. Even though millet remained the predominant staple of rural populations, rice was considered a critical national staple because the numbers of urban residents eating rice had increased (including workers on the state payroll).[29] As a result, the supply and price of rice was politically important. In fact, Mali had to import rice to meet the growing demand after 1968.[30]

The Office produced 40 percent of the total national consumption of rice.[31] In fact, while overall food production decreased in Mali between 1960 and the mid-1980s, the Office increased its rice production during a brief period in the mid to late 1970s.[32] In 1968–69 the total recorded production of rice paddy was forty-five thousand tons. It went up to seventy-five thousand tons in 1972–73 (probably due more to increased yields than to closer state control of the harvest).[33] The Office certainly alleviated some of the urban and state need for rice.[34] However, in the actual context of national food supply, its contribution was limited. Ironically, even Office residents who were farming for the nation would rely heavily on food aid when it arrived.

Women's responses to the cereal controls at the Office were marked by specifically gendered claims to the harvest. Women made and gave birth to rice babies. They also made rice bundles conform to their bodies in other ways. It is useful here to recall Mauss's analysis of bodily techniques, specifically his description of the body as a natural tool.[35] During the drought women at the Office followed suit. They turned their own bodies into "pregnant" ones. The underlying success of this technique was that pregnancy was understood to be a normative state for women and, therefore, should have been an unremarkable sight for the guards. Hiding rice was also accomplished by mimicking the act of carrying a child (rice bundle) on the back (see figure 5.1). In this way, women transformed mundane bodily acts, which were perceived as part of normal women's being (pregnancy) or ordinary female behavior (carrying a child) into a subversive act. In a sense the women's bodies became tools for collecting the harvest.[36]

Women at the Office also addressed food concerns by living in and interacting with the very particular space of the project's technological apparatus and altered natural environment. During the years of drought and intensified surveillance the Office environment changed. This altered how women engaged with the agricultural space. Most obviously, their bodies became essential for hiding and keeping as much of the harvest as possible. The setting further changed with the arrival of food aid cereals. These were grown outside the region and were often of poor quality. Women still had to transform them into edible meals. To do so, they drew on their own senses and acquired knowledge of food, taste, and aesthetics.[37]

RICE BABIES AND OFFICE SURVEILLANCE

After the appearance of the Economic Police and other guards, women across the Office began smuggling rice from their household fields. When residents (women and men) were asked about the guards, the memory of hiding rice under their clothes was always mentioned first. Most everyone also emphasized that all women did this; no one woman was isolated in secretly taking rice.[38] Many women like Fatoumata Coulibaly (who told the story about giving birth to rice in the market) laugh now when they recall the practice. A few women also hid rice around their posterior. When a woman was caught carrying rice in this way by a guard it was often preceded by a comment like: "Your butt is not that big; you have something."[39] Most likely, this comment not only underscored the fact that women were extremely thin in these economically lean years but also reveals the gendered and sexualized dimension of struggles over the harvest.

FIGURE 5.1. The technique of carrying a baby on the back, 1973. Courtesy of the Library of Congress. Prints and Photographs Division, lot no. 11515 (17). Photo collection credited to the Food and Agricultural Organization and the World Food Program.

Guards' comments about women's bodies highlight the harsh scrutiny women were under. The words used in these comments also emphasized the women's vulnerability. Other guards questioned women who suddenly appeared pregnant overnight.[40] Nièni Tangara (who moved to the Office in the 1970s to marry her husband and was a young woman at the time) surmised that women hid rice in this way because the guards would not touch a married woman. Yet she admitted that when guards questioned her she quickly gave up her rice out of fear.[41] From other women's accounts it is apparent that the guards physically searched any woman they suspected of smuggling rice. Other women who were afraid of being touched under their clothes hid rice under their head wraps.[42] Another strategy was to wear the biggest garments possible to cover their bodies and the rice baby.[43] Older women were certainly aware of or had previously experienced unwanted physical attention from monitors during the colonial era. The archival record further suggests that complaints by women in this regard continued into the early 1960s.[44] Once the guards were on the lookout for women smuggling rice, the possibilities of unwanted touching and molestation only increased.

Djewari Samaké (who moved to the Office during the colonial era) did not laugh when she discussed the guards. She recounted going into the fields early in the morning with her sister. Like many other women she only went for rice in the company of other women. Djewari's hiding method was to put some rice in a wash basin under a cloth. Then she said that "the guards were always watching for people coming from the fields, if they saw you they would search everywhere including your head, belly, and back."[45] If the guards searched Djewari's basin first and found her rice, she was saved from being touched in this way. Oumou Dembélé remembered that when the guards did find rice in a woman's clothes or under her head wrap, they often violently tossed the cloth back at her. As she told the story she mimicked the guard's action with her own headscarf to highlight her point.[46] It was a time of physical hunger but also one of intimate intrusions by the state into women's bodily space and household labor.

Women were particularly aggrieved by the guards and felt they had the right to the rice they took. In part, women (especially Bamana women) customarily gathered rice from grain fields not just at the Office but across Mali.[47] Leftover grains were considered a surplus, and women had the right to sell what they collected from the ground, or they could use it for home consumption. For this reason Aramata Diarra emphasized that what she did was not "stealing," as rice smuggling was described by male farmers. Aramata only took what rice she could gather from the ground.[48] The gendered language is interesting to note. Men said when they took rice it was like "stealing" from their own fields. Women only described their own actions: cutting, threshing, and bundling the rice out of plain sight. They also described the threats of violence toward women caught with rice. Women clearly felt that the illegitimate actions lay with the guards. While the Office enforced a strict ration on the men, women believed that the guards should at least let them, as *women* with customary rights, glean some rice.

Ordinarily, women acquired grains through a variety of means such as gleaning.[49] These practices, as discussed in previous chapters, were especially important for women without their own fields. Another way that young women acquired small amounts of millet or rice was to set some aside for themselves when they winnowed for the household.[50] When women performed labor in a man's fields during the harvest, they were also generally paid with a measure of grain.[51] In all these cases, women earned or claimed relatively small quantities of grain, usually no more than a calabash's worth. Women seem to have considered the rice smuggling as a form of such customary rights. In fact, some guards did let women pass once in a while without searching them.

With the arrival of the guards, women could not always be sure that their customary rights would be upheld. In a lot of cases, gleaned rice became confiscated rice. Rice that women earned by doing work in their husband's or other men's fields was also subject to confiscation. During the harvest, even during the difficult years of drought and heavy surveillance, men in charge of fields would call women from their own and other households to help bring in the crop. In return, women were paid with a calabash of rice (similar to the colonial-era practice). The militarized production and control of the harvest did not regularly allow for these otherwise typical exchanges. In these cases, women did not have written permission from an Office staff member to carry their earnings (i.e., the rice).[52] They only had the consent of the man in charge of that field. The problem was that he no longer legally controlled his household's production. As noted elsewhere in this book, the state monopoly over the harvest undercut farmer (male) profits.[53] It is also clear that enforcing state control over production disturbed household dynamics and the gendered distribution of the harvest.

Women needed rice not just to eat but to sell in exchange for necessities like salt, fish, other sauce ingredients, or soap and clothes for the family. In previous decades many women had become adept at trading cotton and rice for other foodstuffs (see chapter 3). As the government controls tightened, women struggled just to keep some rice to sell. In addition, a significant number of male farmers saw absolutely no profit for at least a decade under the Traoré regime.[54] The families of these men must have relied heavily on women's efforts. Even households with cash on hand found it difficult to purchase other cereals like millet from outside the Office. The barriers that kept rice from leaving the Office also stopped millet from circulating.[55]

The need to sell rice for other foodstuffs added labor to women's already weighty duties. Women rose at 2 or 3 a.m. to collect rice that they had set aside the day before, and often they left the fields to head directly to the markets, which were sometimes located a great distance away from the fields. Women undertook these trips knowing that even if they successfully hid the rice as they left the fields, they could be stopped en route to the market and searched. This happened to Assane Coulibaly one day when she was lucky enough to find a ride to Macina on a wagon. That same day she lost all her rice.[56] In 2010, it was still a vivid memory. Assane Pléah, who lived just opposite from the major Office market town of Niono, transported her rice by wading through the water of the canal that separated her town from Niono. This task was made easier because the water levels in the canal had dropped due to the drought. As Assane crossed, she pushed a wash basin containing her rice covered with a cloth. In this way, she hoped to miss the guards who focused most of their attention on the fields and roads.[57] Now the canals, which were in many ways symbolic of state control over production (colonial and postcolonial), could enable women to circumvent the cereal controls and unwanted searches.

During the harvest, male farmers were officially permitted to take rice from their fields once a week. The formal process was as follows: the male head of the household requested permission from his monitor for the amount of rice he wished to take. Typically, the staff member allotted the man about half of the quantity that he requested. Farmers tended to have very exact memories of getting these paper permissions for rice. Moussa Coulibaly explained that every Friday his family was allotted only two sacks of rice.[58] Given the overburdened size of most households—because they were hosting hungry guests—even two large sacks of rice would not feed the family for the whole week. Once a man had the paper in hand, he had to show it to the guard who would allow the farmer to pass. He and other accompanying family members then had to hand thresh their "ration."[59]

Generally, every town had at least four guards. Two men were stationed in town, and two more watched at strategic points along the roads from the fields and leading out of town.[60] There were more residents than guards, but the guards constituted a considerable force. For example, there were 408 guards working at the Office in 1982. This number did not even include the Economic Police guarding the frontiers of the Office. In Macina, there were more guards proportionally than in any other region, suggesting that some areas were more known for rice smuggling. Not incidentally, the Macina sectors of Kokry and Kolongotomo were the poorest areas of the Office.

The guards were stationed at the Office for the duration of the harvest. In some years this period could be translated into as many as eight months out of the year. This was the case for the 1982–83 agricultural season when several threshing machines were in disrepair. For this reason, the collection of the harvest that year went into May.[61] During this time, the guards were tasked with making sure all rice was collected by the Office (save the household ration). The state-controlled harvest was further overseen by teams of workers who threshed, packed, and transported the rice. Teams of at least twenty-one Office workers did this work (the teams rotated throughout the Office during the harvest). This additional state presence included one man to operate the thresher, sixteen men to feed paddy into the machine, two men to sew and repair sacks, one repair worker, and one transport driver.[62] The threshers only processed the rice for sale outside the Office. Farmers had to hand thresh what they ate.

In the face of such controls, women had to exert extra energy in their efforts to provide for the family. Women's already physically intense labor routine was even more challenging with the addition of nighttime trips to the fields. Moreover, this additional effort was not always successful: sometimes women were able to hide rice to eat or sell, and sometimes it was taken from them by the guards. It was a heavy load to cook and work in the field during the day and then to rise early the next day to take rice from the fields. On those days women also had to hand thresh the rice in the field without the guard noticing.[63] This task was especially tiring because the women most in need of rice were likely to be chronically hungry. When the guards confiscated rice from a woman it often meant that she and her family did not eat that night.[64]

While women ensured much of the household grain and food supply, both men and women circumvented the strict controls. It was not a coordinated effort per se, but both husband and wife knew the other was sneaking rice from the fields (as well as cowives, sisters, and brothers-in-law). Often women did not tell their husbands which evening they planned to go into the field.[65] This way, women controlled how they would use the rice they took. It was perhaps easier for women to conceal their gleaned rice than it was for their husbands to "steal" it. When men went into the fields at night, they did not necessarily go to great lengths to hide what they took. Certainly they were on the lookout for guards. They also walked through the fields to avoid the roads.[66] At the same time, some men took a cart with them so as to collect a large amount of rice.[67] Other men simply carried the rice in a bundle from the field and only occasionally tried to hide it under their hats and cloth head coverings (which lent themselves to concealing only small amounts of rice).[68]

Men had another option for augmenting the store of rice at home. The worker who drove the threshing machine often negotiated in the fields with farmers to record a smaller harvest. Thus, he allowed the farmer to take some rice sacks home in addition to the ration. In exchange for this assistance, the farmer gave the operator one or two sacks for himself.[69] Kadja Coulibaly, the wife of a former thresher operator, knew that their household stores in Kokry were always full of rice precisely because her husband engaged in this kind of negotiation.[70] It is worth noting here that wives of workers continued to benefit from their position vis-à-vis the institution during these years, especially in relation to food rations and provisioning. Another important observation to be made is that men could take more rice home after one day of negotiation with the thresher operator than any one person could take from a clandestine trip to the field. The smaller amounts women took fed the family for a day. This meant that women likely made more trips and were at a greater risk of being searched.

Even when women could avoid being searched, the guards were a part of daily life. Office guards did not carry guns, but they nevertheless physically represented the coercive surveillance of the state over Office residents. Guards watched the fields both day and night. When men demonstrated how they "stole" rice from the fields at night, they mimicked looking over their shoulder as they cut and threshed rice.[71] People felt as if the guards could be anywhere at any time. Alarmingly, these men could even enter your house to confiscate rice. This was Maïssa Sountura's unfortunate experience one day when she was carrying a rice bundle to her house in Kouyan-Kura. She told me that the guard followed her all the way from the field. She ran to her house hoping to escape him, but the guard eventually showed up at her house to take the rice.[72] Under the guise of food security (or food sovereignty), the state intervened in household labor negotiations, violated women's bodies, and invaded people's domestic space.

THE PROMISE OF TECHNOLOGY
AND THE PROBLEM OF MALNUTRITION

From the perspective of the Traoré administration, the Office's large-scale irrigation infrastructure was a success in the 1970s. The years when rice production numbers dramatically rose were held up for observers as evidence. In fact, international funders like the World Bank advised even greater investment in improving the irrigation works. They also encouraged increased mechanization as a means of maximizing the size of the harvest. When the Malian government approached the World Bank for funding in 1978, the official report actually credited increased mechanical threshing with the earlier dramatic increase in production. Ultimately, the World Bank supported the request to improve these capacities.[73] The funding was expected to reinforce national food security. In practice, the technical apparatus had always emphasized production for sale to the broader region (and specifically to its coastal neighbors), which undermined local food needs.

Many mechanical elements of the scheme, such as those for field preparation and harvesting, were actually in decline in the 1970s.[74] This meant greater manual labor was needed, especially given the Office's high production estimations. In fact, mechanization disrupted local food autonomy and did little to alleviate women's daily labor. Hand threshing during the 1970s and 1980s marked an abrupt shift for women from the 1960s. Despite the financial and social problems of the Keita era, by the early 1960s more and more of the rice kept at home for consumption had been processed by a machine.[75] This technological aid was despite the fact that other mechanical services were already in decline by that time. In fact, in these years the Office encouraged farmers to prepare their fields using plows, in part because the machines for this purpose were expensive for farmers but also for the Office. However, the harvesting and threshing machines that had been purchased in the 1950s remained in operation.

It was as the presence of the military guards increased that more rice was threshed by hand. Or, in any case, women felt like they threshed more by hand. Certainly, women (and men) manually processed all rice taken in secret. In 1965 (only five years into independence), Office administrators reported to the Keita government that farmers routinely hand threshed rice surpluses, which marked the shift. They suggested that it was done to avoid administrative controls and sell their harvest for higher prices on the black market. A report from the same year blamed these farmers for the underwhelming results of the Office's cotton and rice production; the logic in the argument being that the Office was productive but that too much of the harvest was sold on the black market. Following the 1965 report and similar assessments of farmer subterfuge, the Economic Police force at the Office was reinforced, which only lead to more hand threshing.[76]

The resistance of farmers to government efforts to direct the sale of their harvest by controlling the harvesting machines was not new. Since the Office opened, men and women resisted selling all of their production to the institution because it consistently offered poor prices in comparison to local markets. It was when the Office liberalized the market in the late colonial era that women began to process less cotton by hand and thresh less rice by hand. Most likely this was due in part to the increasing number of processing machines at the Office during the 1950s. Farmers in those years could also sell most of the Office processed harvest outside the institution's marketing infrastructure. The trend toward increased mechanical processing was only gradually reversed after the Keita government declared a monopoly over the Office harvest in 1960 (lifted in 1968 and reinstated in 1974 as a response to the drought).[77]

Just when women had become accustomed to the option of mechanical threshing, the number of working machines began to dwindle. In the long decade of drought and increased militarization in the 1970s, Office residents threshed everything they ate. They also threshed whatever amount they sold on the black market (including confiscated rice).[78] Women (and men) simply remember that there were no machines at this time. In any case, there were no machines to help process rice for home consumption. One farmer, Seyba Coulibaly, attributed this to the fact that when the ground was wet the machines could not get to the field.[79] Indeed, the farmers who hand threshed smuggled rice were aided by the frequent delay of the harvest and threshing machines in reaching Office fields.[80] This was partly due to the disrepair of the roads and the increasingly poor state of the Office's equipment yard.[81] It was the moment that the Traoré government was claiming that irrigation technology and mechanization were successfully addressing the problem of food sovereignty.

Certainly, farmers hand threshed when necessary, but they also appreciated the mechanical option. In general, the women I spoke to were enthusiastic about the arrival of the machines at the Office because some of their labor burden would be eased. They explained to me that women could "relax a little," or "were not so tired."[82] In 1972, several officials even noted the existence of illegal (nonstate or private) threshing machines in the Office zone. These private machines made it easier for farmers to process rice they took from the fields.[83] Women, too, would have benefited from these circulating machines because hand threshing had come to occupy so much of their labor time. Moreover, their language about resting their bodies speaks to how they experienced and measured these years in terms of their physical labor and well-being.

Officials claimed that the privately operated machines were detrimental to the nation's food supply. They routinely cited the clandestine sale of rice from the Office as a problem for the institution's financial health and its effectiveness at producing rice. They may have overstated (perhaps deliberately) the importance of rice sold on the black market. For example, after the harvest in 1976, Office staff recorded that threshing machines processed 51,520 tons of paddy. Officially, farmers hand threshed only 3,638 tons total for their ration. That same year, the Office reported that the overall amount of the harvest that had been processed mechanically was up. It is unlikely that hand threshing for consumption and clandestine sale greatly surpassed the recorded three or four thousand tons or approached anything close to the amount processed by a machine.[84] At the same time, women (and men) spent a considerable amount of their labor time processing rice paddy by hand so they could eat.

Then in the early 1980s, a few officials began to voice concerns about the time that women spent threshing rice by hand. The primary worry was the impact of this labor on local diets. In 1981, Victor Douyon reported four deaths related to malnutrition in Sabula (near Kolongotomo). The victims in Sabula were diagnosed with beriberi, an illness caused by a lack of B vitamins.[85] Beriberi is often related to the consumption of large amounts of white rice (lacking in nutrients from processing).[86] After an investigation into their deaths, Douyon concluded, not surprisingly, that the problem resulted from Sabula residents' overreliance on rice in the diet. During these years, most Office residents were similarly dependent on rice for most of what they ate. It was perhaps for this reason that the situation in Sabula was so alarming. Following the deaths, the Office sent six tons of millet (high in B vitamins) to bring down mortality rates in Sabula. The nearby Catholic Mission in Kolongotomo also sent powdered milk, rice powder, sugar, and vitamin B supplements.[87]

These extreme cases of malnutrition arose in spite of women's considerable labor and effort to produce and prepare food. In a brief report on the situation, Douyon suggested a couple of long-term solutions. For one, he advocated adding millet to local diets. He also called for increased gardening by women in town. Both suggestions further underscored the lack of diversity in diets not only in Sabula but across the Office. Only a few years earlier, the Office director had tried to repress such activities because they detracted from overall rice production. With these constraints in mind, it is no surprise that residents were short of food other than rice. Moreover, Sabula was just one town out of many that were negatively impacted by the suppression of diverse food cultivation activities.

Finally, Douyon suggested that the town acquire a mechanical rice thresher to ease women's labor burdens. He thought this would allow them more time to garden.[88] The obvious question is: why were farmers not allowed to use the threshers for their ration in the first place? Around the same time, women's development experts emphasized the desire and need among women in Africa for grinding mills. The mills, too, were meant to relieve women of some of the burden of processing millet and other cereals.[89] In Sabula, it was also not clear how residents could purchase more millet, given the strict cereal controls enforced by the Economic Police. Women would likely have appreciated a mechanical thresher, as women's labor time was critical for producing food, but a mechanized solution simply sidestepped deeper structural issues. At the Office, it is clear that despite all the talk of "food sovereignty" the trappings of improved farming were not meeting local dietary needs.

FOOD AID AT THE OFFICE AND "THE RED THING"

The severe droughts of the 1970s and the 1980s across the Sahel generated a virtual industry on famine, food aid, and appropriate international intervention.[90] Most analysts at the time supported the need for temporary food aid, even drawing attention to the slow arrival of such assistance in the initial years of the 1970s crisis. Great attention was also brought to bear on the Sahelian climate. Other scholars have cautioned that an emphasis on rainfall and the environment obscures the social and political exchanges involved in what people had to eat. Rather, a broader view of food production, the economy, and the state is necessary to understand the underlying causes of food shortage and famine.[91] As it will become clear, the distinction that Malian officials made between the areas of need and the Office, as a region of drought-resistant production, distorted the reality of daily hunger for Office residents.

Following the malnutrition scare in Sabula, the broader Sahel region experienced a second acute drought and subsequent widespread food shortages. The level of food aid sent to the Sahel increased beginning in the 1980s as this second food crisis emerged. For example, between 1982 and 1985 the World Food Programme (WFP) shipped thirty-one thousand tons of cereals to Mali.[92] It was a significant boost to the food supply, although the Office alone regularly commercialized more than forty thousand tons of rice. In this period, food aid from the WFP would increasingly become a staple of the Office foodscape. However, most of the aid sent directly to the Office was reserved for new settlers, and longtime residents had to travel outside the Office villages to receive assistance. This was the case even for families that had only been settled for a year or two. When grain stores were empty, a village chief sent a couple of men to purchase cereals for town families. These representatives purchased mostly red millet from government warehouses in administrative centers such as Macina or Ségou. This particular type of food aid marked the way many women and men remembered these years.

The sale of WFP commodities was an institutional irregularity. Generally donated food was not meant to be commercialized because it was part of WFP emergency aid operations. Both Mali and Burkina Faso were cited for these practices in a WFP review for the period 1983–85. The same report highlighted the problems of requiring people to travel to distribution centers to receive their aid. In these cases most people sold some of their food aid to pay the transportation costs, meaning less overall aid for many. It also resulted in an economic strain on already struggling farmers.[93] What the WFP evaluators missed was that people resented paying for aid that to them could hardly be considered food. Conversely, the WFP report recorded that "WFP emergency food aid was generally very well accepted."[94] Of course communities facing immediate starvation accepted food aid in any form.

Even though Office residents suffered from hunger, the project still offered refuge to northerners facing famine. In fact, hundreds of Tuareg pastoralists traveled to the Office to request settlement as farmers in order to receive food assistance. For example in June 1985, 175 people traveled from the Niafounké region along the Nampala road to camp about thirty kilometers from the Office center of Dogofiry. The group, which comprised some thirty-eight families, reached out to Office officials to request settlement and assistance. The Office settled them in the northern Office zone of N'Debougou. By the time these needy families arrived just outside of Dogofiry, seventeen adults from their original party were dead due to starvation.[95] In exchange for the agreement to settle at the Office, they were offered transportation to their new village, immediate food aid, and wood for fuel and construction. Despite being fatigued, young men from this group of families were called upon to construct the lodgings.[96]

For the leaders of this Tuareg group the decision to settle at the Office must have been difficult. For decades prior to and then during the drought, nomads and pastoralists resisted administrative pressure to take up farming. Initially, the French colonial government pursued a policy of "sedentarization," and this policy was revived during the Great Sahel Drought. The policy move was often justified by claims that overgrazing had contributed to drought and desertification. However, nomads and pastoralists often understood the policy simply as intrusive state intervention.[97] In moving to the Office, members of this Tuareg group were expected to immediately begin learning the techniques of irrigated rice cultivation. They were guaranteed food aid during the cultivation season but were also expected to produce a harvest that same year. Perhaps in the hopes of an eventual return to pastoralism, the group's leader, Mohamed Ag Moustapha, negotiated for four or five young men to continue caring for their remaining animals.[98]

Seen from the outside, Office residents had enough rice. Famine, as it was defined politically, did not strike them.[99] However, as the cases of beriberi in Sabula suggest, severe hunger afflicted most everyone living there. Like other Malians, many Office residents survived the lean months on outside food donations. Certainly, hunger and malnutrition among Office residents were less easily identifiable than the extreme cases of starvation that afflicted people in the regions just north of the scheme. However, what constituted starvation and need was highly political.[100] Most obviously, the bulk of the rice harvest supplied the urban civil servants and military or was shipped to the nation's urban markets. Yet, there was great need elsewhere. New settlers seeking refuge from highly visible drought conditions received cereals, milk, meat, and cooking oil from WFP donations.[101] Meanwhile, the majority of Office residents were largely expected to survive on extremely minimal rations of rice. In fact, the administration (and many Malians in regions neighboring the Office) believed Office residents to be better off than most. For example, when Moussa Coulibaly, who farmed in the Kokry sector, asked the administration for food assistance, he remembered being told that "there was water" where he had settled and that he was supposed to "work for rice." The staff member he approached even called him "lazy."[102] The insult was especially offensive given that the whole project relied on the manual labor of farmers like Moussa.

In effect, the Office operated much like a food-for-work scheme during these years. The food-for-work framework for organizing aid was common in drought-affected countries. In neighboring Niger a major food-for-work scheme intended to promote reforestation, and erosion control gained nationwide acclaim: moreover, its predominantly female workforce was lauded. In fact, the project was promoted by the government as the epitome of drought relief and a national development project. In practice, the food women received only offered a bare minimum of food security.[103] The food-for-work approach betrayed the underlying concern of both state officials and international advisors that food aid had the potential to negatively impact production.[104] In fact, WFP officials worried that the generalized distribution of food aid in Mali in the mid-1980s had hampered the ability of farmers to recruit outside labor to work during the agricultural season.[105] What is clear from the archival and oral record is that these years for rural Sahelian men and women were characterized by hard physical labor in exchange for inadequate amounts of food. Longtime residents of the Office suffered from a persistent lack of food and from acute hunger. As was the case in Sabula, many Office residents ate rice (or red millet, which will be further discussed in this chapter) and not much else. When representatives from a medical mission based in Holland visited the Office in 1981, one nurse noted that kwashiorkor and beriberi, two diseases resulting from malnutrition, were prevalent.[106] This hunger and malnutrition had far-reaching and debilitating effects.[107]

The diet of Office residents was grim for more than a decade. Most towns were short on basics like salt and millet, not to mention nutritious ingredients like shea butter, baobab leaf powder, and meat. Government-organized cooperatives in each town were charged with buying salt, sugar, and other necessary items from OPAM and then selling the goods to residents. OPAM frequently failed to distribute goods to cooperatives due to inadequate supplies. This was a fact reported by Mme. Correze, who led an investigating team sponsored by the Agricultural Ministry and an outside donor group in 1981. Even officials from the ministry who were critical of the team's findings admitted that OPAM was slow to acquire and distribute basic foodstuffs.[108] In fact, the records for this period from the scheme's Division du Paysannat list only one distribution of thirty tons of sugar (which provided little nutritional value) to Office cooperatives in 1980.[109] For the most part, people survived on their normal ration and what men and women smuggled out of the fields. If they wanted to eat anything other than rice, women had to sell some. When the rice was gone, only red millet was left.

Cooking during the Famine of the Red Millet, as many residents called the hunger-stricken years, was a particular challenge for women. Older women had grown up either with the experience of cultivating dry-season foods, or with some knowledge of cooking wild foods like the grains fonio and jéba. However, fewer of these local and often appetizing grain alternatives were accessible at the Office, as rice fields occupied most land. Elsewhere, wild food collection constituted roughly 80 percent of people's food supply during the food crisis.[110] While some Office households succeeded in provisioning its stores in rice, other households were short on grains, especially in the months preceding the next harvest. During some parts of the year families might even go for three or four days without eating.[111] In Sabula, where Douyon reported the severe cases of beriberi, Oumou Sow pounded calabashes to eat. Dried calabashes were commonly used as vessels for food storage, but they were not usually used for food. Yet with the pounded and ground remains of a calabash Oumou made couscous (or basi) for the meal.[112] It was a sign of a severe shortage: Oumou had to destroy some of her store of household goods just to have something to eat.

It was only in 1976 that the World Food Aid Program began to give assistance to the Office (two years after the worst period of the early 1970s drought).[113] From 1976 until 1987 the kinds of food goods donated remained consistent. Donor countries sent grains, vegetable oil, canned meat or fish, and powdered milk. In 1976 and 1977 the Office received millet, and in subsequent years, the institution received mostly ground corn.[114] Initially, these goods were sold at set prices, probably to workers with cash to spare. In 1981 the prices for these foodstuffs were as follows: 1,200 francs Malien (FM) for one sack of corn (50 kg); 480 FM for one liter of cooking oil; 362.66 FM for one tin of canned chicken; 28.26 FM for one can of fish; and 28.26 FM for 3 kg of powdered milk.[115] Few farmers remember purchasing (or even seeing) any of these goods.

Beginning in 1983, this international aid was integrated into the Office recruitment program. That year, the Division du Paysannat was reorganized and charged with two major tasks: (1) the recruitment of new families and (2) the distribution of food aid to those new families. New settlers were promised food assistance for a brief installation period.[116] This included the standard selection of what was, by then, corn, canned meat or fish, vegetable oil, and powdered milk. In 1983–84 the Office provided food aid to 306 new households. As had been the case in the 1970s, an increasing number of households petitioned to join the Office during the drought years of the 1980s.[117] The total number of families receiving aid more than tripled from 306 to 1,005 in the following year (1984–85). The number increased again to 1,879 families for 1985–86. In fact, the numbers of families receiving aid each month steadily increased between April and October every subsequent year.[118] These were the months when most households would have run out of their grain stores from the previous harvest.

The basic food supplies donated by the WFP mostly came from the United States and several European countries.[119] By and large the United States provided the bulk of the corn flour, vegetable oil, and powdered milk. The United States also sent millet to the Office and other government distribution centers, which was what most farmers remember purchasing outside the Office. The canned fish and meat arrived mostly from Holland with additional shipments from Germany, Norway, Sweden, Denmark, Canada, and the CEE (Central and Eastern Europe).[120] Humanitarian aid efforts have already been criticized by scholars for promoting assistance supplied only from the outside.[121] One problem with the standard variety of food assistance in Mali was that these commodities were not readily incorporated into local diets or meal preparation. For example, powdered milk is not necessarily an easy substitute for fresh milk. To use the powdered milk, women needed a good supply of clean water, which was not always available at the Office during these years. These provisions also did not necessarily fulfill local health or nutritional needs. Likewise, the cereals provided through this program were appreciated by recipients to ward off starvation but were believed to be the primary cause of digestive illnesses.

One example of a more locally oriented response comes from Markala. Few people in Markala (the old workers' town) had direct access to rice during these years because they were not working in Office fields. Mariam N'Diaye Thiam, who was then working with women in Markala and the surrounding rural areas for the Ministry of Rural Development, mentioned that rice did not come out of the Office. In other words, Office rice could not be purchased in Markala, even though the town was located just on the periphery of where the rice had been grown.[122] As such, towns just outside the Office were severely short of food. Several international agencies distributed food aid in Markala. However, this assistance was not enough. Soon, with the backing of Thiam and some international female aid workers, a group of Markala women created their own cereal bank. They all contributed small amounts of grain or cash to create a stock for the women to draw from in the case of famine.[123] In this case, when women took anything from the cereal bank, they could be assured of getting something of quality that they knew how to cook.

Broadly speaking, international food aid from this period is best described as unpredictable.[124] This was also true for food donations sent to the Office. For example, as of September 1983, only oil, canned fish, canned chicken, and powdered milk had been distributed to the new settlers. The Office warehouses had received no corn from the WFP. The cereal donation finally arrived, and it was distributed in October. One must ask: What did those families eat in place of the corn from April or May until September?[125] While the holdup may have originated in the United States, at the port of Dakar, or with the Malian administration in Bamako, the delay itself was characteristic of the unevenness of outside food assistance. In a similar instance, neither oil nor powdered milk were available for distribution until the last months of the year in 1985.[126] Cooking without oil must have especially difficult for women when there was little else to season food or to provide vital nutrients. Food aid destined for Mali between 1984 and 1985 was in fact substantially delayed in Dakar. A World Food Program assessment of the institution's operations in the Sahel confirms that a significant amount of food aid arrived in Mali only five to seven months after its arrival at Dakar. Part of the problem was that materials for the Manantali dam construction project in Mali had been given priority over the shipments of food aid.[127] The dam was part of Mali's long-term strategy to ensure water resources and an investment in technological development, but here it was an impediment to immediate relief. Again in 1986, powdered milk was not available for several months, and in 1987, cooking oil was not distributed until the October–December period.[128] The international donations were not reliable sources for all the ingredients women needed for cooking.

Moreover, these distributions went to a limited number of people. The peak for the number of settlers receiving food aid from the Office was not quite two thousand in 1985–86. The total population of the Office was near sixty thousand; the best WFP aid simply did not reach the vast majority of families, most of whom were also unofficially hosting a large number of drought refugees. By all accounts these groups also badly needed protein-rich and other nutritious foods. At least food aid from the

WFP included meat, milk, and oil in addition to cereals. Beginning in the 1930s, the Office promised to provide new families with food assistance. To receive this help, families had to arrive just before the start of the new agricultural season and be in need. This logic continued to hold sway with officials during the 1980s when they distributed international food aid only to new settlers. They repeated a pattern that left the vast majority of farmers without reliable assistance. The policy was also based on the assumption that once farmers were installed, they just needed to work to grow rice (thereby negating any need for help even during an era of severe rationing).

WFP officials were even concerned that food aid shipments were repeatedly distributed to more people than had been anticipated, which spread the donations thin. The 1986 assessment reveals as much: "In practice at village level, in a country which has suffered a terrible drought and where most of the rural population lacks food and purchasing power, food is distributed to all; a village chief or local administrator cannot select the worst-hit families."[129] The need was simply much greater than the resources. One further problem was that the WFP operations in the Sahel did not have much oversight, making it harder to estimate which families were most in need. In some cases, state officials (like Douyon) kept detailed records about the receipt and distribution of aid. A few administrators even wrote down the number of families receiving aid in a village and the number of food aid bags delivered, but such records were often not communicated to the WFP. Calculating the amount of aid needed was a problem, worse still were the cases when not all aid shipments were distributed on time or at all.[130]

Predating the 1986 WFP assessment Douyon (from the Division du Paysannat) openly recognized that more Office families needed assistance. In 1984, he proposed that four thousand previously settled households also receive assistance in the coming year. He suggested that helping with food needs would make it easier for them to pay off some of their debt (an ongoing concern) to the Office.[131] This last point was probably how Douyon hoped to persuade his superiors. His division was poorly staffed. It also had little power, but his records offer a picture of a man who genuinely wanted to improve the poor living conditions of the project's residents. From the existing records of the WFP distributions at the Office, it does not appear that Douyon's suggestion was directly implemented. However, the next year the Banque Nationale de Developpement Agricole (BNDA), in cooperation with the Office, began offering agricultural loans for equipment that included food assistance. From the record of meetings between the BNDA and farmers it does not appear that the loan scheme was very successful. This was in large part because the Office (or an outside aid organization) would make all of the purchases for farmers. This proposal was not ideal. For example, the man who received a loan of oxen would have little recourse if their new animals died; the farmer would still have to repay the loan.[132]

Perhaps only the wives of workers for the Office (not farmers) received regular distributions of food like canned meat and fish. Kadja Coulibaly (whose husband was a thresher operator) cooked with a lot of food aid ingredients. She explained to me that what she called the "fish in a box" was so salty she had to soak it for two hours before cooking it. Even though the canned goods required special handling, Kadja appreciated having access to these goods. She remembers that this food was given to her, but it may have been the case that her husband paid a reduced price for the goods out of his wages.[133] Her husband similarly could have purchased rice at half price because of his employment.[134] In any case, because of these connections, Kadja's household was more easily provisioned than any of the drought refugees or Office farm families in desperate need of food. Nevertheless, Kadja and many other women shared the experience of having to transform what were often unfamiliar or unappetizing ingredients into edible meals.

Many women and men remember only receiving or paying for red millet. Locally the red millet was sometimes called the "red thing." It was widely understood that this millet was unhealthy, of poor quality, and unappetizing. For one, women recalled that when they opened a sack of red millet it always had a terrible smell.[135] After people ate it they were often afflicted with stomachaches or diarrhea (symptomatic of several digestive illnesses and other viruses).[136] Reflecting on the "red thing," Sekou Salla Ouloguem quoted the Bamana saying, "Mogo t'i fa fen bla i balo fenyen" (to keep from starving you will eat something that will make you ill). He also blamed a cholera outbreak from this era on the "red thing."[137] This millet was clearly widely associated with ill health and a last recourse against starvation. In recounting these memories many Malians signaled a serious local critique of food aid during the Great Sahel Drought.

Some women also remember receiving what they called *magnomugu*. It was a kind of cereal powder that they used to make couscous. Magnomugu came in big sacks that to women resembled the kind used to transport cement. When women opened the sacks, they found insects burrowed in with the grain powder.[138] It is not surprising that women who were accustomed to cleaning and processing their own grains right before cooking were shocked to be given powder that was unclean and were still expected to cook with it. Magnomugu was perhaps the ground corn that a few people received in food aid. In any case, it was poorly packaged and, like the red millet, required adept handling to make it edible. Even when distribution centers offered local types of food (like sweet potato), its quality was generally assessed as "rotten."[139]

Survivors of the drought at the Office spoke most vividly about red millet. More than the fact that it made people ill, its color made food visually unappetizing. Most women agreed that it could only be used to make couscous. If you used it to make toh it would be red "like blood."[140] Some women did make toh and dege with the food aid millet. To diminish the red color and turn the cooked millet dark brown, a woman could add potash to the cooking pot.[141] The meal still made most who ate it sick to their stomach, but it was more palatable. One donated cereal that was white in color was described as "not as bad to eat as the red thing." One likely reason was that the grain's color made it look more like food one could eat. This kind of millet came up in conversation in contrast to the red millet, and Sekou Salla Ouloguem emphasized the *white* (not red) color.[142] Clearly, the distribution of food aid, which from the perspective of most donors was the "aid" action, was incomplete until women made food that people could physically stomach eating.[143]

Women like Assane Coulibaly who were fortunate enough to have rice year-round traded their red millet for other foodstuffs like fish.[144] Interestingly, Assane traded her food aid millet to nearby Bozo families. In the 1930s, the Office botanist Roberty documented that while fishing was the main occupation for Bozo settlements, these communities also grew a kind of red millet that was resistant to river flooding.[145] It was likely that an aesthetic familiarity for red millet among Bozo families was why Assane was even able to trade the otherwise detested "red thing." At any rate, the red color was perhaps not as distasteful to Assane's Bozo trading partners. For her, the red millet was best used to trade for better things to eat. Thus, her family got to eat fresh fish. At a time when the water levels were low, fish was a luxury.[146]

Simply put, people understood the red millet to be famine food. Bintu Dieunta, who was from a Bozo family and who did cook with it, told me that it was "for people with hunger."[147] Many other people said that where red millet originated from it was considered animal or horse food, meaning it was not for human consumption.[148] It must have been awful to consciously eat something that was not perceived as food for human beings. Nevertheless, for those in grave need, it did ward off starvation.

Women even took the broth from cooked red millet in exchange for helping another woman to pound grains.[149]

There is a historical precedent among farmers in the region for associating a type of red-colored millet with hunger and famine. One type of millet that was a particularly hardy and quick-growing plant that Bamana and other farmers cultivated in the first decades of the century in the event of a poor rain (or likely poor harvest) was a red millet that also gave people stomachaches.[150] Aïssata Mallé, whose family came from the Koutiala region in the colonial era, remembered that her father sometimes grew a kind of red millet. If he had food made from this grain with him in the field, he would hide it so that other farmers did not notice him eating it.[151] Such antisocial behavior around food was extreme and speaks to the desperation of her father for growing and eating it. The foreign red millet similarly evoked feelings of revulsion and shame among those who ate it in the 1970s and 1980s. It was not anything to be openly shared but endured for the sake of survival. The more women could do to make it edible, the more their families could actually eat and could at least partially lessen their hunger pangs.

Stories about famine both in the past and those circulating about the Famine of the Red Millet similarly highlight the shame associated with food and eating during such crises. They also draw out tensions surrounding the ideals of hospitality in moments of ecological crisis. In 1923 Moussa Travélé published the story "During the Famine." It likely represents the social memory of the 1913–14 famine. Even if the story predates that particular crisis, the early twentieth-century famine likely influenced its circulation. In the story, a man and his son are eating when a starving stranger approaches them. The father and son invite the stranger to share their meal. However, the son soon notices that the stranger is eating an inordinately large portion of the food, an observation he shares with his father. The audience is then told that the father is equally dismayed that the stranger is eating more than his share. At the close of the story the narrator asks the audience who they think was the first to ask the stranger to leave, the father or the son.[152] Indeed, the stranger appears to have violated a social norm. Yet the question certainly invites a discussion of proper social behavior toward those in need, especially in times of food shortages. One of the potential answers to the narrator's query is that neither the father nor the son should deny the stranger hospitality. However, the tale foregrounds the fact that social breakdown is likely to occur during times of ecological crisis and food shortage.

A similar story of food denial was related in interviews about the Famine of the Red Millet for this book. In this story, a wife denies food to her husband, explaining to him that the meal she had prepared was for their child. The wife then hides the bowl of food only to secretly eat it herself that night. A woman who wanted to emphasize how terribly hunger affected people related this story. This act was especially shameful for a woman but understandable because of the exceptional circumstances of the famine. At the same time, it signifies extreme social disorder, since many people also repeatedly held up the ideal woman as one who cooks to share with others, including strangers.[153]

Indeed, women also prided themselves on producing good food to share even when they had few resources. The wild grains and other foods that women have historically cultivated from the bush for the lean periods were no longer found in most areas of the Office; however, women in other areas continued to collect and prepare these foods into the 1970s and 1980s. At the Office it was even unlikely that women could trade rice, let alone red millet, for fonio or any other of these foods because the women who found these foods likely prepared them immediately. The extreme poverty of famine food during these years was further evident in the fact that there was often no sauce to accompany the rice, couscous, or toh made from red millet. Even families who had access to WFP donations did not receive vegetables in any form (even canned). Wherever possible women at least added some salt to the water for taste as the toh was cooking. According to Aïssata Mallé, you would go to a person you thought was selling salt. You would not say the word kogo (salt) but ask, "Do you have something?" If that person said yes, you would give them some money and hide the salt. To get money for this purchase Aissata sometimes would walk a great distance to cut wood and then sell it in the Niono market.[154] Her recollection suggested that the limited black market was a more efficient way to get basic cooking ingredients. The gravity of the situation for women is perhaps understood in context with the Bamana saying: "It is for the husband to supply the to . . . and for the woman to supply the sauce."[155] Despite these women's physical labors and efforts to secure rice or transform red millet in these years, they had a hard time making any sauce at all.

HUNGER MIGRANTS: CREATING LOCAL FOOD AID

The problems of food at the Office notwithstanding, the scheme became a refuge of sorts and host to thousands of migrants and new settlers. In this way, the Office became a source of regional food aid. This was made possible at the Office mainly through women's labor.[156] It is no coincidence that the population of the project rose dramatically beginning in 1969 when some northern regions were already suffering from poor rains. The official population census recorded 30,356 residents for the 1969–70 year. Even though many areas began to recover in 1974 after several years of poor harvests, the entire decade was characterized by food shortages across much of the country.[157] During this time, the Office population steadily rose. By 1980–81 the total was 58,150 people.[158] In only ten years, the population had almost doubled.

In 1981, the largest populations were in the sectors closest to the paths of many northern migrants: Niono, N'Debougou, Molodo, Kourouma, and Dogofiry. Residents in the Macina area suggested that migrants frequently settled in other areas of the Office because there was no (paved) road to Kokry and there was less vehicular traffic there than in the other areas.[159] It is a plausible theory. Whatever the case, residents of Molodo-Centre remember that so many women came to work in the Office fields during the famine years that their sheer numbers would have been impossible to count.[160] Women made up a large proportion of the migrants, but at times whole families migrated together.[161] Many of these migrants arrived from regions at some distance to the north the Office. At the same time, women and men from nearby regions also traveled to the Office. For example, in Kokry, people even came from the Monipé region, where in the past Office residents had purchased millet.[162]

Many of these women and men stayed with relatives who were Office farmers. Women winnowed in the fields of their family members in exchange for rice to sell or to try to take home. Migrant women also helped Office women thresh and pound grains for daily cooking. Oumou Dembélé, a longtime female resident in the Niono sector, recalled that as soon as some women left to take the rice back home, more would come to take their place.[163] Some women and men who had no relations at the scheme also came to do similar work (men would cut the rice paddy). They all came in the hopes of earning enough rice to help alleviate food shortages at home to avoid possible starvation. Sekou Salla Ouloguem remembered that there were so many migrants that they could not all find hosts.[164] One household in Molodo-Centre was reputed to have hosted as many as one hundred people during an especially difficult agricultural year. Hawa Diarra explained the overwhelming number of migrants in similar terms: they could not feed them all. Most of these migrants shared the popular perception that there was no famine at the Office. Indeed, Hawa told me that during these hard times they did not experience "famine" because of the rice.[165] Of course, there is merit in her statement. Office farmers grew a lot of rice and were able to support hundreds of people from their limited rations. Mariam N'Diaye Thiam, who worked at the time for the Ministry of Rural Development in Markala and towns on the periphery of the Office, recalled that back then the Office seemed like an "îlot de prosperité" (island of prosperity) by comparison.[166]

People who lived in towns close to the Office also traveled there just to try to buy some rice. In Boky-Were, people most often went to the commercialization center at Kolongotomo where prices were cheaper than in individual Office towns.[167] Elsewhere, people secretly purchased rice in smaller project towns. For example, in Molodo-Bamana, men arrived from all over by bike carrying with them an empty canvas or other bag. Upon arrival the visitor purchased rice from a farmer and hid it in the bag. This way they hoped to avoid the guards and successfully carry their food purchase home.[168]

Perhaps in response to the numbers of visitors and migrants departing with rice, a few administration officials suggested formalizing the informal aid that farmers already provided to family members. As early as 1973, it was apparent that Office farmers wanted to and did send rice to family members outside of the project area. Douyon at the Division du Paysannat proposed in a letter to the director that families who were not indebted to the institution be given permission to send rice from their family ration to outside relations. He reasoned that a family of up to ten people could spare 330 kilograms of rice (roughly one year's ration for one person at the Office). He further calculated that a household with between eleven and thirty members could spare 600 kg (a ration for approximately two people). Finally, Douyon thought that families with thirty or more members could spare up to 1,000 kg (a ration for around three people). These calculations in no way approached the numbers of family members and other visitors coming to the Office. Also many families who were indebted were clearly already helping to feed others. From the archival record, it is not clear whether or not families ever received the right to send rice to their families as Douyon recommended. However, his interest in such a program implied that people were already supporting their relations with the rice that they had. It further suggested that administrators were fully aware of these actions.

Even as the Office indirectly addressed the problem of food shortages in neighboring regions, the high numbers of people living at the Office exacerbated the hunger of Office residents. The official rice ration (that Douyon used in his calculations) was based on the formal number of residents in the household and not the large number of visitors that a given family hosted during the harvest. The rise in the official census numbers reflected some of this migration. However, most farmers agreed that their ration never covered all of the people living with them. When these migrants were at the Office they were fed, but their numbers stretched the ration thin. This greater number of mouths to feed increased the need for secret trips to the fields.[169] For almost two decades the Economic Police and Office guards jointly exerted strict control over the harvest. The result was a new seasonality of predominantly female migration and of smuggling rice babies.

The Great Sahel Drought was a prolonged moment of hunger and ecological crisis. From the perspective of rural Malians it was also a moment of state failure, despite the claims made about successful irrigation technology and its promise of food sovereignty. Their stories from this period resonate with earlier ecological myths, and for women especially, the story of Muso Koroni. Their circulation also recalls the prevalence of stories relating to food, women's cooking, and famine from the early twentieth-century regional corpus. Together these stories offer collected knowledge about women's central role in surviving ecological crises.

The stories repeated by women and men about red millet and women smuggling rice babies are new collective memories. My reading of the rice baby stories finds two distinct interpretations. The first reading grapples with the overwhelming fear of bodily intrusion. We should take seriously the fact women are still telling stories about guards following them into their homes to take away their food in the interest of food security. Indeed, the persistence of women's gleaning is also suggestive of women's ongoing concerns over their precarious access to land and resources. The second reading considers the ways that women framed their smuggling as akin to gleaning: it is clear they saw their actions as a means to claim rights over agricultural production. Moreover, their transformation of cloth into a container for rice recalls the Koue-Bamana story about women's bodily sacrifices in the midst of famine. In feigning pregnancy, childbirth, or childbearing in their smuggling women also claimed their fruits of the harvest as a widely understood female task, one that resonated with women's customary roles in ensuring the well-being of society. The rice babies carried on women's backs have additional meaning. A woman's back was a ritual space associated with reproduction. In the first decades of the twentieth century, when a woman lost a child, she was required to make an offering before she could carry another baby on her back. The ritual offering was to ensure the long life of other children and future births. During the drought at the end of the century, social reproduction depended urgently on women's embodied and symbolic labors.[170] In reengineering their food production to respond to extreme hunger they drew on embodied ritual resources. Their artificially pregnant bodies and rice babies signaled severe and widespread hunger, but these images were also emblematic of women's determination to ensure rural survival during one of the worst environmental crises of the twentieth century in West Africa.

FIGURE C.1. Food security motto for the Molodo zone, 2010. Photo by author.

Conclusion

A SIGN hangs above one of the administrative compounds for the Office that reads Pour Une Veritable Securite Alimentaire au Mali et Dans le Reste de la Sous Region, or "For real food security in Mali and the region." It is the current motto for the Molodo project region (see figure C.1).[1] The staff members no doubt believe that they have an important role in realizing this goal. However, the slogan was a hard-fought reality for Office residents in the 1970s and 1980s, when they supported thousands of hungry migrants, as well as their own households, on their meager ration and illicitly obtained rice. Office residents who farmed in the 1940s also remember fulfilling this role in the past for their neighbors—those who had been hit by poor millet harvests—when selling rice without official permission. This history of local food security in rice was ensured through both women's and men's labor. During the Great Sahel Drought, it was a task that especially drew on deep-rooted associations between women's bodies, agricultural and human fertility, and women's obligations to provide regional hospitality. It also exposed women's bodily vulnerability to hunger and state intrusion. For women, food security was heightened as a matter of the body.

For much of the twentieth century, women at the Office made sure that food security extended beyond the provision of rice to include diverse and nutritious ingredients, familiar textures, and enticing smells in the daily meals that they prepared. When the concept of food security is expanded to include the sensorial aspects of how food tastes and feels in the body, its nutritional content, and the character of the bodily labor necessary to make daily meals, it is women who are at the center of the story. They labored in cotton, millet, and rice fields, but they also produced—or acquired at market—shea butter, soumbala, tegedege, namugu, and other tasty, nutrient-rich ingredients. Women at the Office also retained the texture of familiar meals when making it with rice, and they changed the displeasing color of food aid millet to make it more palatable and familiar. Dramatic changes in the environment and the food resource landscape made the task of engineering food production a consistently challenging one. Yet the aesthetics of food mattered on a very basic level, especially for men and women living on a novel and unpredictable development scheme where being watched was a quotidian experience. Familiar and satisfying food was necessary for survival, and it was also central to the celebration of the harvest. In the absence of women, food crises signaled broader political and social instability.

Following the prolonged drought of the 1970s and the food shortages of the 1980s, the Office administration established a program for women's development. The sudden emergence of women on the Office development agenda was due in part to pressure from international aid organizations and experts.[2] Strikingly, the projects that took hold in the 1990s and after replicated what women at the project had been doing for decades: market gardening. In conjunction with the allotment of small women's communal gardens, a few newly formed women's groups received solar dryers from a local development organization called Alphalog.[3] Women in these groups processed the onions they grew in their gardens into durable dried foodstuffs for sale in regional markets. From the Niono area women became particularly well known for their onions and tomatoes, even in the Bamako market.[4] With respect to the solar dryers, a gendered distinction in the introduction of technologies is apparent in their distribution because women at the Office were still expected to engage in sauce production and the manufacture of foodstuffs rather than large-scale rice production. A small number of women were allotted rice fields after 1997, but the emphasis in women's development remains on market gardening. Where women are understood in relation to food, technology, and development in these projects is squarely in the domain of sauce cultivation and not the wide array of technological work that supports food security. Moreover, women found the solar dryers of little use, and the few examples women took me to see were rarely in use. Women had little say in the design of the dryers and found that they did not help women to improve their overall market sales. Indeed, they were more impressed with stories about the wealthy woman at the Office who owned her own rice thresher, a machine long associated with the project and male workers.[5] Yet, none of the women sought to upend gender roles with relation to food; rather they sought to choose and even to claim the machines with the potential to aid in their "women's work." No wonder the nickname they gave to the new threshing machines (which also winnowed) was "Muso na aka chira," or "The old woman and her broom." These machines began to proliferate in the same years that women were newly targeted for development, but the new machines were owned and operated by men.[6] Rather than claim the new machine as women's technology, the nickname criticized it for sweeping away one of women's money-making avenues now that the machine winnowed in their place.

The gendered bias in relation to technology at the Office also remained in place after women's groups were recruited to do the work of rice planting in men's fields beginning in the 1990s. The work of transplanting rice has always been labor intensive, and women complain of poor compensation. Without their labor, the men's rice fields would not benefit from the rice-planting technique that many farmers now employ in the hopes of gaining higher yields at harvest time. In light of women's history at the Office, the gendered labor of transplanting rice begs for comparison with the physically demanding bodily labor of smuggling rice babies in prior decades.[7] Women are acutely aware of how much physical labor they have expended and continue to exert in ensuring the survival of communities at the project and wider region. It is striking that while women's communal groups are now offered solar dryers (and millet grinders) in the name of development, women's physical labor remains central to agricultural success at the Office. In several interviews, women repeated to me the simple phrase "an segenna" (we are tired). Their insistence on such fatigue was certainly a critique of the project in that their labors have not always been rewarded with greater economic benefits (for the women or their families). At the same time, it is also an insistence on the recognition of their physical labors and expertise. Indeed, over the course of the twentieth century, women have consistently displayed the social, economic, and political value of their embodied work. Yet they are also interested in shaping the technological possibilities for their work and, more broadly, the economic situation of their communities.

∽

In my final visit to see Fatouma Coulibaly and Nièni Tangara in Kolony (km26), I offered to take a photo of the two women. I visited Nièni's extensive and well-cared-for garden and observed cooking in her compound. Nièni was Fatouma's daughter-in-law and listened in on several conversations I had with Fatouma. Over several months, we had long discussions about women, food, and technology at the Office. We talked about Fatouma's commission of a small mortar and pestle from a local blacksmith, and I had the chance to see the final product. She had ordered it because the wooden mortar and pestle wears down with use, and she wanted one that would not need to be replaced. When I remarked that it was heavy, she explained to me that is why she had ordered a small one made from metal.[8] It was an example of Fatouma's willingness to experiment with new materials and tools for cooking. We also discussed the fragility of new imported eating vessels made of plastic, the installation of electricity in her terra cotta home, and even her desire for a refrigerator. Like many other women, Fatouma is actively contemplating the meanings and forms of the current reengineering of the foodscape.

FIGURE C.2. Fatouma Coulibaly (left) and Nièni Tangara with a moto in Kolony (km 26), 2010. Photo by author.

When I suggested the photo, Fatouma insisted that she sit on top of a moto (see figure C.2). It figured into her self-styling for the souvenir of our time together. She wore an outfit featuring fish, in homage to the fish widely found at the Office and of course her fish sauce. In fact, she made sure to point out the food-related design to me. Fatouma belonged to a women's Muslim association in Kolony and decided she also wanted to hold her prayer beads to signify her faith. When she was ready, she also chose to pose with Nièni, the daughter-in-law who joined her in interviews with me. Fatouma might have meant to include Nièni in our interview sessions as a way to impart knowledge to a younger woman under her supervision, but Nièni actively participated with her own memories and interpretations of women's labor. What stood out the day of the photo was Fatouma's insistence that a young man in the compound go find a moto for them to pose with. She, like most older women, does not drive a moto but has ridden on the back of one driven by young male relatives. Fatouma had also seen young women like my interview assistant, Aissata, drive motos even in rough rural terrain (I, like Fatouma, was a passenger). Fatouma laughed as she grabbed the handles, a gesture perhaps to show her awareness of the new modern amenities at the Office. I also understood that Fatouma was presenting herself as technologically savvy, even with machines more often associated with men (or younger people) in rural Mali. It is an image that speaks to rural women's awareness that they are drivers of technological change. Even though women are still expected to provide difficult labor and confine themselves to women's technologies that now include solar dryers and grain grinders, it is a picture of how Fatouma sees herself and younger women at the center of African technological history.

Notes

INTRODUCTION

1. "La sauce aux petits poissons," in *Anthologie des chants mandingues: Cote d'Ivoire, Guinée, Mali*, trans. and ed. Kaba Mamadi (Paris: Harmattan, 1995), 81. Mamadi recalls hearing this and other women's songs in the 1930s and early 1940s but suggests their composition likely predated the period. The territory of the contemporary Republic of Mali covers a region conquered by the French in the late 1880s and 1890s. The new territory was referred to as the western Soudan, and it was briefly administered as Haut-Sénégal et Moyen Niger in 1900. In 1904 the colony was reorganized as Haut-Sénégal-Niger and finally renamed Soudan Français in 1920. For simplicity, I refer to this territory during the colonial era only as the French Soudan, and when referencing the long twentieth century, I simply refer to the territory as Mali.

2. See the tale "La marmite," recorded and translated by Charles Monteil in *Contes soudanais* (Paris: Ernest Leroux, 1905), 26–27.

3. See, for example, the stories "Le gros singe et la femme" and "La fille modeste et les phenomènes (ou la bonne et la mauvaise fille)," in Moussa Travélé, *Proverbes et contes bambara: Accompagnés d'une traduction française et précédés d'un abrégé de droit coutumier* (Paris: Librarie Orientaliste Paul Geuthner, 1923), 163–65, 205–13; Moussa Travélé, "Baba qui ne mange pas des harciots," in *Petit dictionnaire français-bambara et bambara-français* (Paris: Librairie Paul Geuthner, 1913), 265–68; and "Le gourmand" and "L'hyène" in Monteil, *Contes soudanais*, 25–26, 57–61.

4. Marcel Mauss, "Techniques of the Body," in *Incorporations*, ed. Jonathan Crary and Sanford Kwinter (New York: Zone, 1992), 454–77.

5. I employ the term "bush" as a translation of *kúngo* (*koungo*). See Gérard Dumestre, *Dictionnaire bambara-français: Suivi d'un index abrégé* (Paris: Éditions Karthala, 2011). See also Travélé, *Petit dictionnaire*, 197. The term refers to the natural world outside the village or cultivated fields; bush is understood as "wild" or "untamed" land.

6. Téréba Togola, "Memories, Abstractions, and Conceptualization of Ecological Crisis in the Mande World," in *The Way the Wind Blows: Climate,*

History, and Human Action, ed. Roderick J. McIntosh, Joseph A. Tainter, and Susan Keech McIntosh (New York: Columbia University Press, 2000), 187.

7. Naminata Diabate, *Naked Agency: Genital Cursing and Biopolitics in Africa* (Durham, NC: Duke University Press, 2020), 1–12.

8. Diabate, 4.

9. In making this claim about the inseparability of women from their contextual and shifting natural world I draw upon work by ecofeminist scholars who have challenged romantic notions of women as natural managers of the environment: this recent work recognizes not only the material relationships between the environment and human health but also the symbolic and cultural relationships between social identity and specific culturally claimed environments. See Stacy Alaimo, *Undomesticated Ground: Recasting Nature as Feminist Space* (Ithaca, NY: Cornell University Press, 2000); and Christina Holmes, *Ecological Borderlands: Body, Nature, and Spirit in Chicana Feminism* (Urbana: University of Illinois Press, 2016).

10. I take the notion of the "mindful body" as one in which emotion and the senses influence the feeling and state of bodily health or wellness. See Nancy Scheper-Hughes and Margaret M. Lock, "The Mindful Body: A Prolegomenon to Future Work in Medical Anthropology," *Medical Anthropology Quarterly (New Series)* 1, no. 1 (1987): 6–41.

11. Donna J. Haraway, "Otherworldly Conversations, Terran Topics, Local Terms," in *Material Feminisms,* ed. Stacy Alaimo and Susan Hekman (Bloomington: Indiana University Press, 2008), 157–87.

12. Emphasis in the original. Marica-Anne Dobres, "Archaeologies of Technology," *Cambridge Journal of Economics* 34 (2010): 108.

13. For a discussion of women's work, gendered constraints, and social expectation in Mali, see Adame Ba Konaré, *Dictionnaire des femmes célèbres du Mali (des temps mythico-légendaires au 26 Mars 1991); précédé d'une analyse sur le rôle et l'image de la femme dans l'histoire du Mali* (Bamako, Mali: Éditions Jamana, 1993), 27–28.

14. For an elaboration of this foundational critique, see Chandra Talpade Mohanty, "Under Western Eyes: Feminist Scholarship and Colonial Discources," *Boundary* 2, no. 1 (1984): 333–58; and Arturo Escobar, *Encountering Development: The Making and Unmaking of the Third World* (Princeton, NJ: Princeton University Press, 1995). Malian historian Adame Ba Konaré similarly criticized historical narratives that flattened African women into generic victims of patriarchy and offered little to the scholar or student of women's history. See Ba Konaré, *Dictionnaire des femmes,* 5.

15. Ester Boserup, *Woman's Role in Economic Development* (London: Earthscan, 1989), see especially 3–24, 68–69.

16. For a critical analysis of "technology transfer" as a concept in relation to Africa, see Clapperton Chakanetsa Mavhunga, "Introduction: What Do

Science, Technology, and Innovation Mean from Africa?," in *What Do Science, Technology, and Innovation Mean from Africa?*, ed. Clapperton Chakanetsa Mavhunga (Cambridge, MA: MIT Press, 2017), 2–9.

17. For a discussion of the international focus on women's wood fuel collection and their role as environmental managers, see Richard A. Schroeder, *Shady Practices: Agroforestry and Gender Politics in The Gambia* (Berkeley: University of California Press, 1999), 4–15. For a critical discussion of international discourses representing women as natural managers of the environment see Melissa Leach and Cathy Green, "Gender and Environmental History: From Representation of Women and Nature to Gender Analysis of Ecology and Politics," *Environment and History* 3, no. 3 (1997): 345–53.

18. Jane Guyer, "Female Farming in Anthropology and African History," in *Gender at the Crossroads of Knowledge*, ed. Micaela di Leonardo (Berkeley: University of California Press, 1991), 257–77.

19. Chéibane Coulibaly, *Politiques agricoles et stratégies paysannes au Mali de 1910 à 2010: Mythes et réalités à l'Office du Niger*, 2nd ed. (Paris: L'Harmattan Mali, 2014), 30–32.

20. The influential French agronomist Auguste Chevalier was openly critical of Bélime's project to irrigate the Niger River in several publications, including one in an edited collection of responses to Bélime's original proposal. See Emile Bélime et al., *Les irrigations du Niger: Discussions et controverses* (Paris: Comité du Niger, 1923). See also Pierre Herbart, *Le chancre du Niger* (Paris: Gallimard, 1939). Historians have also pointed to the persistent poverty among early Office settlers and the debt incurred by male farmers. See Amidu Magasa, *Papa-commandant a jeté un grand filet devant nous: Les exploités des rives du Niger, 1902–1962* (Paris: François Maspero, 1978); Jean Filipovich, "Destined to Fail: Forced Settlement at the Office du Niger, 1926–45," *Journal of African History* 42, no. 2 (2001): 239–60; and Monica van Beusekom, *Negotiating Development: African Farmers and Colonial Experts at the Office du Niger, 1920–1960* (Portsmouth, NH: Heinemann, 2002).

21. David Arnold, "Europe, Technology, and Colonialism in the 20th Century," *History and Technology* 21, no. 1 (2005): 96.

22. See, for example, Sara Berry, *Fathers Work for Their Sons: Accumulation, Mobility, and Class Formation in an Extended Yorùbá Community* (Berkeley: University of California Press, 1985); Barbara M. Cooper, *Marriage in Maradi: Gender and Culture in a Hausa Society in Niger, 1900–1989* (Portsmouth, NH: Heinemann, 1997); Karen Tranberg Hansen, ed., *African Encounters with Domesticity* (New Brunswick, NJ: Rutgers University Press, 1992); Rachel Jean-Baptiste, *Conjugal Rights: Marriage, Sexuality and Urban Life in Colonial Libreville, Gabon* (Athens: Ohio University Press, 2014); Meredith McKittrick, "Forsaking Their Fathers? Colonialism, Christianity, and Coming of Age in Ovamboland, Northern

Namibia," in *Men and Masculinities in Modern Africa*, ed. Lisa A. Lindsay and Stephan F. Miescher (Portsmouth, NH: Heinemann, 2003), 33–51; Kenda Mutongi, *Worries of the Heart: Widows, Family, and Community in Kenya* (Chicago: University of Chicago Press, 2007); Heidi J. Nast, *Concubines and Power: Five Hundred Years in a Northern Nigerian Palace* (Minneapolis: University of Minnesota Press, 2004); Emily Lynn Osborn, *Our New Husbands Are Here: Households, Gender, and Politics in a West African State from the Slave Trade to Colonial Rule* (Athens: Ohio University Press, 2011); Elizabeth Schmidt, *Peasants, Traders, and Wives: Shona Women in the History of Zimbabwe, 1870–1939* (Portsmouth, NH: Heinemann, 1992); and Lynn M. Thomas, *Politics of the Womb: Women, Reproduction, and the State in Kenya* (Berkeley: University of California Press, 2003).

23. Emily Burrill, *States of Marriage: Gender, Justice, and Rights in Colonial Mali* (Athens: Ohio University Press, 2015); Ba Konaré, *Dictionnaire des femmes*, 51–53, 57–62; Richard Roberts, *Litigants and Households: African Disputes and Colonial Courts in the French Soudan, 1895–1912* (Portsmouth, NH: Heinemann, 2005); and Stephen R. Wooten, "Colonial Administration and the Ethnography of the Family in the French Soudan," *Cahiers d'Études Africaines* 33, no. 131 (1993): 419–46.

24. In particular, work by scholars in feminist technology studies and African STS have influenced my own work. See Francesca Bray, *Technology and Gender: Fabrics of Power in Late Imperial China* (Berkeley: University of California Press, 1997); Ruth Schwartz Cowan, *More Work for Mother: The Ironies of Household Technology from the Open Hearth to the Microwave* (New York: Basic Books, 1983); Judith A. McGaw, "Reconceiving Technology: Why Feminine Technologies Matter," in *Gender and Archaeology*, ed. Rita P. Wright (Philadelphia: University of Pennsylvania Press, 1996), 52–75; Ruth Oldenziel and Karin Zachman, "Kitchens as Technology and Politics: An Introduction," in *Cold War Kitchen: Americanization, Technology, and European Users*, ed. Ruth Oldenziel and Karin Zachman (Cambridge, MA: MIT Press, 2009), 1–29; Joy Parr, "What Makes Washday Less Blue? Gender, Nation, and Technology Choice in Postwar Canada," *Technology and Culture* 38, no. 1 (1997): 153–86; Libbie Freed, "Networks of (Colonial) Power: Roads in French Central Africa after World War I," *History and Technology* 26, no. 3 (2010): 203–23; Gabrielle Hecht, *Being Nuclear: Africans and the Global Uranium Trade* (Cambridge, MA: MIT Press, 2014); Pauline Kusiak, "'Tubab' Technologies and 'African' Ways of Knowing: Nationalist Techno-Politics in Senegal," *History and Technology* 26, no. 3 (2010): 225–49; Marianne de Laet and Annemarie Mol, "The Zimbabwe Bush Pump: Mechanics of a Fluid Technology," *Social Studies of Science* 30, no. 2 (2000): 225–63; Clapperton Chakanetsa Mavhunga, *Transient Workspaces: Technologies of Everyday Innovation in Zimbabwe* (Cambridge, MA: MIT Press,

2014); Toluwalogo Odumosu, "Making Mobiles African," in *What Do Science, Technology, and Innovation Mean from Africa?*, ed. Clapperton Chakanetsa Mavhunga (Cambridge, MA: MIT Press, 2017), 137–50; and Abena Dove Osseo-Asare, *Atomic Junction: Nuclear Power in Africa after Independence* (Cambridge: Cambridge University Press, 2019).

25. For an overview of African histories of technology, see David Serlin, "Confronting African Histories of Technology: A Conversation with Keith Breckenridge and Gabrielle Hecht," *Radical History Review* 127 (2017): 87–102; and Laura Ann Twagira, "Introduction: Africanizing the History of Technology," *Technology and Culture* 61, no. S2 (April 2020): S1–19.

26. Laura Ann Twagira, "'Robot Farmers' and Cosmopolitan Workers: Divergent Masculinities and Agricultural Technology Exchange in the French Soudan (1945–68)," *Gender and History* 26, no. 3 (2014): see especially 465–66. This pattern is mirrored in the scholarship as most studies of the Office treat science, technology, and work at the project as masculine. See, for example, Van Beusekom, *Negotiating Development*; Christophe Bonneuil, "Development as Experiment: Science and State Building in Late Colonial and Postcolonial Africa, 1930–1970," *Osiris* 15 (2000): 258–81; Isaïe Dougnon, *Travail de Blanc, travail de Noir: La migration des paysans dogon vers l'Office du Niger at au Ghana (1910–1980)* (Paris: Karthala-Sephis, 2007); Magasa, *Papa-commandant*; and Emil Schreyger, *L'Office du Niger au Mali 1932 à 1982: La problématique d'une grande entreprise agricole dans la zone du Sahel* (Paris: L'Harmattan, 1984).

27. Birama Diakon, *Office du Niger et pratiques paysannes: Appropriation technologique et dynamique sociale* (Paris: L'Harmattan Mali, 2012), 34–41, 43–44.

28. Diakon, 102–12.

29. Suzanne Moon, "Place, Voice, Interdisciplinarity: Understanding Technology in the Colony and the Postcolony," *History and Technology* 26, no. 3 (2010): 190.

30. Winther looked at this question for contemporary Africa in her study of rural electrification in Zanzibar. See Tanja Winther, *The Impact of Electricity: Development, Desires and Dilemmas* (New York: Berghahn, 2008), see especially chapters 9 and 10.

31. The foundational work on the relationship between society and technology is the 1987 edited collection (now in a revised version): Wiebe E. Bijker, Thomas P. Hughes, and Trevor Pinch, eds., *The Social Construction of Technological Systems: New Directions in the Sociology of History and Technology* (Cambridge, MA: MIT Press, 2012). For key works on gender and technology, see Michael Adas, *Machines as the Measure of Men: Science, Technology, and Ideologies of Western Dominance* (Ithaca, NY: Cornell University Press, 1989); Francesca Bray, "Gender and Technology," *Annual Review of Anthropology* 36 (2007): 37–53; Wendy Faulkner, "The

Technology Question in Feminism: A View from Feminist Technology Studies," *Women's Studies International Forum* 24, no. 1 (2001): 79–95; and Nina E. Lerman, Ruth Oldenziel, and Arwen P. Mohun, eds., *Gender and Technology: A Reader* (Baltimore, MD: Johns Hopkins University Press, 2003).

32. For an overview of recent literature on "users" see Nelly Oudshoorn and Trevor Pinch, "Introduction: How Users and Non-users Matter," in *How Users Matter: The Co-construction of Users and Technology*, ed. Nelly Oudshoorn and Trevor Pinch (Cambridge, MA: MIT Press, 2003), 1–25. Diakon's work on male plow users in Mali is a good example of the "users" approach, bridging African history and the history of technology. See Diakon, *Office du Niger et pratiques paysannes*. Jenna Burrell has also notably expanded the user framework in her study of internet café users in urban Ghana. See Jenna Burrell, *Invisible Users: You in the Internet Cafés of Urban Ghana* (Cambridge, MA: MIT Press, 2012).

33. Oudshorn and Pinch, *How Users Matter*, 5.

34. Judy Wajcman, *Technofeminism* (Cambridge, UK: Polity, 2004), 3–9.

35. Lynn M. Thomas, "Historicising Agency," *Gender and History* 28, no. 2 (August 2016): 324–39.

36. Sarah C. Brett-Smith, *The Silence of the Women: Bamana Mud Clothes* (Milan: 5 Continents Editions, 2014).

37. Barbara E. Frank, "Marks of Identity: Potters of the Folona (Mali) and Their 'Mothers,'" *African Arts* 40, no. 1 (2007): 30–41.

38. Shadreck Chirikure, "The Metalworker, the Potter, and the Pre-European African Laboratory," in *What Do Science, Technology, and Innovation Mean from Africa?*, ed. Clapperton Chakanetsa Mavhunga (Cambridge, MA: MIT Press, 2017), 63–77.

39. Eugenia W. Herbert, *Iron, Gender, and Power: Rituals of Transformation in African Societies* (Bloomington: Indiana University Press, 1993). Other historians of precolonial Africa have also established gender as central to understanding ritual and technology. See, for example, Christine Saidi, "Pots, Hoes, and Food: Women in Technology and Production," in *Women's Authority and Society in Early East-Central Africa* (Rochester, NY: University of Rochester Press, 2010), 128–46. For Mali, Sara Brett-Smith has similarly argued that the production of sculpture by male blacksmiths is entangled with human sexuality as a creative but potentially dangerous force. See Sarah C. Brett-Smith, *The Making of Bamana Sculpture: Creativity and Gender* (Cambridge: Cambridge University Press, 1994).

40. Historian Ruth Oldenziel first made this point for the gendered definition of technology that emerged in the nineteenth and twentieth centuries in the United States. See Ruth Oldenziel, *Making Technology Masculine: Men, Women and Modern Machines in America 1870–1945* (Amsterdam: Amsterdam University Press, 1999).

41. Ralph A. Austen and Daniel Headrick, "The Role of Technology in the African Past," *African Studies Review* 26, no. 3/4 (1983): 174.

42. Margaret Ehrenberg has argued that women's childcare needs prompted one of the earliest human technologies—the sling. See Margaret Ehrenberg, *Women in Prehistory* (London: British Museum, 1989), 46–48.

43. I draw here from Londa Schiebinger's work on gendered ideology and reproductive metaphors in the development of Western biological sciences. See Londa Schiebinger, *Nature's Body: Gender in the Making of the Modern Science* (New Brunswick, NJ: Rutgers University Press, 2004).

44. I have elaborated on this point in relation to grain grinders elsewhere. See Laura Ann Twagira, "Machines That Cook or Women Who Cook? Lessons from Mali on Technology, Labor, and Women's Things," *Technology and Culture* 61, no. S2 (April 2020): S77–103.

45. See Magasa, *Papa-commandant*.

46. At the end of the Second World War, international political pressure pushed France and several other colonial powers to outlaw forced labor. See Frederick Cooper, *Decolonization and African Society: The Labor Question in French and British Africa* (Cambridge: Cambridge University Press, 1996), 25–56, 182–202.

47. Coulibaly, *Politiques agricoles et stratégies paysannes*; Diakon, *Office du Niger et pratiques paysannes*; and Van Beusekom, *Negotiating Development*.

48. Brian Larkin, *Signal and Noise: Media, Infrastructure, and Urban Culture in Nigeria* (Durham, NC: Duke University Press, 2008), 16–47.

49. In addition to multiple colonial-era critiques, in the 1970s sociologist Amidu Masa harshly criticized the postcolonial practices of the Office du Niger. See Magasa, *Papa-commandant*. More recent work continues to critique the Office's ongoing problems in assuring local food security. See Nicolette Larder, "Possibilities for Alternative Peasant Trajectories through Gendered Food Practices in the Office Du Niger," in *Postcolonialsm, Indigeneity and Struggles for Food Sovereignty: Alternative Food Networks in Subaltern Spaces*, ed. Marisa Wilson (New York: Routledge, 2017), 106–26.

50. Libby Freed similarly observed that colonial infrastructure projects like colonial roads in French Central Africa were meant to extend the imperial reach in those territories. See Freed, "Networks of (Colonial) Power," 205–10.

51. As Freed points out for roads, technologies of landscape order were distinct in colonial Africa. They did not become unremarkable elements of the modern landscape as theorized by other historians. Rather, they were clearly visible and imbued with political meaning. In making this point, Freed was responding to Paul Edwards, who previously asserted that in the so-called modern world people become so accustomed to the technological infrastructures of daily life that they eventually appear to be a part of the natural environment. See Freed, 218; and Paul N. Edwards, "Infrastructure and Modernity: Force, Time, and Social Organization in the History of Sociotechnical Systems," in *Modernity and Technology*, ed. Thomas J. Misa, Philip Brey, and Andrew Feenberg (Cambridge, MA: MIT Press, 2003), 185–225.

52. Mavhunga, *Transient Workspaces*. See also Jennifer Hart, *Ghana on the Go: African Mobility in the Age of Motor Transportation* (Bloomington: Indiana University Press, 2016); Mavhunga, "Introduction"; Jamie Monson, *Africa's Freedom Railway: How a Chinese Development Project Changed Lives and Livelihoods in Tanzania* (Bloomington: Indiana University Press, 2009); and Winther, *Impact of Electricity*.

53. Frederick Cooper, "Modernizing Bureaucrats, Backward Africans, and the Development Concept," in *International Development and the Social Sciences: Essays on the History and Politics of Knowledge*, ed. Frederick Cooper and Randall Packard (Berkeley: University of California Press, 1997), 70.

54. Emil Schreyger, *L'Office du Niger au Mali*, 2, 136–45.

55. Modibo Keita had already been involved in the affairs of the Office du Niger as a West African depute in the French parliament. For a detailed chronology of this political history, see Tony Chafer, *The End of Empire in French West Africa: France's Successful Decolonization?* (New York: Berg, 2002).

56. Gabrielle Hecht, "Introduction," in *Entangled Geographies: Empire and Technopolitics in the Global Cold War*, ed. Gabrielle Hecht (Cambridge, MA: MIT Press, 2011), 1–12.

57. Twagira, "Robot Farmer and the Cosmopolitan Worker," 464–65.

58. Hecht, "Introduction," 5.

59. Allen Isaacman and Barbara Isaacman characterize this enthusiasm for dam construction in Africa as the "dam revolution." See Allen F. Isaacman and Barbara S. Isaacman, *Dams, Displacement, and the Delusion of Development: Cahora Bassa and Its Legacies in Mozambique, 1965–2007* (Athens: Ohio University Press, 2013), 7–10.

60. See, for example, Djibril Aw and Geert Diemer, *Making a Large Irrigation Scheme Work: A Case Study from Mali*, Directions in Development (Washington, DC: World Bank, 2005); Van Beusekom, *Negotiating Development*; Bonneuil, "Development as Experiment"; Mamdou Diawara, "Development and Administrative Norms: The Office du Niger and Decentralization in French Sudan and Mali," *Africa* 81, no. 3 (2011): 434–54; Filipovich, "Destined to Fail"; Marcel Kuper, Jean-Philippe Tonneau, and Pierre Bonneval, eds., *L'Office du Niger, grenier à riz du Mali: Succès économiques, transitions culturelles, et politiques de développement* (Paris: Karthala, 2003); and Schreyger, *L'Office du Niger au Mali*.

61. This point has been made by many scholars who have noted the emergence of multiple practices and discourses of development in the twentieth century. For example, see Sara Berry, *No Condition Is Permanent: The Social Dynamics of Agrarian Change in Sub-Saharan Africa* (Madison: University of Wisconsin Press, 1993); Frederick Cooper and Randall Packard, "Introduction," in Cooper and Packard, *International Development and the Social Sciences*, 1–41; Andrea Cornwall, Elizabeth Harrison, and Ann

Whitehead, "Gender Myths and Feminist Fables: The Struggle for Interpretive Power in Gender and Development," *Development and Change* 38, no. 1 (2007): 1–20; James Ferguson, *Anti-politics Machine: Development, Depoliticization, and Bureaucratic Power in Lesotho* (Minneapolis: University of Minnesota Press, 1994); James Ferguson, *Global Shadows: Africa in the Neoliberal World Order* (Durham, NC: Duke University Press, 2006); Timothy Mitchell, *Rule of Experts: Egypt, Techno-Politics, Modernity* (Berkeley: University of California Press, 2002); Richard Rottenburg, *Far-Fetched Facts: A Parable of Development Aid* (Cambridge, MA: MIT Press, 2009); James C. Scott, *Seeing Like a State: How Certain Schemes to Improve the Human Condition Have Failed* (New Haven, CT: Yale University Press, 1998); and Helen Tilley, *Africa as a Living Laboratory: Empire, Development, and the Problem of Scientific Knowledge, 1870–1950* (Chicago: University of Chicago Press, 2011).

62. Van Beusekom, *Negotiating Development*.

63. In the 1920s preventing famine was increasingly perceived as an obligation of the colonial state across the French Empire. See Yan Slobodkin, "Famine and the Science of Food in the French Empire, 1900–1939," *French Politics, Culture, and Society* 36, no. 1 (2018): 52–75. See also Vincent Bonnecase, *La pauvreté au Sahel: Du savoir colonial à la mesure internationale* (Paris: Éditions Karthala, 2011), 29–34.

64. James Fairhead and Melissa Leach, *Misreading the African Landscape: Society and Ecology in a Forest-Savanna Mosaic* (Cambridge: Cambridge University Press, 1996), see especially chapter 1; James Fairhead and Melissa Leach, "Dessication and Domination: Science and Struggles of Environment and Development in Colonial Guinea," *Journal of African History* 41, no. 1 (2000): 35–54; Jeremy Swift, "Desertification: Narratives, Winners and Losers," in *The Lie of the Land: Challenging Received Wisdom on the African Environment*, ed. Melissa Leach and Robin Mearns (Portsmouth, NH: Heinemann, 1996), 74–77. This French misreading of the West African landscape fits a larger pattern of misleading environmental assessments of Africa in the twentieth century by outside experts. See Melissa Leach and Robin Mearns, "Environmental Change and Policy: Challenging Received Wisdom in Africa," in *Lie of the Land: Challenging Received Wisdom on the African Environment*, ed. Robin Mearns and Melissa Leach (Portsmouth, NH: Heinemann, 1996), 1–33.

65. Monica M. van Beusekom, "From Underpopulation to Overpopulation: French Perceptions of Population, Environment, and Agricultural Development in French Soudan (Mali), 1990–1960," *Environmental History* 4, no. 2 (1999): 198–219.

66. In this respect the Office du Niger was not unlike other development projects in Africa featuring large-scale dams. See Emmanuel Kwaku Akyeampong, *Between the Sea and the Lagoon: An Eco-social History of the Anlo of Southeastern Ghana, c. 1850–Recent Times* (Athens: Ohio

University Press, 2001); Isaacman and Isaacman, *Dams, Displacement, and the Delusion of Development*. See also Laura Ann Twagira, "Peopling the Landscape: Colonial Irrigation, Technology, and Demographic Crisis in the French Soudan, ca. 1926–1944," *PSAE Research Series* 10 (2012): 1–29.

67. Emmanuel Kreike, *Environmental Infrastructure in African History: Examining the Myth of Natural Resource Management in Namibia* (Cambridge: Cambridge University Press, 2013), 21–23.

68. Kreike, 24–25.

69. A 1935 decree declared vast tracts of "empty" land under the management the colonial state. Tor A. Benjaminsen, "Natural Resource Management, Paradigm Shifts, and the Decentralization Reform in Mali," *Human Ecology* 25, no. 1 (1997): 130–31.

70. In making this point I draw from the work of Carola Lentz, who has discussed the sociopolitical import of land claims in West Africa. See Carola Lentz, *Land, Mobility, and Belonging in West Africa* (Bloomington: Indiana University Press, 2013).

71. Amartya Sen, *Poverty and Famines: An Essay on Entitlement and Deprivation* (Oxford: Clarendon, 1981).

72. Shipton also emphasized the need to study how people actually survive food crises. Specifically, he suggested that memories of past food shortages illuminate the complexities of poverty, age, gender, and myriad other issues that influence who has access to food. See Parker Shipton, "African Famines and Food Security: Anthropological Perspectives," *Annual Review of Anthropology* 19 (1990): 353–94.

73. See the tale "L'ami du lion," in Monteil, *Contes soudanais*, 126–35.

74. Diana Wylie, *Starving on a Full Stomach: Hunger and the Triumph of Cultural Racism in Modern South Africa* (Charlottesville: University Press of Virginia, 2001).

75. Two important works are Richard W. Franke and Barbara Chasin, *Seeds of Famine: Ecological Destruction and the Development Dilemma in the West African Sahel* (Montclair, NJ: Allanheld, Osmun, 1980); and Michael Watts, *Silent Violence: Food, Famine and Peasantry in Northern Nigeria* (Berkeley: University of California Press, 1983).

76. However, several historical studies dating to the same period called for a longer-term examination of famine in Africa. See, for example, Jill R. Dias, "Famine and Disease in the History of Angola C. 1830–1930," *Journal of African History* 22, no. 3 (1981): 349–78; Joseph C. Miller, "The Significance of Drought, Disease, and Famine in the Agriculturally Marginal Zones of West-Central Africa," *Journal of African History* 23, no. 1 (1982): 17–61; Megan Vaughan, *The Story of an African Famine: Gender and Famine in Twentieth-Century Malawi* (Cambridge: Cambridge University Press, 1987); and Watts, *Silent Violence*.

77. Bonnecase, *La pauvreté au Sahel*, see especially chapter 5.

78. Rosamond L. Naylor, ed., *The Evolving Sphere of Food Security* (Oxford: Oxford University Press, 2014); and Christopher B. Barrett and Daniel G. Maxwell, *Food Aid after Fifty Years: Recasting Its Role* (New York: Routledge, 2005).

79. Priscilla Parkhurst Ferguson, "The Senses of Taste," *American Historical Review* 116, no. 2 (April 2011): 372. On taste and consumption in Mali, see Mamadou Diawara and Ute Röschenthaler, "Green Tea in the Sahel: The Social History of an Itinerant Consumer Good," *Canadian Journal of African Studies* 46, no. 1 (2012): 39–64; and Gérard Dumestre, "De l'alimentation au Mali," *Cahiers d'Études Africaines* 36, no. 144 (1996): 689–702.

80. I draw this insight from Tanja Winther, who examined the flavor aspects of cooking with wood fuel or electric cookers in contemporary Zanzibar. See Winther, *Impact of Electricity*, 197–208.

81. I draw here from Pierre Bourdieu's work on taste and how the construction of acceptable or superior taste generates and reinforces social values and distinctions, which in the case of Mali incorporated daily food sharing and the organization of harvest festivals as essential aspects of rural hospitality and respectability. See Pierre Bourdieu, *Distinction: A Social Critique of the Judgement of Taste* (Cambridge, MA: Harvard University Press, 1984).

82. On the cultural meanings of food and as a measure of social well-being, see Jon Holtzman, *Uncertain Tastes: Memory, Ambivalence, and the Politics of Eating in Samburu, Northern Kenya* (Berkeley: University of California Press, 2009).

83. For more on the role of international NGOs in Mali during these years, see Gregory Mann, *From Empires to NGOs in the West African Sahel* (Cambridge: Cambridge University Press, 2015).

84. Benedetta Rossi, *From Slavery to Aid: Politics, Labour, and Ecology in the Nigerien Sahel, 1800–2000* (Cambridge: Cambridge University Press, 2015). Strikingly, following the drought, women in Niger became the face of national development as the predominant participants in the landscape-rehabilitation scheme known as the Keita Project. See especially chapters 5 and 6.

85. Misty L. Bastian, "The Naked and the Nude: Historically Multiple Meanings of Oto (Undress) in Southeastern Nigeria," in *Dirt, Undress, and Difference: Critical Perspectives on the Body's Surface*, ed. Adeline Masquelier (Bloomington: Indiana University Press, 2005); Janice Boddy, *Wombs and Alien Spirits: Women, Men, and the Zar Cult in Northern Sudan* (Madison: University of Wisconsin Press, 1989); Bray, *Technology and Gender*; Marie Grace Brown, *Khartoum at Night: Fashion and Body Politics in Imperial Sudan* (Stanford, CA: Stanford University Press, 2017); Naminata Diabate, *Naked Agency*; Marisa J. Fuentes, *Dispossessed Lives: Enslaved Women, Violence, and the Archive* (Philadelphia: University of Pennsylvania Press,

2016); Heidi Gengenbach, "Boundaries of Beauty: Tattooed Secrets of Women's History in Magude District, Southern Mozambique," *Journal of Women's History* 14, no. 4 (2003): 106–41; Holtzman, *Uncertain Tastes*; Julie Livingston, *Debility and the Moral Imagination in Botswana* (Bloomington: Indiana University Press, 2005); Jennifer L. Morgan, *Laboring Women: Reproduction and Gender in New World Slavery* (Philadelphia: University of Pennsylvania Press, 2004); Schiebinger, *Nature's Body*; Paul Stoller, *Embodying Colonial Memories: Spirit Possession, Power, and the Hauka in West Africa* (New York: Routledge, 1995); and Thomas, *Politics of the Womb*. Other scholars not specifically focusing on gender have also examined the body and embodiment in West Africa. See Eric Silla, *People Are Not the Same: Leprosy and Identity in Twentieth-Century Mali* (Portsmouth, NH: Heinemann, 1998); and Rudolph T. Ware, *The Walking Qur'an: Islamic Education, Embodied Knowledge, and History in West Africa* (Chapel Hill: University of North Carolina Press, 2014).

86. On oral history, see Luise White, Stephan F. Miescher, and David William Cohen, eds., *African Words, African Voices: Critical Practices in Oral History* (Bloomington: Indiana University Press, 2001). On oral history and gender, see Gail Hershatter, *The Gender of Memory: Rural Women and China's Collective Past* (Berkeley: University of California Press, 2011); and Yasmin Saikia, *Women, War, and the Making of Bangladesh: Remembering 1971* (Durham, NC: Duke University Press, 2011). On participant observation see Trevor H. J. Marchand, *The Masons of Djenné* (Bloomington: Indiana University Press, 2009). Between 2009 and 2010, I lived in several Office du Niger towns and villages to conduct interviews with women and men living at the project, as well as with longtime project staff. However, I first observed and participated in daily cooking activities, including the seasonal production of shea butter, in 2000. In that year I began service as a Peace Corps volunteer in the village of Kalaké-Bamana to the south of Ségou.

87. Susan Geiger, *TANU Women: Gender and Culture in the Making of Tanganyikan Nationalism, 1955–1965* (Portsmouth, NH: Heinemann, 1997), 14.

88. Tamara Giles-Vernick, "Lives, Histories, and Sites of Recollection," in *African Words, African Voices: Critical Practices in Oral History*, ed. Luise White, Stephan F. Miescher, and David William Cohen (Bloomington: Indiana University Press, 2001), 194–213.

89. Barbara M. Cooper, "Oral Sources and the Challenge of African History," in *Writing African History*, ed. John Edward Philips (Rochester, NY: University of Rochester Press, 2005), 191–215.

90. Luise White, *Speaking with Vampires: Rumor and History in Colonial Africa* (Berkeley: University of California Press, 2000), 30.

91. White, 8–9, 30. Moreover, as Toyin Falola and Fallou Ngom suggest for oral and written literatures more broadly, many of the stories I consulted had increasingly circulated in the twentieth century in both oral and textual forms. See Toyin Falola and Fallou Ngom, "Introduction: Orality,

Literacy and Cultures," in *Oral and Written Expressions of African Cultures* (Durham, NC: Carolina Academic Press, 2009), xvii–xxxviii.

92. By oral tradition, I mean the oral transmission of information about past events, heroic figures, and their exploits, as well as broad social concerns, in the form of myths, legends, stories, and songs. They are a means of reproducing and even reworking a community's history and identity. On oral tradition and its historical uses, see Harold Scheub, "A Review of African Oral Traditions and Literature," *African Studies Review* 28, no. 2/3 (1985): 1–72. See also Stephen Belcher, *Epic Traditions of Africa* (Bloomington: Indiana University Press, 1999); David C. Conrad, "'Bilali of Faransekila': A West African Hunter and World War I Hero According to a World War II Veteran and Hunters' Singer of Mali," *History in Africa* 16 (1989): 41–70; Cooper, "Oral Sources and the Challenge of African History"; and David Henige, "Oral Tradition as a Means of Reconstructing the Past," in *Writing African History*, ed. John Edward Philips (Rochester, NY: University of Rochester Press, 2005): 169–90.

CHAPTER 1: MAKING THE GENEROUS COOKING POT, CA. 1890–1920

1. Masked hyena characters frequently parody human society, especially moments of human imprudence or naivety, and in their performances they often adopt the comportment of a thief. Dominique Zahan, *Antilopes du soleil: Arts et rites agraires d'Afrique noire* (Vienna: Edition A. Schendl, 1980), 30.

2. The version of this Bamana tale collected by Charles Monteil does not explicitly name shea butter as the substance in the story, but it was a common cooking fat (butter or oil) produced and consumed in farming communities across the wider West African region. For more on the mythical and symbolic value of shea butter in societies from the Mande cultural region, see Germaine Dieterlen, *Essai sur la religion Bambara* (Paris: Presses Universitaires de France, 1951), 18, 38.

3. A marsh is a likely space to find an object imbued with special powers because many marshes, natural wells, and other water sources held sacred value in the region. See Dieterlen, *Essai sur la religion Bambara*, 24.

4. This tale and the name of the pot are from Charles Monteil's translations of a story he collected while serving as a colonial administrator during the years 1897–99 in Médine and 1901–2 in Djenné. Some of the meaning was likely lost from the original tale, which Monteil even suggests in the introduction to his collection of Soudanese stories. However, the theme of generosity appears in several other stories collected by Monteil and in other story collections from the region. Charles Monteil, *Contes soudanais* (Paris: Ernest Leroux, 1905), 1–14, 57–61.

5. Old women are untrustworthy characters in many of the folktales collected from the region, representing a negative gender trope and anxieties over their potentially dangerous power.

6. This story was collected and published by Moussa Travélé, who served as primary language interpreter in the French Soudan. Travélé was also a researcher and intellectual. He wrote and published a Bambara grammar guide (1910) and a French-Bambara dictionary (1913) in addition to publishing his 1923 collection of regional proverbs and tales, for which he wrote an ethnographic introduction to Malinke and Bambara societies. Moussa Travélé, "Trois personnes rapides," in *Proverbes et contes bambara: Accompagnés d'une traduction française et précédés d'un abrégé de droit coutumier* (Paris: Librarie Orientaliste Paul Geuthner, 1923), 57–58.

7. Several regional collections of stories and songs centrally feature food, famine, and agricultural work. See, for example, Pascal Baba F. Couloubaly, *Une société rural bambara à travers des chants de femmes* (Dakar: Editions IFAN, Université de Dakar, 1990); Veronika Görög and Abdoulaye Diarra, *Contes bambara du Mali* (Paris: Publications Orientalistes de France, 1979); Monteil, *Contes soudanais*; and Travélé, *Proverbes et contes bambara*. For a review of the symbolic values of food and hunger in contemporary West African literature, see Shirwin Edwin, "Subverting Social Customs: The Representation of Food in Three West African Francophone Novels," *Research in African Literatures* 39, no. 3 (2008): 39–50.

8. Travélé, *Proverbes et contes bambara*, 58.

9. For a gendered discussion of hunters as heroes, see Mary Jo Arnoldi, "Wild Animals and Heroic Men: Visual and Verbal Arts in the Sogo Bo Masquerades of Mali," *Research in African Literatures* 31, no. 4 (2000): 63–75.

10. I draw from multiple sources in narrating and interpreting the Muso Koroni myth. See Dieterlen, *Essai sur la religion Bambara*; Germaine Dieterlen, "The Mande Creation Myth," *Africa: Journal of the International African Institute* 27, no. 2 (1957): 124–38; Barbara G. Hoffman, "Gender Ideology and Practice in Mande Societies and in Mande Studies," *Mande Studies* 4 (2002): 1–20; and Dominique Zahan, *The Bambara* (Leiden, The Netherlands: E. J. Brill, 1974). Adame Ba Konaré identifies the figure as Musokoroni Kuntyé and explains that the female deity is also associated with sexuality, infidelity, and jealousy. See Adame Ba Konaré, *Dictionnaire des femmes célèbres du Mali (des temps mythico-légendaires au 26 Mars 1991); précédé d'une analyse sur le rôle et l'image de la femme dans l'histoire du Mali* (Bamako, Mali: Éditions Jamana, 1993), 99–100. Germaine Dieterlen and other anthropologists participating in the Marcel Griaule missions to the French Soudan collected multiple accounts of the Mande creation myth and religious practices over several decades in the early twentieth century. While aspects of Dieterlen's work have been rightly criticized, I agree with Ralph Austen that her ethnographic accounts contain local belief and ritual practices, especially for the central Mali region historically occupied by Bamana and Dogon communities. See Ralph A. Austen, "The Problem of the Mande Creation Myth," in *Mande Mansa: Essays in Honor of David C. Conrad*, ed. Stephen Belcher, Jan Jansen, and Mohamed N'Daou (Zürich: Lit, 2008), 37.

11. Margaret Ehrenberg, *Women in Prehistory* (London: British Museum, 1989), 77–90. See also Judith A. Carney and Richard Nicholas Rosomoff, *In the Shadow of Slavery: Africa's Botanical Legacy in the Atlantic World* (Berkeley: University of California Press, 2009), 15–18; and Christopher Ehret, *The Civilizations of Africa: A History to 1800* (Charlottesville: University of Virginia Press, 2002), 35–39, 61–64.

12. The anthropologist Solange de Ganay similarly collected Bamana stories that credit the mother of the mythical Ciwara with the planting of wild seeds. Subsequently, the Ciwara taught humans the techniques of cereal cultivation. See Zahan, *Antilopes du soleil*, 59.

13. The Muso Koroni myth also betrays tensions in regional gender ideologies. In Barbara Hoffman's interpretation, Muso Koroni's innovation of circumcision and excision instituted chaos because the natural state of the human body was dual gendered. It is only through procreation and cooperation that adults in society are made complete, thereby restoring social balance. See Hoffman, "Gender Ideology," 16. Kassim Kone, examining linguistic evidence, has further argued that gender in Mali is flexible in specific performance and social contexts despite ideologies of strict gender stratification. Kassim Kone, "When Male Becomes Female and Female Becomes Male in Mande," *Wagadu* 1 (Spring 2002): 21–29. On Muso Koroni's association with circumcision, see also Ba Konaré, *Dictionnaire des femmes*, 100.

14. Zahan, *Antilopes du soleil*, 35–41. See also Stephen Wooten, *The Art of Livelihood: Creating Expressive Agri-culture in Rural Mali* (Durham, NC: Carolina Academic Press, 2009), especially chapter 5. I also draw from my interview about farming songs with performers Madu Saré (singer), Kadja Coumaré (singer), Bading Traoré (Ciwara mask maker), and Issa Saré (dancer) in Markala-Kirango, June 3, 2010.

15. Zahan, *Antilopes du soleil*, 41, 78. Stephen Wooten further notes that prayers given during Ciwara performances are given both for a successful agricultural season and for multiple healthy births during the year. See Stephen R. Wooten, "Where Is My Mate? The Importance of Complementarity: A Bamana Headdress (Ciwara)," in *See the Music, Hear the Dance: Rethinking Art at the Baltimore Museum of Art*, ed. Frederick John Lamp (Munich: Prestel, 2004), 168. For a comparative study of women's fertility, ideology, and spirituality in Africa, see Iris Berger, "Fertility as Power: Spirit Mediums, Priestesses and the Pre-colonial State in Interlacustrine East Africa," in *Revealing Prophets: Prophecy in Eastern African History*, ed. David M. Anderson and Douglas H. Johnson (Athens: Ohio University Press, 1995).

16. Stacy Alaimo and Susan Hekman, "Introduction: Emerging Models of Materiality in Feminist Theory," in *Material Feminisms*, ed. Stacy Alaimo and Susan Hekman (Bloomington: Indiana University Press, 2008), 3.

17. One influential example is Carolyn Merchant, *The Death of Nature: Women, Ecology, and the Scientific Revolution*, reprint ed. (New York: Harper One, 1989).

18. Hoffman, "Gender Ideology," 6.
19. For women potters in Mali, see Barbara E. Frank, *Mande Potters and Leather Workers: Art and Heritage in West Africa* (Washington, DC: Smithsonian Institution Press, 1998); and Barbara E. Frank, "Marks of Identity: Potters of the Folona (Mali) and Their 'Mothers,'" *African Arts* 40, no. 1 (2007): 30–41.
20. For a critical elaboration of the predominant view of male dominance in Mali, see Hoffman, "Gender Ideology," 3.
21. Hoffman, 16.
22. Obioma Nnaemeka, "Mapping African Feminisms," in *Readings in Gender in Africa*, ed. Andrea Cornwall (Bloomington: Indiana University Press, 2005), 31–32.
23. Tellingly, the first mythical women extracted oil from the shea nuts to accompany the consumption of the tree's fruits. See Dieterlen, *Essai sur la religion Bambara*, 18.
24. Dieterlen, "The Mande Creation Myth"; Dieterlen, *Essai sur la religion Bambara*, 1–33. Marcel Griaule and Germaine Dieterlen, "L'agriculture rituelle des Bozo," *Journal de la Société des Africanistes* 19, no. 2 (1949): 209–11. While many accounts describe Faro as both male and female, Adame Ba Konaré identifies Faro as a female deity whose bodily curves left their imprint on the course of the Niger River. See Ba Konaré, *Dictionnaire des femmes*, 98–99.
25. Ethnographers and botanists have documented that farmers across the western Soudan region do not cut the balanzan when they clear land for fields. Its branches, leafless during the rainy season, do not hinder the growth of crops under its canopy; the trees also provide animal forage, wood fuel, medicines, food products, and tannins for leather production. More recently, scientists have documented the species' role in restoring soil fertility, a key concern for farmers. See Guy Roberty, *Les associations végétales de la vallee moyenne du Niger* (Bern, Switzerland: Verlag Hans Huber, 1946), 96–97; Dieterlen, *Essai sur la religion Bambara*, 24; G. E. Wickens, *Role of Acacia Species in the Rural Economy of Dry Africa and the Near East* (Rome: Food and Agricultural Organization of the United Nations, 1995). See also J. Th. Broekhuyse and Andrea M. Allen, "Farming Systems Research on the Northern Mossi Plateau," *Human Organization* 74, no. 4 (Winter 1988): 331; and Birama Diakon, *Office du Niger et pratiques paysannes: Appropriation technologique et dynamique sociale* (Paris: L'Harmattan Mali, 2012), 71.
26. The term "Sahel" refers to the belt of land extending from the western regions of Senegal and Mauritania to the east along the southern border of the Sahara. The same region is also often referred to as the Sudan (or Soudan).
27. In fact, paleoclimatologists characterize the region not by its rainfall patterns but by its climatic anomalies. See Roderick J. McIntosh, "Social

Memory in Mande," in *The Way the Wind Blows: Climate, History, and Human Action*, ed. Roderick J. McIntosh, Joseph A. Tainter, and Susan Keech McIntosh (New York: Columbia University Press, 2000), 145–57; and Téréba Togola, "Memories, Abstractions, and Conceptualization of Ecological Crisis in the Mande World," in McIntosh, Tainter, and McIntosh, *The Way the Wind Blows*, 181–85.

28. Mande broadly refers to several farming societies, including the Bamana, that are understood to share common cultural practices and a common historical and mythical past. Mande also constitutes a language group with speakers of related languages in Mali and several other West African countries including Guinea, Senegal, Sierra Leone, and the Ivory Coast.

29. Roderick McIntosh and his coauthors define social memory as "a concept to describe the ways by which communities curate and transmit both past environmental states and possible responses to them." See Roderick McIntosh, Joseph A. Tainter, and Susan Keech McIntosh, "Climate, History, and Human Action," in McIntosh, Tainter, and McIntosh, *The Way the Wind Blows*, 24. See also Togola, "Memories, Abstractions, and Conceptualization"; and McIntosh, "Social Memory in Mande."

30. Togola, 185–86. Michael Gomez offers an alternative interpretation of the legend, suggesting it represents a political shift from a religious atmosphere of tolerant coexistence to one in which African practices were suppressed in favor of Islam. See Gomez, *African Dominion: A New History of Empire in Early and Medieval West Africa* (Princeton, NJ: Princeton University Press, 2018), 38.

31. Dieterlen, "The Mande Creation Myth," 127.

32. McIntosh, "Social Memory in Mande," 160–68.

33. In this analysis I draw inspiration from Melissa Leach and Robin Mearns, whose foundational work challenges tropes about the African environment. See Melissa Leach and Robin Mearns, "Challenging Received Wisdom in Africa," in *The Lie of the Land: Challenging Received Wisdom on the African Environment*, ed. Melissa Leach and Robin Mearns (Portsmouth, NH: Heinemann, 1996), 9–16.

34. William B. Cohen, "Imperial Mirage: The Western Sudan in French Thought and Action," *Journal of the Historical Society of Nigeria* 7, no. 3 (1974): 417–45. The observations of early European travelers of agricultural riches in West Africa are substantiated by Judith Carney. See Carney and Rosomoff, *In the Shadow of Slavery*.

35. Trade along the North-South axis of the Soudan has a long history that the Segu Kingdom reoriented along the Niger River during the late eighteenth and early nineteenth centuries. See chapter 2 in Richard Roberts, *Warriors, Merchants, and Slaves: The State and the Economy in the Middle Niger Valley, 1700–1914* (Stanford, CA: Stanford University Press, 1987), 210–11.

36. In the last two decades of the century the French military sponsored missions to explore the course of the Niger River and establish political relations with the chiefs of polities along the river. See, for example, Jean Gilbert Nicomède Jaime, *De Koulikoro à Tomboctou sur la cannonière "Le Mage"* (Paris: Les Libraires Associés, 1894).

37. B. Marie Perinbam, *Family Identity and the State in the Bamako Kafu, c.1800–c.1900* (Boulder, CO: Westview, 1997), 16–18, 114. See also Pierre Viguier, *Sur les traces de René Caillié: Le Mali de 1828 revisité* (Versailles: Éditions Quae, 2008), 65.

38. Perinbam, *Family Identity*, 21. Arabic speakers and European visitors to the western Soudan used the term "Bambara" (or "Bamana") to refer to a diverse group of agriculturalists. In some cases Bambara only referred to people who were non-Muslim or beer drinking. In other cases Bambara meant speakers of Mande languages. See Jean Bazin, "A chacun son Bambara," in *Au coeur de l'ethnie: Ethnies, tribalism et État en Afrique*, ed. Jean-Loup Amselle and Elikia M'Bokolo (Paris: Paris Éditions de la Découverte, 1985).

39. Ethnicity in the Middle Niger region is generally attributed to groups who dominated particular ecological niches, such as herding, fishing, or farming. However, ethnic boundaries were not historically rigid or always clearly defined (see note 38). For example, in the nineteenth century the Segu Kingdom promoted the emergence of a new fishing community made up of ethnically diverse slaves. In exchange for state service, members of the new Somono group were granted specific privileges over Bozo fishers. See Richard Roberts, "Fishing for the State: The Political Economy of the Middle Niger Valley," in *Modes of Production in Africa: The Precolonial Era*, ed. Donald Crummey and C. C. Stewart (Beverly Hills, CA: SAGE, 1981). B. Marie Perinbam has also noted that the founders of the Bamako Kafu changed ethnic identity over time, as it was common for migrating families to take on new cultural identities as they moved to new regions. She also notes that in the nineteenth century, the Segu state promoted a general association with a Bamana identity. See Perinbam, *Family Identity*.

40. The present work refers to the Bamana Empire and capital as Segu. For the same city and region under French control the modified name Ségou is used. See Mungo Park, *Travels in the Interior Districts of Africa [1799]*, ed. Kate Ferguson Marsters (Durham, NC: Duke University Press, 2000), 189, 201. As with each of the travel narratives cited here, Park is cautiously read as a primary source and early observer of agriculture and foodways in the Western Soudan. Many scholars have pointed out that his travel narrative was heavily edited, especially with an interest in promoting the work of antislavery societies in England. See Kate Ferguson Marsters, "Introduction," in *Travels in the Interior Districts of Africa*, ed. Kate Ferguson Marsters (Durham, NC: Duke University Press, 2000).

41. René Caillié, *Journal d'un voyage à Tombouctou et à Jenné, dans l'Afrique centrale*, 2 vols. (Paris: P. Mongie, 1830). See also Viguier, *Sur les traces de René Caillié*.

42. David Robinson, *The Holy War of Umar Tal* (Oxford: Oxford University Press, 1988). See also chapter 3 in Roberts, *Warriors, Merchants, and Slaves*.

43. The Segu Empire controlled a large but shifting territory between roughly 1712 and 1861. A rival empire called the Caliphate of Hamdullahi emerged in the Macina around 1816.

44. Cohen, "Imperial Mirage," 438–39.

45. Eugène Mage, *Voyage dans le Soudan occidental (Sénégambie-Niger)* (Paris: Hachette, 1868).

46. Mage, 81, 170.

47. Mage, 49, 70, 81, 87, 127.

48. Most notably the empire established by Samori Touré in the late nineteenth century and neighboring city-state Sikasso. For more on the French conquest of both southern states, see chapters 1 and 2 in Brian J. Peterson, *Islamization from Below: The Making of Muslim Communities in Rural French Soudan, 1880–1960* (New Haven, CT: Yale University Press, 2011).

49. Roberts, *Warriors, Merchants, and Slaves*, 144–45. See also chapter 5 in Martin Klein, *Slavery and Colonial Rule in French West Africa* (New York: Cambridge University Press, 1998).

50. Commander Frey in 1888, cited by Klein, *Slavery and Colonial Rule*, 77.

51. Jaime, *De Koulikoro*, 401.

52. Despite the emerging sense of the region as marginal, the early colonial government relied heavily on agricultural production in the Soudan to feed its army. Much like the previous warrior states, the French military demanded labor, grains, and other goods from conquered populations. To supply the African troops, French military leaders established villages of captured ex-slaves to grow grain for its African troops. These ex-slaves also provided labor in towns along the French supply line known as the *route de ravitaillement*.

53. The regional trade economy in the nineteenth century had been supported by slave labor and the slave trade. Over the course of that century both intensified. See Klein, *Slavery and Colonial Rule*, 134–37. On the intensification of slavery in the nineteenth century, see chapter 3. See also Richard Roberts, *Litigants and Households: African Disputes and Colonial Courts in the French Soudan, 1895–1912* (Portsmouth, NH: Heinemann, 2005), 99. In 1903 the governor-general of French West Africa issued a decree abolishing enslavement and the exchange of people. However, most administrators hoped to foster compromise between enslaved communities and their masters in order to maintain control over agricultural production. The decree did not directly abolish the institution of slavery, though it was often later interpreted to do so. See Klein, *Slavery and Colonial Rule*, 134–37, 78–79.

54. In fact, French administrators made a concerted effort to keep slaves in the Macina from leaving the region. See Klein, 179–85.
55. See chapter 4 in Roberts, *Litigants and Households*.
56. "Exode des serviteurs," n.d., 1907 folder, Correspondance affaires politiques cercle de Ségou et Sansanding 1891–1909, Archives nationales du Mali (hereafter ANM) 1E 219.
57. Roberts, *Litigants and Households*, 105.
58. Mademba Sy, the French-appointed ruler of Sansanding, reported raids on Marka and Bamana towns near Sokolo by northern Moors. See 1907 folder, Correspondance affaires politiques cercle de Ségou et Sansanding 1891–1909, ANM 1E 219.
59. Pawning is the practice of providing the labor of a dependent (often a daughter) as collateral for a debt. The pawn provides labor in the household of the creditor until the debt is repaid. The pawn's labor serves as an interest payment and does not count toward payment of the debt. Often young girls were married into the households into which they were pawned. Martin Klein and Richard Roberts have outlined the important role of female slaves and pawns to domestic and agricultural labor in the late nineteenth and early twentieth centuries. Distinctions between a young wife, a female pawn, and a female slave were hard for Europeans to grasp during these years as their roles in the household overlapped, and their status often shifted through marriage. The administrator Henri Ortoli documented pawning in the French Soudan in the 1930s. See Henri Ortoli, "Le gage des personnes au Soudan français," *Bulletin de l'Institut Français d'Afrique Noire* 1 (1939): 313–24. See also Roberts, *Litigants and Households*, 114–18; Martin A. Klein, "Women in Slavery in the Western Sudan," in *Women and Slavery in Africa*, ed. Claire C. Robertson and Martin A. Klein (Portsmouth, NH: Heinemann, 1997); Martin A. Klein and Richard Roberts, "The Resurgence of Pawning in French West Africa during the Depression of the 1930s," *African Economic History* 16 (1987): 23–37. Missionaries working in neighboring Mossi regions similarly worried about women's status and servitude. See Marie-André du Sacré-Coeur Sr., "La femme Mossi, sa situation juridique," *Ethnographie* 33–34 (1937): 15–33.
60. It is not clear from the record whether Koita had been enslaved in Sansanding. After 1903, Mademba Sy would have been prohibited by law from labeling Koita a slave in official reports. However, it is probable that Koita had been enslaved, because Sansanding was one of the major Marka production centers characterized by slave plantations. See 1905 folder, Correspondance avec le Fama de Sansanding Cercle de Ségou 1890–1917, ANM FA 1E 220. Other cases in the same folder record a few pending cases against litigants accused of enslaving children. Mademba Sy had himself been a large-scale slave owner, many of whom became his wives. See Richard Roberts, "The Case of Faama Mademba Sy and the Ambiguities of Legal Jurisdiction in Early Colonial French Soudan,"

in *Law in Colonial Africa*, ed. Kristin Mann and Richard Roberts (Portsmouth, NH: Heinemann, 1991), 110, 16, 18.

61. Roberts, *Litigants and Households*, 109, 117.

62. When I visited the National Archives of Mali in the fall of 2009, a set of records relating to the 1913–14 famine was missing ("Renseignement sur la famine," 1915, ANM FA 1D 52). I later found these documents when I requested the record "Ravitaillement production agricole-prévisions situation alimentaire 1926 à 1934," ANM FR I 1D 2879. The other documents in this second file recorded food shortages and famine during the 1930s, and it is likely that administrators responding to the 1930s food shortages consulted responses made by the earlier administration. The archivist retained a copy of these records in the ANM FR 1D 2879 file to preserve the record of historical consultation by the 1930s administration. During the 1913–14 famine, the lieutenant-governor sent 394,000 tons of millet and 76,000 tons of rice to the Tombouctou, Niafounké, Bandiagara, Goumbou, Sokolo, Mopti, Djenné, and Kayes cercles (*cercle* is the French term for the subregional level of administration). An additional 123,000 tons of millet, 72,000 tons of rice paddy, 6,000 tons of groundnut, and 2,000 tons of corn were shipped to furnish seed stores. See the table attached to the document "Extrait du rapport d'ensemble sur la situation économique de la colonie du Haut-Sénégal-Niger pendant l'année 1914," in a folder marked "Achats pour les cercles (mil, riz, paddy, arachides, et maïs)." See also Lieutenant-Gouverneur du Haut Sénégal et Niger to the Gouverneur Général, February 5, 1914 (letter no. 102), ANM FA 1D 52, found in ANM FR I 1D 28791.

63. The French spelling of Tombouctou (Timbuktu) is employed here for the colonial era. See Lieutenant-Gouverneur du Haut Sénégal et Niger to the Gouverneur Général, June 11, 1914 (letter no. 412 AS), ANM FA 1D 52, found in ANM FR I 1D 28791. Cumulative reports from 1913 estimate as many as half a million deaths in the wider region due to famine. See Yan Slobodkin, "Famine and the Science of Food in the French Empire, 1900–1939," *French Politics, Culture, and Society* 36, no. 1 (2018): 52.

64. Lieutenant-Gouverneur to the Gouverneur Général, June 11, 1914.

65. In general, famines provoked widespread movements of people in search of food. See Klein, "Women in Slavery in the Western Sudan," 210–11.

66. Banankourou residents also requested Christian prayers for rain, and when some rain did fall in 1913, followers of the Catholic mission clashed with other community members over whose prayers brought the rain. See diary records from the White Sister missionaries for Banankourou (Région du Soudan) in the volumes 1912–1914, Fonds des Soeurs Blancs (FSB), Archives de la Société des missionaires d'Afrique (hereafter MAFR).

67. "Extrait du rapport d'ensemble sur la situation économique de la colonie du Haut-Sénégal-Niger pendant l'année 1914," ANM FA 1D 52, found in ANM FR I 1D 28791.

68. Djenebu Coulibaly, interview by author, Kolony (km 26), August 7, 2010. Coulibaly is referring to Bamana practice in particular and *jigine* is a Bamana term for "granary." She also described granaries as round mud structures whose foundations were dug slightly into the ground. Her description is reinforced by research conducted in the 1950s at the Office. M. Holstein, who conducted various research projects at the scheme for the Office de la recherche scientifique outre-mer (ORSTOM), recorded the layout of typical houses at the project in 1952. His renderings of typical house plans included ten or more granaries of varying sizes, probably in Mossi towns. See M. Holstein materials in folder 26: Documents divers, Office du Niger Archives (hereafter AON) 118/26.

69. Djenebu Coulibaly, interview, August 7, 2010.

70. Grenier de réserve 1915, ANM FR 1R 180.

71. Fada N'Gourma is located in present-day Burkina Faso. In 1914 the region was part of the French Soudan, and Gourma farmers in the area were also severely impacted by drought and food shortages from 1913 to 1914.

72. Grenier de réserve 1915.

73. "Extrait du rapport d'ensemble."

74. Réorganisation de l'agriculture 1913, ANM FA 1R 217; Arrêts décrets circulaires réorganisant le Service de l'agriculture 1913, ANM FA 1R 220. The decree was issued under Governor General Ponty.

75. See Caillié, *Journal d'un voyage*; Mage, *Voyage dans le Soudan*; and Park, *Travels in the Interior Districts of Africa*.

76. Réorganisation de l'agriculture 1913.

77. Denise Savineau, *La famille en A.O.F.: Condition de la femme* (Dakar: Gouvernement de l'A.O.F., 1938), 62.

78. Peanut farming in many areas competed with the cultivation of a local plant closely resembling the peanut: *voandzu*, or the Bambara groundnut. European observers often mistook one cultivar for the other, but peanut cultivation had reached western and northwestern Bambara territory by the eighteenth century. See Zahan, *Antilopes du soleil*, 48.

79. For more on Travélé see note 6 in this chapter. "L'orphelin (Conte de femme)," in Travélé, *Proverbes et contes bambara*, 123–25.

80. The term cited here for this wild grain was employed by women in Koue-Bamana; the spelling is the one offered by Aïssata Kassonke.

81. Most likely calabashes and baskets.

82. Recounted by Kadja "Ma" Coulibaly in Kadja "Ma" Coulibaly, Fatoumata Guindo, Fatoumata Zaré Coulibaly, and Lalafacouma Tangara, interview by author, Koue-Bamana, May 26, 2010.

83. The analysis in this section draws from important insights from Elias Mandala and Megan Vaughan relating to gender and the availability of food and its distribution. See Megan Vaughan, *The Story of an African Famine: Gender and Famine in Twentieth-Century Malawi* (Cambridge: Cambridge University Press, 1987); and Elias Mandala, *The End of*

Chidyerano: A History of Food and Everyday Life in Malawi, 1860–2004 (Portsmouth, NH: Heinemann, 2005).

84. A calabash is a large fruit that when hollowed, dried out, and halved serves as an organic bowl or container.

85. The colonial administration pressured farmers in the French Soudan to sell agricultural products to government markets following the war in Europe (1914–18). These demands are recorded for the Ségou region in several archival records: Ravitaillement correspondance cercle de Ségou 1914–1920, ANM FA 1Q 177; Rapport d'inspection sur le ravitaillement de la colonie de l'A.O.F., 1916, ANM FA 1Q 190; Ravitaillement des grains 1913–1919, ANM FR 1Q 77.

86. In 1915, Bintu might have cultivated groundnuts, peanuts, or both of these. See note 78 in this chapter.

87. The seasonality of the leaves used to make the sauce for *basi* (pronounced like "bashi") and, thus, the seasonality of *basi* was pointed out by Hawa Fomba in Kalaké-Bamana. She explained that the plant whose leaves were used in the sauce grow low to the ground and are harvested just after the rainy season. See Hawa Fomba, interview by author, Kalaké-Bamana, September 11, 2010.

88. The French administrator-ethnographer Henri Labouret observed that women in West Africa sometimes added pounded manioc to cooking fonio, which added more variety to the diet. See Henri Labouret, *Paysans d'Afrique Occidental* (Paris: Gallimard, 1941), 196.

89. On the nutritional value of leafy greens, nuts, roots, and other products cultivated from forested areas, see M. B. Nordeide et al., "Nutrient Composition and Nutritional Importance of Green Leaves and Wild Food Resources in an Agricultural District, Koutiala, in Southern Mali," *International Journal of Food Sciences and Nutrition* 47 (1996): 455–68. The authors point out that the spice *soumbala* is particularly rich in protein, fats, and energy-providing nutrients. Further, the common preparation practice of fermenting seeds and other raw products has been shown to increase the nutritional value of such raw foodstuffs.

90. Labouret, *Paysans d'Afrique Occidental*, 197–203.

91. Diana Wylie, "The Changing Face of Hunger in Southern African History 1880–1980," *Past and Present* 122 (1989): 160–62.

92. Labouret, *Paysans d'Afrique Occidental*, 197–203.

93. Fatoumata Coulibaly, Kadja Mallé, Mariam Mallé, Aminata Dembélé, and Kadja Koné, interview by author, Kouyan-Kura, May 27, 2010.

94. Guy Roberty to the Head of Laboratory Services, August 31, 1935 (letter no. 13), Correspondances, Dossier botanique (suite), AON 93/3. Roberty also highlighted a tree called Zeguéné found in the northern regions, along with cotton, ricin, and curcas oil. In interviews women recall collecting shea nuts for several months on a household rotation. Hawa Coulibaly, Mariam Sango, Djéné Mariko, Adam Coulibaly, and Nyé Diarra, interview by author, Molodo-Bamana, April 19, 2010.

95. Labouret, *Paysans d'Afrique Occidental*, 192.
96. Daouda Coulibaly and Almamy Thiènta, interview by author, Sokolo, April 5, 2010. During the interview, the two elderly men recalled how their mothers prepared sauces.
97. A large, round organic container with a small hole in the top, similar to a calabash or gourd, the baarakolo was made by drying the plant's large fruit and hollowing it out.
98. The cover allowed women to transport meals to the fields protected from the elements. It also helped to keep food warm. Hawa Fomba, informal conversation with author, Kalaké-Bamana, October 2, 2010.
99. Hawa Diarra, interview by author, Nara, April 30, 2010; Djewari Samaké, interview by author, Kolony (km 26), June 2, 2010.
100. In the months following the grain harvest, much of women's labor time was occupied with artisanal industry and foodstuffs manufacture. Women picked cotton and began carding and spinning yarn. Women also made potash by burning dried millet stalks; the potash was then used to make soap with shea oil. Women dyers cultivated indigo and other plants used to dye cloth. Fisherwomen processed dried and smoked fish for the household and similarly made cooking oil and soap from parts of certain fish. In these same dry months, women collected wood fuel for the year and made cooking oils and other nutritious ingredients. Much technical innovation went into these and other manufactures. Kono Dieunta, interview by author, Kokry, May 3, 2010; Rokia Diarra, interview by author, Nara, April 29, 2010; see also Agricultural calendar, n.d., Agriculture divers 1914–1931, ANM, FR II 1R 16; "Rapport du botaniste pour 1934," Dossier botanique, AON 30/10; Charles Monteil, *Le coton chez les noirs* (Paris: Émile Larose, 1927), 51–53; M. S. Martin et al., "Characterization of Foliar-Applied Potash Solution as a Non-selective Herbicide in Malian Agriculture," *Journal of Agricultural, Food, and Environmental Sciences* 2, no. 1 (2008): 1–8; and Jacques Daget, "La pêche dans le Delta central du Niger," *Journal de la Société des Africanistes* 19, no. 1 (1949): 52–56.
101. Mage, *Voyage dans le Soudan occidental*; Caillié, *Journal d'un voyage à Tombouctou et à Jenné*; and Louis Gustave Binger, *Du Niger au Golfe de Guinée par le pays de Kong et le Mossi*, 2 vols. (Paris: Hachette, 1892). Indeed, several women interviewees commented that across West Africa people offered visitors water, but only in Mali did they also offer food. According to them, this was especially the case when visitors passed by when others were eating. The visitor is always called to join with the phrase *nyaduminke* or "come eat." This last point was especially emphasized in one interview. Aïssata Mallé and Assane Pléah, interview by author, Kouyan-N'Péguèna, May 28, 2010.
102. The importance of sharing food with guests, especially for women, was reinforced in an informal conversation with the animatrice Mme. Tamboura Fatoumata Guindo on April 14, 2010.

103. Daouda Bouaré and Bintu Dembélé, interview by author, Sokorani, June 1, 2010. My research assistant Aïssata Kassonke believed this practice to have originated specifically from the Ségou region. Dege may also have been prepared by women for men leaving on long trips for hunting or other activities. Several times between 2000 and 2001, I was offered dege by Sekou Diarra, my elderly male hunter host in Kalaké-Bamana.

104. Aïssata Mallé and Assane Pléah, interview, May 28, 2010. Brian Peterson links this practice with an enduring culture of servile labor relations. His informants in southern Mali equated this task and posture with women's ongoing service status. See Peterson, *Islamization from Below*. However, Aïssata and Assane stressed that this action was a way of displaying *mogoya*, or respect. In the same interview the two women explained to me that this is why the elderly now ask young children to offer water to guests: the children learn *mogoya* by serving others.

105. Park, *Travels in the Interior Districts of Africa*, 192–93.

106. Monteil, *Contes soudanais*, 25–26.

107. Women who were cowives of the same man also rotated evenings with their husband according to the cooking schedule. Food and sex were very much intertwined, and good food meant that the woman who prepared the meal was more likely to benefit from sex (from bearing children for the lineage to improved relations with the husband). See Jack Goody, *Cooking, Cuisine, and Class: A Study in Comparative Sociology* (New York: Cambridge University Press, 1982); and Tanja Winther, *The Impact of Electricity: Development, Desires and Dilemmas* (New York: Berghahn, 2008), 195–96.

108. In an essay on women's status among the Mossi, Marie-André du Sacre-Coeur emphasizes domestic hierarchies between cowives and jealousy as a major cause of discord. See Du Sacré-Coeur, "La femme Mossi."

109. Elsewhere Paul Stoller explored how women in Niger demonstrated displeasure or protested household decisions by purposefully preparing unappetizing food. See Paul Stoller, *The Taste of Ethnographic Things: The Senses in Anthropology* (Philadelphia: University of Pennsylvania Press, 1989), 15–34.

110. James McCann rightly argues that regional West African diets have constituted a cuisine or distinctive foodway. The qualities distinguishing West African cuisine included the taste, texture, and visual presentation of dishes. McCann emphasizes staple starch dishes; my research also points to specific cooking oils and spices that characterized regional cuisines. See James C. McCann, *Stirring the Pot: A History of African Cuisine* (Athens: Ohio University Press, 2009), 7–8, 109–36.

111. Labouret, *Paysans d'Afrique Occidental*, 192–93.

112. Comment from Kadiatou Traoré during interview by author with Sékou Coulibaly and Kadiatou Traoré, Nyamina, May 31, 2010.

113. Comment from Traoré.

114. Djewari Samaké, interview by author, Kolony, March 29, 2010; Hawa Coulibaly, interview by author, Fouabougou, March 30, 2010; Aïssata Mallé and Assane Pléah, interview, May 28, 2010; Fatoumata Coulibaly et al., interview, May 27, 2010. Henri Labouret similarly described food preparation as a shared endeavor across much of French West Africa, especially the labor of pounding grains. See Labouret, *Paysans d'Afrique Occidental*, 191.

115. Couloubaly collected women's songs between 1981 and 1984 in N'Tiola, a village to the southwest of Bamako in the Koutiala region.

116. Couloubaly, *Une société rural bambara*, 66–67. Similarly, Jacques Dubourg observed that the male head of a Mossi household was charged with distributed daily grain allowances to the woman cooking. See Dubourg, "La vie des paysans mossi: Le village de Taghalla," *Cahiers d'Outre-Mer* 10, no. 40 (1957): 295.

117. See the diary entry for Toma on February 12, 1928, 2 vols. (Région du Soudan), Fonds des Soeurs Blancs (FSB), MAFR. Toma is in the region of contemporary Burkina Faso that was previously administered by the French Soudan.

118. Emile Bélime also observed the collection of a grain that he called *paikin* (which looked like fonio) in the region immediately north of Molodo. "Le Riz," n.d., Dossier Bélime, AON 28A; "Le Soudan Nigérien," n.d., Dossier Bélime, AON 28A.

119. Kadja "Ma" Coulibaly, Fatoumata Guindo, Fatoumata Zaré Coulibaly, and Lalafacouma Tangara, interview, May 26, 2010.

120. While fini is sold in the Niono market, jéba is no longer found near Niono. Djenebu Coulibaly, interview, August 7, 2010.

121. Daouda Coulibaly and Almamy Thiènta, interview, April 5, 2010.

122. Aïssata Mallé, Salimata Samaké, Assan Pléah, Mariam Sall, and Sitan Mallé, interview by author, Kouyan-N'Péguèna, April 12, 2010.

123. Like fini, gokun can still be found in rural markets. Moussa Diawara, Aminata Tangaré, and Hawoyi Diawara, interview by author, Boky-Were, May 7, 2010.

124. Moctar Coulibaly was the dugutigi (chief) of Sirakora and from the Koutiala region called the lean season "koala." Interview by author with Moctar Coulibaly, Mariatou Traoré, and Arouna Coulibaly, in Sirakoro, April 28, 2010. Some of these fruits were consumed raw, and others were transformed into beverages. For example, one small pungent fruit from the *béré* tree was extremely bitter. However, once the seeds were removed and soaked, the bitter taste was diluted. After the seeds were soaked, they were then cooked into a beverage. See Nianzon Bouaré and Harouna Bouaré, interview by author, Molodo-Bamana, April 16, 2010.

125. Assane Coulibaly, interview by author, Sirakoro, April 28, 2010; Alimata Dembélé, Moussa Coulibaly, and Nana Dembélé, interview by author, Koutiala-Kura, May 3, 2010; Kadja "Ma" Coulibaly, Fatoumata Guindo,

Fatoumata Zaré Coulibaly, and Lalafacouma Tangara, interview, May 26, 2010; and Nianzon Bouaré and Hawa Coulibaly, interview by author, Molodo-Bamana, May 29, 2010.

126. Aïssata Mallé and Assane Pléah, interview, May 28, 2010.

127. Women in Kouyan-N'Péguèna recalled that during end of the season parties, food, and dolo were in separate spaces. See Kadja "Ma" Coulibaly et al., interview, May 26, 2010. The White Father Catholic missionaries stationed in Ségou observed similarly animated harvest festivals featuring an abundance of beer and food. See, for example, the diary entry for May 16, 1909, describing a festival in Zogofina in the mission diary for Notre Dame de Ségou II, 1908–1935, MAFR.

128. *Dugutigi* (town caretaker) is often translated as "chief" but is more akin to a trustee for the *dugu* or town.

129. Nianzon Bouaré and Hawa Coulibaly, interview, May 29, 2010.

130. The fruit drink, likely fermented, was prepared especially for this season. See Moussa Diawara, Aminata Tangaré, and Hawoyi Diawara, interview, May 7, 2010.

131. Couloubaly, *Une société rural bambara*, 14–16.

132. Roberts, *Warriors, Merchants, and Slaves*, 28–29.

133. Roberts, 40.

134. Roberts, 33. See also, Jean Bazin, "Genèse de l'état et formation d'un champ politique: Le royaume de Segu," *Revue Française de Science Politique* 38, no. 5 (1988): 709–19.

135. Studies of precolonial alcohol consumption tend to associate drinking such beverages produced from grain harvests with communal welfare and sharing the harvest. See Deborah Fahy Bryceson, "Alcohol in Africa: Substance, Stimulus, and Society," in *Alcohol in Africa: Mixing Business, Pleasure, and Politics*, ed. Deborah Fahy Bryceson (Portsmouth, NH: Heinemann, 2002), 5–6.

136. Bryceson, "Alcohol in Africa."

137. Deborah Fahy Bryceson, "Changing Modalities of Alcohol Usage," in Bryceson, *Alcohol in Africa*, 24. Ethnographic research has focused largely on social relations around alcohol consumption. See also Bryceson, "Alcohol in Africa," 10.

138. Dolo is reputed to be 40 percent alcohol. See Perinbam, *Family Identity*, 112. However, research on precolonial alcohol suggests that most drinks were fermented and thus low in alcohol content. Often children consumed various forms of fermented drinks and porridges, which could often constitute a large portion of the overall grains consumed. In areas where water supplies were unsafe, fermented drinks were safe refreshments. See Bryceson, "Alcohol in Africa," 7–8.

139. Park, *Travels in the Interior Districts of Africa*, 193.

140. Park, 191.

141. Roberts, *Warriors, Merchants, and Slaves*, 52.

142. The Bambara identity was ambiguous in early outside accounts of the region (including Arabic accounts), and being identified as Bambara was often imbued with multiple meanings: an ethnicity, a religion, a culture, and a language. Bazin, "A chacun son Bambara."

143. Perinbam, *Family Identity*, 112.

144. See Correspondance avec le Fama de Sansanding Cercle de Segou 1890–1917, folder 1902, ANM FA 1E 220.

145. Ouagadougou is spelled "Waghadougou" in the Binger text. See Binger, *Du Niger au Golfe*, 497. Ougadougou, a region targeted by Office du Niger recruiters, is located in present-day Burkina Faso.

146. See the diary entry for December 22, 1912, from Diaires, Mission de Ségou, 1912, MAFR. The fathers also recorded administrative efforts to curb dolo production and consumption in several diary entries from the 1920s and 1930s.

147. "La culture du cotonnier," n.d., Dossier Bélime, AON 28.

148. See Fabrication d'alcool de mil, March 12, 1898, folder 603, Archives nationales d'outre mer (hereafter ANOM) FM GEN/60; "Régime des alcools en Afrique occidentale française," n.d., Direction des affaires politiques, Régime des alcools de 1890 à 1940, ANOM FM 1 AFFPOL/397.

149. See Ravitaillement des grains 1913–1919, folder 1916–1917, ANM FR 1Q 77.

150. Ouahigouya was located in Upper Volta, now Burkina Faso, and it was heavily targeted by recruiters for the Office du Niger.

151. Situation du cercle de Ouahigouya en fin Novembre 1934, Ravitaillement production agricole-previsions situation alimentaire 1926 à 1934, ANM FR I 1D 2879.

152. François Sorel, "L'alimentation des indigènes en Afrique occidentale française," in *L'alimentation indigène dans les colonies françaises, protectorats et territoires sous mandat*, ed. Georges Hardy and Charles Richet (Paris: Vigot Frères, 1933), 165.

153. M. Simon to the Lieutenant-Governeur of Haut-Sénégal et Niger, 1911, Agriculture correspondances 1910–1920, ANM FA 1R 27. Only a fragment of the original letter has been preserved, and portions of the letter are illegible or torn. Underlined portions are from the original.

154. See the research collected for the industrial possibilities of bénéfing in the following folder: Bénéfing 1912–1914, ANM 1R 167.

155. Bamako Commandant du Cercle to the Lieutenant-Governor of the Soudan, March 8, 1912, Bénéfing 1912–1914, ANM 1R 167.

156. Rapport du botaniste pour 1934, Dossier botanique, AON 30.

157. Guy Édouard Roberty later became the director of the French colonial scientific research center ORSTOM (Office de la recherche scientifique et technique d'outre-mer) based in Dakar.

158. See the research reports in the following file: Cultures vivrières, Introduction de plantes à Soninkoura, AON 85.

159. Office du Niger Botanist (Guy Roberty) to Le Chef (director) du Service Agronomique, January 28, 1936, Dossier mil, AON 87. In later decades, researchers would similarly document a diversity of millet types grown by Mossi farmers. See Dubourg, "La vie des paysans mossi," 304–5.

160. Women's role in seed selection has been established for rice cultivation in Guinea and other southern regions of West Africa. Judith Carney argued that women rice growers in West Africa selected seeds and as such have long been major agricultural innovators. While millet is not an exclusively female crop, it is likely that women also played a major role in its seed selection. See Judith A. Carney, *Black Rice: The Origins of Rice Cultivation in the Americas* (Cambridge, MA: Harvard University Press, 2001), 49–50.

161. Brahmin Fané, Madu Saré, Kadja Coumaré, Bading Traoré, and Issa Saré, interview by author, Markala-Kirango, June 3, 2010. The masculine characterization of farming by the performers was based on their knowledge and interpretation of Bamana farming songs, which emphasize youthful and especially masculine competition during work. Stephen Wooten also discusses the specifically male obligation to produce grain for the household and the complementary task of women to produce the sauce. See Wooten, *The Art of Livelihood*, 71–80.

162. Mariam Thiam, "The Role of Women in Rural Development in the Segou Region of Mali," in *Women Farmers in Africa: Rural Development in Mali and the Sahel*, ed. Lucy E. Creevey (Syracuse, NY: Syracuse University Press, 1986), 76–77. Sr. Marie-André du Sacré-Coeur noted a similar gendered dynamic food production among Mossi farmers. See Du Sacré-Coeur, "La femme Mossi," 21.

163. Maria Grosz-Ngaté, "Hidden Meanings: Explorations into a Bamanan Construct of Gender," *Ethnology* 28, no. 2 (1989): 171–72.

164. "Note sur la participation du H.S.N. au ravitaillement de la métropole 1916–1918," June 16, 1917, Bureau des affaires economiques, ANM FA 1Q 191.

165. [Jean François] Vuillet, "La culture du riz dans la Vallée du Niger," n.d., AON 38/1.

166. Bintu Traoré, Mariam Doumbia, and Fanta Sogoba, interview by author, Molodo-Centre, April 15, 2010.

167. Grosz-Ngaté, "Hidden Meanings," 77, 171.

168. Assane Coulibaly, interview, April 28, 2010.

169. Hawa Coulibaly, Mariam Sango, Djéné Mariko, Adam Coulibaly, and Nye Diarra, interview by author, in Molodo-Bamana, April 19, 2010; Office du Niger Botanist (Guy Roberty) to Le Chef du Service des Laboratoires, n.d., Correspondance, Dossier botanique (suite), AON 93/3. Roberty recorded that women cultivated n'tioko for food, but he classified the plant as a weed. French ethnographer Viviana Paques similarly observed women's extensive gardening in the 1950s. See Viviana Paques, *Les Bambara* (Paris: Presses Universitaires de France, 1954), 27.

170. Richard Roberts, "The Coercion of Free Markets: Cotton, Peasants, and the Colonial State in the French Soudan, 1924–1932," in *Cotton, Colonialism, and Social History in Sub-Saharan Africa*, ed. Allen Isaacman and Richard Roberts (Portsmouth, NH: Heinemann, 1995), 232.

171. Monteil, *Le coton chez les noirs*, 22–43. See also Richard Roberts, *Two Worlds of Cotton: Colonialism and the Regional Economy in the French Soudan, 1800–1946* (Stanford, CA: Stanford University Press, 1996).

172. Monteil, *Le coton chez les noirs*, 26, 42–3, 49–50, 59–61; Roberts, "Coercion of Free Markets," 226–31. See also Roberts, *Two Worlds of Cotton*, 51–59. During the first three decades of the twentieth century, women increasingly made money from the sale of cotton yarn and cloth. Women either sold cotton yarn or employed weavers to produce cloth from their yarn; the resulting cotton cloth bands were then sold. See Roberts, "Coercion of Free Markets," 235–43.

173. See Roberts, *Two Worlds of Cotton*, 76. See also Emil Schreyger, *L'Office du Niger au Mali 1932 à 1982: La problématique d'une grande entreprise agricole dans la zone du Sahel* (Paris: L'Harmattan, 1984), 7–16.

174. E. Fossat, "Le coton du Soudan," October 1898–August 1899, Mission d'études au Soudan Français (1898–1901), ANM série géographique Soudan français III/4.

175. Roberts, "Case of Faama Mademba Sy."

176. Roberts, *Two Worlds of Cotton*, especially chapters 4, 5, and 6. Private business interests did not always cooperate with French colonial government in this regard. See also Roberts, "Coercion of Free Markets," 223–26.

177. Monteil, *Le coton chez les noirs*, 88–96; Roberts, *Two Worlds of Cotton*, 76–93; and Roberts, "Coercion of Free Markets," 224–25.

178. Roberts, 223.

179. Roberts, *Two Worlds of Cotton*, 118–44.

180. Roberts, 229–32, 236.

181. *Rapport annuel sur la culture du cotonnier en A.O.F. 1928*, folder 4: Rapports concernant la culture cotonnière avant 1930, AON 377. In the 1920s Emile Bélime was the director of the colonial Textile and Hydrology Service for the AOF. On the point of colonial research looking to local agricultural and other knowledge, Helen Tilley argues that British knowledge production about Africa was more rooted in investigations of local knowledge than previous studies of imperial science have recognized. See Helen Tilley, *Africa as a Living Laboratory: Empire, Development, and the Problem of Scientific Knowledge, 1870–1950* (Chicago: University of Chicago Press, 2011).

182. Emile Bélime, quoting R. H. Forbes in "Le Soudan Nigérien," n.d., Dossier Bélime, AON 28A,

183. Nianzon Bouaré and Hawa Coulibaly, interview, May 29, 2010.

184. Women's work harvesting cotton was widespread in West Africa. For example, Andrew Zimmerman noted that women in coastal Togo cultivated

and sold cotton in the first decades of the twentieth century. See Andrew Zimmerman, "A German Alabama in Africa: The Tuskegee Expedition to German Togo and the Transnational Origins of West African Cotton Growers," *American Historical Review* 110, no. 5 (2005): 1362–98.

185. Roberts, "Coercion of Free Markets," 241–42; Roberts, *Two Worlds of Cotton*, 51–59, 93–105.

186. Much of the labor problem at Diré was due to the scheme's reliance on forced labor. In addition, the cotton harvests at Diré were consistently poor in the 1920s. See Roberts, *Two Worlds of Cotton*, 121, 28–35.

187. Lieutenant-Governor of the Soudan to the Commandant de Cercle Goundam, June 11, 1924, Correspondance 1924, Entretien, culture, égrenage du coton 1914–1925, Agriculture divers 1914–1931, ANM FR II 1R 16.

188. Correspondance 1924, Égrenage du coton 1914–1925, Agriculture divers 1914–1931, ANM FR II 1R 16. See also Roberts, *Two Worlds of Cotton*, 131–33.

189. Lieutenant-Governor of the Soudan to the Commandant de Cercle Goundam. On the politics of introducing a colonial currency in French West Africa, see Mahir Şaul, "Money in Colonial Transition: Cowries and Francs in West Africa," *American Anthropologist* 106, no. 1 (2004): 71–84.

190. Fatoumata Coulibaly, Kadiatou Mallé, and Maïssa Sountura, interview by author, Kouyan-Kura, April 14, 2010.

191. The origin of the workers in question is unclear from these documents.

192. E. Thiriet, *Au Soudan français: Souvenirs 1892–1894, Macina—Tombouctou* (Paris: André Lesot, 1932), 64.

193. Conversation led by Aïssata Khassonké with women in Baba Kassamba-ra's household in Markala regarding the purchase of a mortar and pestle, June 3, 2010. I was also present during this exchange.

CHAPTER 2: BODY POLITICS, TASTE MATTERS,
AND THE CREATION OF THE OFFICE DU NIGER, CA. 1920-44

1. Nemabougou folder 22, Enquête sur la situation de la colonisation, fiches individuelles des colons, 1938, ANOM FM 1TP/735.

2. Family no. 2, Nemabougou folder 22, Enquête sur la situation de la colonisation, fiches individuelles des colons.

3. Family no. 32, Nemabougou folder 22, Enquête sur la situation de la colonisation, fiches individuelles des colons.

4. Enquête sur la situation de la colonisation, fiches individuelles des colons. Surveys were carried out in Bamako-Koura, Bediambougou, Dar Salam, Demabougou, Nara, Nemabougou, and Sangarébougou.

5. Family no. 14, Nemabougou folder 22, Enquête sur la situation de la colonisation, fiches individuelles des colons.

6. Nemabougou folder 22, Enquête sur la situation de la colonisation, fiches individuelles des colons.

7. Bélime was the first director of the Office, and he remained associated with the project in the minds of most officials and farmers long after his

departure in 1943. One farmer's cooperative in Kolony (km 26) was even named for Bélime. Mamadou "Seyba" Coulibaly, interview by author, Kolony (km 26), April 1, 2010. See also Georges Spitz, *Sansanding: Les irrigations du Niger* (Paris: Société d'Editions Géographiques, Maritimes, et Coloniales, 1949), 66.

8. Emile Bélime et al., *Les irrigations du Niger: Discussions et controverses* (Paris: Comité du Niger, 1923).

9. For an overview of the mise en valeur policy, see Alice L. Conklin, *A Mission to Civilize: The Republican Idea of Empire in France and West Africa, 1895–1930* (Stanford, CA: Stanford University Press, 1997), 23–37. See also Gary Wilder, "Framing Greater France between the Wars," *Journal of Historical Sociology* 14, no. 2 (2001): 203–5.

10. Monica van Beusekom, *Negotiating Development: African Farmers and Colonial Experts at the Office du Niger, 1920–1960* (Portsmouth, NH: Heinemann, 2002), 2–4; and Richard Roberts, *Two Worlds of Cotton: Colonialism and the Regional Economy in the French Soudan, 1800–1946* (Stanford, CA: Stanford University Press, 1996), 122–23.

11. Auguste Chevalier, a well-known colonial agronomist, signaled this problem in his critique of the project and suggested that even Bélime doubted that irrigation would assure the food supply. See Bélime et al., *Les irrigations du Niger*, 3–10.

12. Van Beusekom, *Negotiating Development*, 7–12.

13. Wilder, "Framing Greater France between the Wars," 198–201.

14. Roberts, *Two Worlds of Cotton*, 159.

15. Emil Schreyger, *L'Office du Niger au Mali 1932 à 1982: La problématique d'une grande entreprise agricole dans la zone du Sahel* (Paris: L'Harmattan, 1984), 67.

16. Spitz, *Sansanding*, 63–70.

17. Koké Coulibaly and Fodé Traoré, interview by author, Markala-Kirango, February 3, 2010. See also Amidu Magasa, *Papa-commandant a jeté un grand filet devant nous: Les exploités des rives du Niger, 1902–1962* (Paris: François Maspero, 1978), 50–81; and Catherine Bogosian, "Forced Labor, Resistance and Memory: The *Deuxième Portion* in the French Soudan, 1926–1950" (PhD diss., University of Pennsylvania, 2002), see especially chapter 7.

18. Van Beusekom, *Negotiating Development*, 40–41.

19. Charrures, demandes de charrures 1927–1930, ANM FR 1Q 19. See also Birama Diakon, *Office du Niger et pratiques paysannes: Appropriation technologique et dynamique sociale* (Paris: L'Harmattan Mali, 2012), chapters 1 and 3.

20. "Colonisation," n.d., Dossier Bélime, AON 28.5. See also Van Beusekom, *Negotiating Development*, 40–41.

21. Michael Adas, *Machines as the Measure of Men: Science, Technology, and Ideologies of Western Dominance* (Ithaca, NY: Cornell University Press, 1989).

22. See the circular in folder 1916, Agriculture Correspondances 1910–1920, ANM FA 1R 27.

23. From the Chef de Service de l'Agriculture to the Gouverneur of Haut-Sénégal et Niger, September 17, 1916 (letter no. 134), in the folder Agriculture les terres à coton 1914–1916, Agriculture divers 1914–1931, ANM FR II 1R 16.

24. Recent climate science does suggest humid and arid phases in the climatological history of West Africa. For an overview, see James L. A. Webb Jr., "Ecology and Culture in West Africa," in *Themes in West Africa's History*, ed. Emmanuel Kwaku Akyeampong (Athens: Ohio University Press, 2006).

25. "Le Canal du Sahel," n.d., Dossier Bélime, AON 28.5.

26. Matthew Connelly, *Fatal Misconception: The Struggle to Control World Population* (Cambridge, MA: Belknap, 2008), 7–8, 46–50.

27. Raymond R. Gervais and Issiaka Mandé, "How to Count the Subjects of Empire? Steps toward an Imperial Demography in French West Africa before 1946," in *The Demographics of Empire: The Colonial Order and the Creation of Knowledge*, ed. Karl Ittmann, Dennis D. Cordell, and Gregory H. Maddox (Athens: Ohio University Press, 2010).

28. The farmers were not wrong in their assessment. The intense farming practices favored by the Office required not only the plow but also a great deal of fertilizer to improve its productivity. See Diakon, *Office du Niger et pratiques paysannes*, 29, 73.

29. Emile Bélime, "Irrigation du Delta Central Nigerien," n.d., Dossier Bélime, AON 28.

30. Van Beusekom, *Negotiating Development*, 86.

31. Governor-General Terrason to the Administration of the French Soudan, 1929, from folder Colonisation européenne et indigène (renseignements généraux 1927–1938), Office du Niger colonisation indigène (1927–1944), ANM FR 1R 33.

32. Terrason to the Administration of the French Soudan.

33. Spitz, *Sansanding*, 48–53.

34. M. Bagot, "Mission d'inspection Office du Niger 1934–1935," ANM, FR 2D 61.

35. Christophe Bonneuil has argued that many of the agricultural settlement schemes that started in the 1930s in Africa were akin to scientific experiments, and he makes special reference to the Office du Niger. Moreover, the Office planners thought that they could ultimately predict the outcome of their scheme. See Christophe Bonneuil, "Development as Experiment: Science and State Building in Late Colonial and Postcolonial Africa, 1930–1970," *Osiris* 15 (2000): 258–81. See also Helen Tilley, *Africa as a Living Laboratory: Empire, Development, and the Problem of Scientific Knowledge, 1870–1950* (Chicago: University of Chicago Press, 2011). Tilley argues that the practice of science in colonial Africa was caught up

in the making of empire, but it also produced knowledge that questioned imperial rule. "Science" is a slippery concept that does not lend itself to easy assumptions about the actual practice in colonial spaces as is the case for "development."

36. See Dossier Bélime. Note in another report for the same year, seventy-six people were reported including sixteen children (instead of fourteen). See Office du Niger rapport sur la colonisation indigène au Soudan Français pendant l'année 1927, Office du Niger colonisation indigène (1927–1944), ANM FR 1R 33.

37. It must, however, be noted that none of the families possessed land titles.

38. Office towns were known by either their distance in kilometers from a central Office marking point or by the town name. Kolony is one town often referred to by its kilometer marker, km 26.

39. J. Lenoir, Inspecteur générale des affaires administratives, rapport no. 3/I.G.AA., February 9, 1945, Office du Niger questions de peuplements, 1944–1945, ANM FR 1R 39.

40. Office du Niger: Note sur les méthodes de colonisation indigène, 1935–1937, ANM FR 1R 63.

41. Gervais and Mandé, "How to Count the Subjects of Empire."

42. See the statistical report "Mouvements demographiques parmi les colons des centres de Kokry et de Niono depuis l'origine de ces centres de colonisation," Office du Niger questions de peuplement 1944–1945, ANM FR 1R 39.

43. Van Beusekom, Negotiating Development, 57–58.

44. Note sur le recrutement dans le Baninko, 1930–1940, folder 1939, Office du Niger, ANM FR 1R 58.

45. Robert Léon to the Direction of the Office du Niger, telegram, April 7, 1939, Organisation administrative des terres irriguées, 1937–1947, Office du Niger, ANM FR 1R 28.

46. "Transfert en colonisation des villages trypanosomés, 1939–1940," Office du Niger organisation administrative des terres irriguées, 1937–1947, ANM FR 1R 28; and Office du Niger documentation, 1937–1946, ANM FR 1R 42.

47. Van Beusekom, Negotiating Development, 91.

48. See the accounts of recruiter propaganda recorded by administrators in Ségou in the folder Enquête auprès de l'Office A/S de recrutement, 1938, Office du Niger documentation, 1937–1946, ANM FR 1R 42.

49. The Moor settlers were replaced by Bamana settlers, the majority of whom were unmarried men. See Report no. 190/O.N., Office du Niger questions de peuplement, 1944–1945, ANM FR 1R 39; and Médecin-Colonel G. Lefrou, "Étude démographique des populations de l'office du Niger," Mission de monsieur le gouverneur général reste à l'office du Niger, 1945, ANM FR III 1D 660, 29.

50. Lefrou, "Étude démographique," 43.

51. See the correspondence between the lieutenant-governor of the French Soudan and Emile Bélime between May and July 1932 in the folder Office du Niger villages voisins de Sotuba par la création des villages composés de volontaires des environs 1932, Office du Niger documentation, 1934–1936, ANM FR 1R 41.

52. Office du Niger: Note sur les méthodes de colonisation indigène 1935–1937, ANM FR 1R 63, 99; Office du Niger note sur le recrutement dans le Baninko 1930–1944, ANM FR 1R 58.

53. Tougan, another region in Upper Volta, was also targeted, and in 1939 forty-seven Samogo families were also sent to the Office. See the following folder: Indigènes originaires du cercle de Tougan demandons à entrer en colonisation dans la région de Kokry, Organisation administrative des terres irriguées 1937–1947, Office du Niger, ANM FR 1R 28.

54. Commandant du Cercle Joseph Rocca-Serra, Extrait du rapport politique annuel 1944 du cercle de Ségou, December 31, 1944, Office du Niger questions de peuplement 1944–1945, ANM FR 1R 39.

55. Rocca-Serra, Extrait du rapport politique annuel 1944 du cercle de Ségou.

56. Lenoir, rapport no. 3/I. G.A.A.

57. Enquête sur la situation de la colonisation, fiches individuelles des colons.

58. Van Beusekom, *Negotiating Development*, 88–89, 153–60.

59. Magasa, *Papa-commandant*, 89–111.

60. Denise Savineau, *La famille en A.O.F.: Condition de la femme* (Dakar: Gouvernement de l'A.O.F., 1938), 93.

61. Savineau, *La famille en A.O.F.*, 92.

62. Note from Dr. Sice of the Office du Niger Health Service, October 25, 1937, Office du Niger note sur le recrutement dans le Baninko 1930–1944, ANM FR 1 R58 FR.

63. Nemabougou folder 22, Enquête sur la situation de la colonisation, fiches individuelles des colons.

64. Savineau, *La famille en A.O.F.*, 94.

65. Djenebu Coulibaly, interview by author, Kolony (km 26), March 27, 2010.

66. Fatouma Coulibaly, interview by author in Kolony (km 26), March 24, 2010; Djenebu Coulibaly, interview, March 27, 2010; Djewari Samaké, interview by author, Kolony (km 26), March 29, 2010; and Moctar Coulibaly, Mariatou Traoré, and Arouna Coulibaly, interview by author, Sirakora, April 28, 2010.

67. Savineau, *La famille en A.O.F.*, 91. According to Moctar Coulibaly, some women in Sirakora did initially cultivate peanut fields far outside of town. See Moctar Coulibaly, Mariatou Traoré, and Arouna Coulibaly, interview, April 28, 2010.

68. Rapport sur la colonisation à Niénébalé Campagne 1929–1930, Colonisation indigène de Niénébalé 1928–1935, ANM FR 1R 76.

69. Richard Roberts points out the poor accounting of managers at the Office du Niger and the fact that Bélime had free reign when it came to

spending, often going over budget for the construction of the dam and irrigation works. See Richard Roberts, *Two Worlds of Cotton: Colonialism and the Regional Economy in the French Soudan, 1800–1946* (Stanford, CA: Stanford University Press, 1996), 233.

70. Programme de travaux d'aménagement du casier rizicole de Boky-Wéré, Office du Niger documentation divers 1933–1934, ANM FR 1R 41. The value of the French West African franc was tied to the metropolitan franc, which fell in value during the First World War. However, the franc was competitive against the US dollar for much of the 1930s in the midst of the global depression. In 1936, 1,000 FR, the amount allotted for a six-month ration per Office family, would have been roughly equivalent to $40 in 2021. The funds budgeted for the five new villages in Kokry in 1936 (1,235,000 FR) would be roughly equivalent to $49,400 today. For more on the economic history of French West Africa, see Jacques Marseille, *Empire colonial et capitalisme: Histoire d'un divorce* (Paris: Albin Michel, 1984).

71. See folder 1938: Office du Niger documentation, 1937–1946, ANM FR 1R 42. See also Van Beusekom, *Negotiating Development*, 90.

72. Rapport Tournée Cercle de Tougan, 1932–1935, ANM FR 1R 17.

73. Koké Samaké, testimony, October 5, 1938, in Enquête auprès de l'office A/S de recrutement, 1938, Office du Niger documentation, 1937–1946, ANM FR 1R 42.

74. Mamourou Coulibaly, testimony, November 25, 1938, in Enquête auprès de l'office A/S de recrutement.

75. Office du Niger: Note sur le recrutement dans le Baninko 1930–1944, folder 194, ANM FR 1R 58.

76. Note du chef du service de santé, no. 358/SS, March 26, 1938, Office du Niger alimentation-colons 1938, ANM FR 1 R51.

77. Van Beusekom, *Negotiating Development*, 124.

78. Moctar Coulibaly, Mariatou Traoré, and Arouna Coulibaly, interview, April 28, 2010.

79. Van Beusekom, *Negotiating Development*, 94.

80. See the January 1937 list of settlers to be evicted along with explanatory notes, Office du Niger Centre de Colonisation de Kokry 1937, Office du Niger Colonisation Indigène, 1927–1944, ANM FR 1R 33.

81. Sékou Coulibaly and Kadiatou Traoré, interview by author, Nyamina, May 31, 2010.

82. Office du Niger alimentation-colons 1938.

83. Office du Niger alimentation-colons 1938.

84. Note du chef du service de santé, no. 358/SS.

85. Magasa, *Papa-commandant*, 95.

86. Enquête auprès de l'office A/S de recrutement, 3.

87. H. Desanti (on behalf of the office of the lieutenant-governor) to the Territorial Government of the AOF, August 17, 1938, Office du Niger alimentation-colons 1938, ANM FR 1R 51.

88. During the investigation the administrators who had approved the contaminated shipment claimed that the millet refused by the labor service was subsequently cleaned and processed before its shipment to Kokry. See Joseph Rocca-Serra, Administrator in Macina to the Commandant de Cercle de Ségou, July 30, 1938, Office du Niger alimentation-colons 1938.

89. Vincent Bonnecase, "When Numbers Represented Poverty: The Changing Meaning of the Food Ration in French Colonial Africa," *Journal of African History* 59, no. 3 (2018): 463–81.

90. The first surveys conducted in the 1920s more specifically focused on a basic ration for workers. See Bonnecase, "When Numbers Represented Poverty," 466–73.

91. See the following folder: Alimentation AOF Soudan, Commission d'enquête dans les territoires d'outre mer, ANOM FM Guernut/100. Discussions of food supply for settlers at the Office in the 1930s were caught up in wider international discussions of undernourishment and malnutrition. Surveys such as those collected during the Guernut commission tended to support the universal assessment of general malnutrition in colonized territories—that it was an economic concern. See Michael Worboys, "The Discovery of Colonial Malnutrition between the Wars," in *Imperial Medicine and Indigenous Societies*, ed. David Arnold (Manchester, UK: Manchester University Press, 1988).

92. See the report for Cercle de Ouahigouya, Peulh and Rimaibe (nord-est), Alimentation AOF Soudan, ANOM FM Guernut/100.

93. Report for Bamako, Alimentation AOF Soudan, ANOM FM Guernut/100.

94. Report for Macina (Région de Dia), Alimentation AOF Soudan, ANOM FM Guernut/100.

95. Vincent Bauzil to the Lieutenant-Governor of the French Soudan, January 11, 1938, Office du Niger alimentation des colons 1938.

96. During the Second World War research on rice consumption and beriberi brought greater attention to malnutrition and the impact of hunger and starvation. See Dana Simmons, "Starvation Science: From Colonies to Metropole," in *Food and Globalization: Consumption, Markets and Politics in the Modern World*, ed. Alexander Nützenadel and Frank Trentmann (New York: Berg, 2008). See also Yan Slobodkin, "Famine and the Science of Food in the French Empire, 1900–1939," *French Politics, Culture, and Society* 36, no. 1 (2018): 55–57. By the late 1950s, French scholarship recognized the nutritional importance of millet but continued to promote intensified rice production in West Africa to assure the food supply in cases of shortage. See Dubourg, "La vie des paysans mossi," 322–324.

97. "Rapport à la commission parlementaire chargée d'une enquête sur l'Office du Niger: Observations sur les travaux scientifiques et agronomiques relatifs à la culture du cotonnier et des plantes vivrières dans la vallée

du Niger par le professeur Chevalier, membre de l'Institut, 17 janvier 1939," p. 23, Office du Niger, Inspection généraux des travaux publics, ANM FM 1 TP/878.

98. See the Inspector Carbou section of Enquête auprès de l'office A/S de recrutement, 3–4. Inspector Carbou is only referred to by his last name in this document. On the popularity of the dorome, or five-franc piece, for daily monetary transactions, see Mahir Şaul, "Money in Colonial Transition: Cowries and Francs in West Africa," *American Anthropologist* 106, no. 1 (2004): 79.

99. The French colonial administration established a town called Macina as an administrative center that served much of the area under the jurisdiction of the Office. This administrative center was located in the region of the same name.

100. Each small pile sold for one dorome, the local name for the five-franc unit of currency. See Djenebu Coulibaly and Djewari Samaké, interview by author, Kolony (km 26), March 22, 2010; Rokia Diarra, interview by author, Nara, April 29, 2010. Mahir Şaul similarly describes the dorome five-franc piece as a key unit of early colonial currency in the Volta region. Mahir Şaul, "Money in Colonial Transition," 79.

101. "Le Soudan Nigérien," n.d., Dossier Bélime, AON 28A.

102. Judith Carney's work on the gendered aspects of rice cultivation in this region is well known. See Judith A. Carney, *Black Rice: The Origins of Rice Cultivation in the Americas* (Cambridge, MA: Harvard University Press, 2001); and Judith Carney and Michael Watts, "Disciplining Women? Rice, Mechanization, and the Evolution of Mandinka Gender Relations in Senegambia," *Signs: Journal of Women in Culture and Society* 16, no. 4 (1991): 651–81.

103. "Culture du riz dans la Vallée du Niger (Vuillet)," n.d., AON 38.1. Jean François Vuillet is the likely author of the first report. See also "Le riz dans la Haute Vallée du Niger," 1934, AON 38.2.

104. "Le riz dans la Haute Vallé du Niger."

105. Danel to the Commandant de Cercle de Mopti, January 5, 1922, Riz correspondance 1904–1919, ANM FA 1R 160.

106. Letter to the lieutenant-governor of Guinée (letter no. 37), January 25, 1927, Riz correspondance 1904–1919, ANM FA 1R 160. The author of the letter is obscured, but the writer is presumably from the office of the lieutenant-governor Henri Terasson. The records in this file are disordered and include material dated past 1919.

107. "Rapport sur la campagne 1927," Service agronomique du coton, Dossier botanique (suite), AON 93/4.

108. "Le riz," n.d., Dossier Bélime, AON 28A.

109. "Culture du riz dans la Vallée du Niger (Vuillet)."

110. On the contrary, scholars have established that farming in the region has been characterized as highly experimental. See Paul Richards,

Indigenous Agricultural Revolution: Ecology and Food Production in West Africa (Boulder, CO: Westview, 1985), 54–59.

111. "Le riz dans la Haute Vallée du Niger."

112. Commandant de Cercle de Djenné to the Kayes Delegate, February 1, 1904, Riz correspondance 1904–1919, ANM FA 1R 160.

113. "Le Soudan Nigérien," n.d., Dossier Bélime.

114. This is a point made by Judith Carney. See Carney, Black Rice, 53–54.

115. "Culture du riz dans la Vallée du Niger (Vuillet)."

116. Report by Administrator Floch in Macina, August 6, 1937, Office du Niger organisation administrative des terres irriguées 1937–1947, ANM FR 1R 28.

117. Emily Burrill, States of Marriage: Gender, Justice, and Rights in Colonial Mali (Athens: Ohio University Press, 2015), 32–34, 99–105. See also chapter 5 in Richard Roberts, Litigants and Households: African Disputes and Colonial Courts in the French Soudan, 1895–1912 (Portsmouth, NH: Heinemann, 2005).

118. Dar Salam folder 18, Enquête sur la situation de la colonisation, fiches individuelles des colons.

119. Dar Salam folder 18.

120. Fatoumata Coulibaly et al., interview by author, Kouyan-Koura, May 27, 2010. The fact that women and men ate together in the first half of the nineteenth century was reiterated in several conversations between the author and Hawa Fomba in Kalaké-Bamana in 2010.

121. Dar Salam folder 18.

122. Fatoumata Coulibaly et al., interview, May 27, 2010.

123. According to Lenoir, Dr. Ethès reported that in some towns the ratio of men to women remained as low as 130:100. See Office du Niger documentation, 1937–1946.

124. Assane Pléah, interview by author, Kouyan-N'Péguèna, May 28, 2010.

125. Van Beusekom, Negotiating Development, 153–60.

126. Koké Coulibaly and Fodé Traoré, interview, February 3, 2010.

127. A canton is one administrative level below cercle.

128. Dembélé noted that he had also been commissioned by a fellow settler to carry some of his belongings back to the Office. Dembélé appears to have accomplished neither the return of his wife nor the transport of his fellow's belongings. See Dembélé's testimony in Affaires du "Nia" du Massabougou et plaintes de Nambolo Dembélé, colon, 1938, Office du Niger organisation administrative des terres irriguées 1937–1947, ANM FR 1R 28.

129. Enquête auprès de l'office A/S de recrutement, 3.

130. By 1945, the Office had established family demographic charts to count the numbers of children (0–8 and 8–15 years of age), adults (15–55 years of age), and the elderly (55 years of age and older). Within each category the total number of children, adults, and elderly members was further

categorized by gender. The charts also recorded the total number of births and deaths for that year. Building on the information already being collected, Lefrou proposed that with more comprehensive recording, the demographic situation in Office towns would inform demographic analysis of French West Africa more broadly. Despite the demographic irregularities, Lefrou suggested that the possibilities for accurate demographic reporting at the Office had great scientific value. See Médecin-Colonel G. Lefrou, "Étude démographique des populations de l'office du Niger," pp. 21, 31, 33, Mission de monsieur le gouverneur général reste à l'office du Niger, 1945, ANM FR 1D 660 III.

131. Djewari Samaké, interview, March 29, 2010.

132. Djenebu Coulibaly and Djewari Samaké, interview, March 22, 2010; Fatouma Coulibaly, interview, March 24, 2010; Hawa Coulibaly, interview by author, Fouabougou, March 30, 2010; Aramata Diarra, interview by author, Fouabougou, March 30, 2010; Fati Kindo, interview by author, Kossouka, May 11, 2010. In the nineteenth century, Maraka merchants developed a local cotton and textile industry using slave labor. In this context, the large-scale cotton industry was directly associated with servile status, setting a precedent for how farmers would likely perceive cotton cultivation at the Office. One major difference is that grain production was the predominant activity on Maraka plantations. See Richard Roberts, *Warriors, Merchants, and Slaves: The State and the Economy in the Middle Niger Valley, 1700–1914* (Stanford, CA: Stanford University Press, 1987); and Roberts, *Two Worlds of Cotton*.

133. Mme. Dagno Adam Bah, interview by author, Markala-Diamarabougou, February 6, 2010.

134. Moctar Coulibaly, Mariatou Traoré, and Arouna Coulibaly, interview, April 28, 2010.

135. Moctar Coulibaly, Mariatou Traoré, and Arouna Coulibaly, interview, April 28, 2010; Hawa Diarra, interview by author, Nara, April 30, 2010; Alimata Dembélé, Moussa Coulibaly, and Nana Dembélé, interview by author, Koutiala-Kura, May 3, 2010; and Mariam "Mamu" Coulibaly, interview by author, Kankan (formerly Sangarébougou), May 4, 2010.

136. Mariam "Mamu" Coulibaly, interview, May 4, 2010; and informal conversation with Tchaka Diallo in Markala-Diamarabougou, February 3, 2010.

137. Enquête auprès de l'office A/S de recrutement.

138. Magasa, *Papa-commandant*, 32–33.

139. Djewari Samaké, interview, March 29, 2010.

140. Maria Grosz-Ngaté, "Monetization of Bridewealth and the Abandonment of 'Kin Roads' to Marriage in Sana, Mali," *American Ethnologist* 15, no. 3 (1988): 501–14.

141. Djenebu Coulibaly, interview, March 27, 2010; Mariam "Mamu" Coulibaly, interview, May 4, 2010; and Fatouma Coulibaly, interview, March 24, 2010.

142. Djewari Samaké, interview by author, Kolony (km 26), June 2, 2010.
143. Ghislaine Lydon, "Women in Francophone West Africa in the 1930s: Unraveling a Neglected Report," in *Democracy and Development in Mali*, ed. R. James Bingen, David Robinson, and John M. Staatz (East Lansing: Michigan State University Press, 2000); and Martin A. Klein and Richard Roberts, "The Resurgence of Pawning in French West Africa during the Depression of the 1930s," *African Economic History* 16 (1987): 23–37.
144. Ortoli, like other administrators, noted a resemblance between this practice and slavery but that it was common for male creditors to marry female pawns. See Henri Ortoli, "Le gage des personnes au Soudan français," *Bulletin de l'Institut Français d'Afrique Noire* 1 (1939): 313–24.
145. Rapports de tournée cercle de Ségou, 1932–1935, ANM 1 E 70 FR.
146. Rapports de tournée cercle de Ségou, 1932–1935. Similar conditions were noted by administrator Fabre in neighboring regions.
147. Coutumiers Mossi 1933–1935, ANM FA 1D 212. This file consists mostly of correspondence related to Catholic missionary concerns in 1933 for pawning in Ouagadougou. The Office would later heavily recruit from this Mossi region. The inclusion of descriptions of pawning at the Niénébalé station of the Office with these missionary observations on pawning in Mossi regions suggests that at least some administrators believed pawns were sent to the Office.
148. It is perhaps the case that some of the women interviewed for this book who came to the Office as young girls were pawns, but few directly indicated this. However, one woman interviewee suggested in a separate conversation that she and her whole family came from a servile status. Other women spoke more freely about how their mothers-in-law were brought with their husbands by force but not necessarily as pawns.
149. "Rapport sur la traite des femmes et des enfants 1933," January 23, 1934, ANM FA 1D 210.
150. Emily Burrill recorded a similar marriage crisis in the 1920s and 1930s in the southern region of Sikasso. See Burrill, *States of Marriage*, 99–105.
151. Folder 1941, Project d'instruction pour le recrutement des colons dans les cercles de Koutiala, de Tougan, et de Ouahigouya, Note sur le recrutement dans le Baninko 1930–1944, Office du Niger, ANM FR 1R 58.
152. Lenoir, Report no. 3/I. G.A.A., February 9, 1945.
153. Dembougou folder 19, Enquête sur la situation de la colonisation, fiches individuelles des colons.
154. My italics. See Koké Samaké, testimony, October 5, 1938, in Enquête auprès de l'Office A/S de recrutement.
155. See Stephen R. Wooten, "Colonial Administration and the Ethnography of the Family in the French Soudan," *Cahiers d'Études Africaines* 33, no. 131 (1993): 419–46.

156. Jane I. Guyer and Samuel M. Eno Belinga, "Wealth in People as Wealth in Knowledge: Accumulation and Composition in Equatorial Africa," *Journal of African History* 36, no. 1 (1995): 91–120.
157. Dar Salam folder 19, Enquête sur la situation de la colonisation, fiches individuelles des colons.
158. Dembougou folder 18 and Dar Salam folder 19, Enquête sur la situation de la colonisation, fiches individuelles des colons.
159. Account from an administration-settler meeting held May 1, 1944, Office du Niger questions de peuplement 1944–1945, ANM 1 R 39 FR.
160. Rapport de l'inspecteur des colonies Pruvost commissaire de gouvernement auprès de l'Office du Niger sur le peuplement des terres aménagées par l'Office du Niger, p. 8, Office du Niger questions de peuplement 1944–1945, ANM FR 1R 39. See also Van Beusekom, *Negotiating Development*, 159.
161. 1944 folder, Office du Niger documentation 1937–1946, ANM FR 1R 42.
162. Married men tended to be in their midthirties and older based on the recollections of my sources. Senior men in an Office household tended to be the oldest married men, meaning that a senior man at the Office was not always elderly.
163. Office du Niger questions de peuplement 1944–1945, p. 5. Men at the Office tended to marry in their early thirties according to several women and men interviewed for this book.
164. Account from an administration-settler meeting held May 1, 1944, Office du Niger questions de peuplement 1944–1945.
165. Djewari Samaké, interview, March 29, 2010.
166. Rapport de l'inspecteur des colonies, 4.
167. 1944 folder, Office du Niger documentation 1937–1946.
168. Djewari Samaké, interview, March 29, 2010. Samaké's oral testimony is doubly significant because she lived in the town closest to the center of the Office's Niono administration where protesters confronted officials.

CHAPTER 3: "WE FARMED MONEY"

1. Mariam "Mamu" Coulibaly, interview by author, Kankan (formerly Sangarébougou), May 4, 2010.
2. The Office administration designated these fields *hors-casier*. *Casier* denoted a particular sector or zone in the Office where farmers sold their crops; *hors casier* meant "outside" of the Office.
3. Farmers frequently protested the amount they were charged for water fees. See Affaires agricoles Office du Niger 1955–1956, ANM FR II 1R 1572. The percentage of farmer harvests collected by the Office in payment for fees is recorded in a letter from Deputé Modibo Keita to the Office du Niger Direction. See Deputé Modibo Keita to the Office du Niger Direction, September 12, 1956, Affaires agricoles Office du Niger 1955–1956, ANM FR II 1R 1572.

4. Dossier confidentiel, Colonisation 1953/6 (affaire Sangaré), AON 111; Affaires agricoles Office du Niger 1955–1956, ANM FR II 1R 1572.

5. Dossier divers, AON 118; Rapports politiques et rapports de tournées cercle de Ségou I (1923–1939), ANM FR 1E 40; Rapports politiques et rapports de tournées cercle de Ségou II (1940–1959), ANM FR 1E 40.

6. Dossier confidentiel, Colonisation 1953/6 (affaire Sangaré), AON 111. The town of Kankan was particularly fraught with political tension after its first chief, Mamadou Sangaré, was evicted in 1943. Sangaré had been a clerk for the colonial government and was encouraged by the first director of the Office, Emile Bélime, to found Sangarébougou as a model town. From the start, Sangaré was a vocal supporter of Bélime. Many early residents of Sangarébougou protested Sangaré's heavy hand. After Bélime was ousted from the colonial administration for his open collaboration with the Vichy regime, Sangaré was also ousted. Following Sangaré's eviction, the town was renamed "Kankan" by the farmers. In subsequent years, Sangaré requested readmission to the Office. He also began to organize for farmers' land rights, a cause then supported by Bélime. Sangaré gained some support from Office farmers but also opposition. In February 1955, when he returned to his former town for a political rally, he and his entourage were attacked by men and women wielding sticks. In fact, Mamu Coulibaly, who was a resident at the time, remembered women in town arguing over the politics of Sangaré's eviction and continued activity in town. For the most part, protests in Kankan revolved around access to resources and grievances over Office policies. See Mariam "Mamu" Coulibaly, interview, May 4, 2010. Documentation of Sangaré's eviction and his relationship to the Office and with Bélime is found in Distribution permis d'occupation où evictions, Affaires agricoles Office du Niger 1955–1956, ANM FR II 1R 1572; see also Emil Schreyger, *L'Office du Niger au Mali 1932 à 1982: La problématique d'une grande entreprise agricole dans la zone du Sahel* (Paris: L'Harmattan, 1984), 5–6.

7. Deputé Modibo Keita (and future President of Mali) even visited the Office in the late 1950s to investigate reports of abuse by the institution against farmers. See Affaires agricoles Office du Niger 1955–1956, ANM FR II 1R 1572.

8. Affaires agricoles Office du Niger 1955–1956. After the war, the Office aggressively (and sometimes forcibly) recruited farmers among the Mossi and Samogo of neighboring Upper Volta. See Amidu Magasa, *Papa-commandant a jeté un grand filet devant nous: Les exploités des rives du Niger, 1902–1962* (Paris: François Maspero, 1978), 97–111.

9. Unfavorable articles on the Office that appeared in African papers in the 1950s were collected and reviewed by the director of the Office, as were petitions to the project administration. Copies of the *L'Essor* for this period and *Le Reveil* for 1945–48 are also available in Archives Nationales du Mali. On this point, Monica van Beusekom notes that political and

labor leaders opposed abusive practices at the Office, not the idea of improving or developing agriculture through irrigation or industrial farming. See Monica van Beusekom, "Individualism, Community, and Cooperatives in the Development Thinking of the Union Soudanaise-RDA, 1946–1960," *African Studies Review* 51, no. 2 (2008): 1–25. See also Dossier confidentiel, Colonisation 1953/6 (affaire Sangaré), AON 111.

10. Affaires agricoles Office du Niger 1955–1956.

11. Markala was home to several sections of the Office that employed wage workers. Some workers continued construction on the dam and irrigation works. Other workers manufactured plows, carts, and other agricultural equipment in the industrial workshop. Office jobs also included driving transport vehicles, tractors, harvesting machines, or other heavy equipment. Additional work included canal and machine maintenance. Inspection générale des travaux publics Office du Niger, 1946/1952, ANOM FM 2TP/104; Inspection générale des travaux publics, Office du Niger, 1952/1957, ANOM FM 3TP/332.

12. In 1947 a general strike was in force across French West Africa. The Office strike and other coordinated protests were closely connected to labor actions by railroad workers and political organizing by the Rassemblement Démocratique Africain (RDA). See Frederick Cooper, *Decolonization and African Society: The Labor Question in French and British Africa* (Cambridge: Cambridge University Press, 1996), 241–47.

13. Bakary Marka Traoré, interview by author, Markala-Diamarabougou, January 23, 2010. When recounting this story Traoré hinted that Aoua Keita was among the leaders of the protest and march to Ségou. Aoua Keita was a leading political and women's rights activist who worked as a midwife for the Office medical services in Markala during this time. See Aoua Kéita, *Femme d'Afrique: La vie d'Aoua Kéita racontée par elle-même* (Paris: Présence Africaine, 1975), 48–82.

14. Office employees were granted one representative of their choosing. Farmers were represented by three to five leading farmers, but these men were chosen by the governor. See Schreyger, *L'Office du Niger au Mali*, 169, 77, 83.

15. Farmers were not formally organized into a union until 1954. See Van Beusekom, "Individualism, Community, and Cooperatives," 5–6.

16. Tony Revillon to Cornut-Gentille Haut-Commissaire de la Republique de l'AOF à Dakar, January 6, 1953, Dossier Confidentiel, Colonisation 1953/6 (Affaire Sangaré).

17. Georges Spitz, *Sansanding: Les irrigations du Niger* (Paris: Société d'Editions Géographiques, Maritimes, et Coloniales, 1949), 63–70.

18. Ravitaillement en Céréales des populations de l'A.O.F., 1944, ANM FR III 1Q 1700.

19. Mme. Koné Mariam Diarra and Tchaka Diallo, interview by author, Markala-Diamarabougou, January 26, 2010.

20. James C. Scott, *Seeing Like a State: How Certain Schemes to Improve the Human Condition Have Failed* (New Haven, CT: Yale University Press, 1998).

21. Fati Kindo, interview by author, Kossouka, May 11, 2010; and Hawa Diarra, interview by author, Nara, April 30, 2010. The fact that the Office cut down trees to make fields was something remarked upon by regional observers. For example, before coming to the Office Hawa Diarra had heard that to have fields "people had to cut trees."

22. Musokura means "older woman," and Musokura Jabate is not to be confused with the mythical Musokura mentioned in chapter 1. Nci is sometimes written as Nji. Banbugu is the written form of the town name that conforms to Bamana orthography and is employed by the historian Catherine Bogosian. See Catherine Bogosian, "Forced Labor, Resistance and Memory: The *Deuxième Portion* in the French Soudan, 1926–1950" (PhD diss., University of Pennsylvania, 2002). Official town signs now use the orthography Bambougou or Bambougouji.

23. The course of the canal passed by Diamarabougou, which was incorporated with Kirango under the French to constitute what is today the town of Markala.

24. Tiesson Dembélé, interview by author, Bambougouji, June 3, 2010. See also Bogosian, "Forced Labor, Resistance and Memory," 84–88. Bogosian cites an interview she conducted with Yah Balla Dembele in Markala, July 15, 2000.

25. Bogosian, "Forced Labor, Resistance and Memory," 84–85.

26. This was a point of criticism raised during the official Mission Reste in 1945. Report by M. Dorche 1945 (Reste Mission), Office du Niger 1952/1957, Inspection générale des travaux publics, ANOM FM 3TP/334.

27. Comment from Sitan Mallé during Aïssata Mallé et al., interview by author, Kouyan-N'Péguèna, April 12, 2010. Sitan's parents were among the first group of farmers in Kouyan-Kura, and her recollection speaks to the early adaption of canals for women's daily work.

28. Bintu Traoré, Mariam Doumbia, and Fanta Sogoba, interview by author, Molodo-Centre, April 15, 2010; Kono Dieunta, interview by author, Kokry, May 3, 2010.

29. Mariam "Mamu" Coulibaly, interview, May 4, 2010; and Kono Dieunta, interview, May 3, 2010.

30. Alimata Dembélé, Moussa Coulibaly, and Nana Dembélé, interview by author, Koutiala-Kura, May 3, 2010; Mariam "Mamu" Coulibaly, interview, May 4, 2010; and Fati Kindo, interview, May 11, 2010.

31. Fatoumata Coulibaly, Kadiatou Mallé, and Maïssa Sountura, interview by author, Kouyan-Koura, April 14, 2010.

32. Interview by author with members of the women's group Sabali. Hawa Coulibaly et al., interview by author, Molodo-Bamana, April 19, 2010. Hawa Coulibaly was president of the women's group and wife of the *dugutigi*.

33. Alimata Dembélé, Moussa Coulibaly, and Nana Dembélé, interview, May 3, 2010.

34. Wassa Dembélé and Kalifa Dembélé, interview by author, San-Kura, May 5, 2010.

35. Counting the number of arm lengths required to draw water from the well was a common method for measuring its depth. See Nianzon Bouaré and Harouna Bouaré, interview by author, Molodo-Bamana, April 16, 2010. Fati Kindo made a similar observation, comparing her parents' home region in Burkina Faso to towns in the Office. See Fati Kindo, interview, May 11, 2010.

36. Moussa Diawara, Aminata Tangaré, and Hawoyi Diawara, interview by author, Boky-Were, May 7, 2010.

37. Assane Pléah, informal conversation with the author, Kouyan-N'Péguèna, April 12, 2010.

38. Hawa Diarra, interview, April 30, 2010.

39. Hawa Coulibaly, interview by author, Fouabougou, March 30, 2010.

40. Sékou Coulibaly and Kadiatou Traoré, interview by author, Nyamina, May 31, 2010; and Nièni Tangara and Mamadou "Seyba" Coulibaly, interview by author, Kolony (km 26), August 7, 2010.

41. Amadou Sow and Sekou Salla Ouloguem, interview by author, Molodo-Centre, April 8, 2010.

42. Kono Dieunta, interview, May 3, 2010; and Assane Coulibaly, interview by author, Sirakoro, April 28, 2010.

43. Bintou Diarra and Aïssata Coulibaly, interview by author, Kolongotomo, May 11, 2010. Both women were rural women development workers employed by the Office du Niger in 2010.

44. For more on the construction of Office roads, see Magasa, *Papa-Commandant*, 24–29.

45. Correspondance et renseignements concernant le réseau routier du Soudan français (1932), Soudan Français (Haut-Sénégal-Niger)/Mali, Routes, circulation, et transports routiers, Inspection générale des travaux publics, ANOM FM 1TP/51.

46. Since the First World War, the administration sought an improved transportation infrastructure for the export of agricultural products in the colony. See Ravitaillement des grains 1913–1919, ANM FR 1Q 77; and Grains limitation centres commerciaux points de traite 1928–1934, ANM FR 1Q 272.

47. Aramata Diarra, interview by author, Fouabougou, March 30, 2010. Vehicle traffic in the French Soudan increased after the war. The colonial government documented significant increases in road travel between 1947 and 1952. See Routes, circulation et transports routiers, Notes sur travaux routiers A.O.F. 1953/1960, Afrique occidentale française, Inspection générale des travaux publics, ANOM 3TP/30.

48. Nianzon Bouaré and Hawa Coulibaly, interview by author, Molodo-Bamana, May 29, 2010.

49. Wassa Dembélé and Kalifa Dembélé, interview, May 5, 2010.
50. Kadja Coulibaly, interview by author, Kokry, May 2, 2010.
51. Tchaka Diallo, interview by author, Markala-Diamarabougou, January 27, 2010; and Aïssata Mallé and Assane Pléah, interview by author, Kouyan-N'Péguèna, May 28, 2010.
52. Assane Coulibaly, interview, April 28, 2010; Moctar Coulibaly, Mariatou Traoré, and Arouna Coulibaly, interview by author, Sirakoro, April 28, 2010; and Mariam "Mamu" Coulibaly, interview, May 4, 2010.
53. During the Second World War, much land prepared by Office workers ended up being abandoned by farmers because of its poor quality. See L'Office du Niger: Note de presentation technique, p. 4, May 15, 1960, AON 1.
54. Moctar Coulibaly, Mariatou Traoré, and Arouna Coulibaly, interview, April 28, 2010; and Mme. Dagno Adam Bah, interview by author, Markala-Diamarabougou, February 6, 2010.
55. Assane Coulibaly, interview, April 28, 2010; Moctar Coulibaly, Mariatou Traoré, and Arouna Coulibaly, interview, April 28, 2010; and Affaires agricoles Office du Niger 1955–1956, ANM FR II 1R 157.
56. Notes et réflexions au sujet du rapport de M. Dumont by the Service de l'Exploitation, c. 1950–1951, AON 138/2.
57. Rokia Diarra, interview by author, Nara, April 29, 2010.
58. Moctar Coulibaly, Mariatou Traoré, and Arouna Coulibaly, interview, May 31, 2010.
59. On historical marriage practices in the region, see Maria Grosz-Ngaté, "Monetization of Bridewealth and the Abandonment of 'Kin Roads' to Marriage in Sana, Mali," *American Ethnologist* 15, no. 3 (1988): 501–14.
60. Tchaka Diallo, interview by author, Markala-Diamarabougou, March 10, 2010.
61. Tchaka Diallo, interview, January 27, 2010.
62. The beauty pageant was one of several events hosted by the Office to foster social life. Tchaka Diallo brought up the pageant and other social events to respond to what he knew were well-circulated scholarly and present-day popular reports about the lack of a vibrant social life in economically struggling towns during the colonial era. See Tchaka Diallo, interview, March 10, 2010; Mme. Dagno Adam Bah, interviews by author, Markala-Diamarabougou, January 25, 2010, and February 6, 2010.
63. Rapport Rossin, January 1957, Réorganisation de l'Office du Niger, AON 138/3. This report included notes from the mission by Inspector General Mazodier. See also Rapport de monsieur le gouverneur de la FOM Romani commissaire du gouvernement, December 1956, Réorganisation de l'Office du Niger, AON 138/5.
64. "Etude Socio-Economique des Exploitations des Colons de l'Office du Niger," IER division d'etudes techniques, 1980, AON 464.
65. Sékou Coulibaly and Kadiatou Traoré, interview, May 31, 2010. The town of Bo only joined the Office after 1958. When the town initially requested

integration with the Office, it was refused entry because of its elevation in relation to the Molodo swamp. See Directeur Général to Chef of the Subdivision of Niono, November 13, 1958, Correspondance directeur general 1957–1960, AON 106.

66. Baba Djiguiba, interview by author, Molodo-Centre, April 10, 2010.
67. From a copy of a newspaper article clipped for the Office administration. See Motion from the Syndicat autonome des agriculteurs et colons Nigeriens, published in *L'Essor*, no. 2737 (March 24, 1958), *L'Essor* articles 1957–1958, AON 156.
68. Moussa Diawara, Aminata Tangaré, and Hawoyi Diawara, interview, May 7, 2010.
69. Simple questions, June 10, 1955, Dossier confidentiel, Colonisation 1953/6 (affaire Sangaré).
70. The size of land allotments was a frequent complaint of farmers who brought grievances to their instructor or the administration. See Affaires agricoles Office du Niger 1955–1956, ANM FR II 1R 1572.
71. Local farmers in and outside the Office protested such restrictions on common land. See Motion from the Syndicat autonome des agriculteurs et colons Nigeriens.
72. Motion from the Syndicat autonome des agriculteurs et colons Nigeriens.
73. Djenebu Coulibaly, interview by author, Kolony (km 26), April 9, 2010; Djewari Samaké, interview by author, Kolony (km 26), April 9, 2010; and Director Georges Peter, "Un exemple d'assistance technique: L'Office du Niger," January 1, 1955 (draft copy), AON 118/2.
74. Direction technique no. 14, unpublished report, Problèmes culturaux et mécanisation agricole de l'Office du Niger, 1955, ANM RFD 253.
75. Rokia Diarra, interview, April 29, 2010.
76. From a copy of a newspaper column clipped for the Office administration. "Echos de Brousse de Niénébalé," *L'Essor*, no. 2515 (June 4, 1957), *L'Essor* articles 1957–1958, AON 156.
77. Monica M. van Beusekom, *Negotiating Development: African Farmers and Colonial Experts at the Office du Niger, 1920–1960* (Portsmouth, NH: Heinemann, 2002), 121–26.
78. Van Beusekom, *Negotiating Development*.
79. Mariam "Mamu" Coulibaly, interview, May 4, 2010.
80. Monica van Beusekom also found that some women maintained small gardens for onion, okra, peanuts, and cotton despite labor time constraints and the problem of accessing fields. See Van Beusekom, *Negotiating Development*, 98.
81. Rokia Diarra, interview, April 29, 2010.
82. Alimata Dembélé mentions that the land right next to the canals was not of the best quality. See Alimata Dembélé, Moussa Coulibaly, and Nana Dembélé, interview, May 3, 2010.
83. Fatouma Coulibaly, interview by author, in Kolony (km 26), March 24, 2010; Djenebu Coulibaly, interview, April 9, 2010; and Djewari Samaké,

interview, April 9, 2010. Some women in their forties might have qualified as older in the women's recollections, but women well into their fifties were often without the help of a daughter-in-law and would have continued to work under the strict supervision of Office staff.

84. Fati Kindo, interview, May 11, 2010. Fati used the Bamana term *ngoyo* for "eggplant" during our interview.

85. Datu is a leafy green plant. See Rapport annuel Kokry-Kolongotomo, 1948–1949, AON 367/8.

86. See Rapport Annuel Kokry-Kolongotomo, 1948–1949. See also Schreyger, *L'Office du Niger au Mali*, 96.

87. Fatoumata Coulibaly and Oumou Sow, interview by author, Sabula, May 10, 2010.

88. See Direction technique no. 14.

89. Hawa Diarra, interview, April 30, 2010.

90. Kono Dieunta, interview, May 3, 2010.

91. Djewari Samaké, interview by author, Kolony (km 26), March 29, 2010.

92. Wassa Dembélé and Kalifa Dembélé, interview, May 5, 2010.

93. "Vade Mecum de l'instructeur de colonisation en centre cotonnier," p. 11, c. 1954, AON 11.

94. René Dumont, Réorganisation de l'Office du Niger, p. 10, unpublished report, ca. 1950, AON 138/2. "Kouia" is the French spelling of "Kouya."

95. Dramane Doumbia cited in Isaïe Dougnon, *Travail de Blanc, travail de Noir: La migration des paysans dogon vers l'Office du Niger et au Ghana (1910–1980)* (Paris: Karthala-Sephis, 2007), 176.

96. Delegation au paysannat, Remy Madier, Organisation des associations cooperatives agricoles, March 1958, AON 132/2; and Mme. Koné Mariam Diarra and Tchaka Diallo, interview, January 26, 2010. Comment by Tchaka Diallo.

97. Dumont, Réorganisation de l'Office du Niger, 6–7.

98. Schreyger, *L'Office du Niger au Mali*, 159–62.

99. Nianzon Bouaré and Harouna Bouaré, interview, April 16, 2010.

100. Hawa Coulibaly et al., interview, April 19, 2010; and Nianzon Bouaré and Harouna Bouaré, interview, April 16, 2010.

101. Emile Schreyger previously examined the demands of the Office labor schedule in Schreyger, *L'Office du Niger au Mali*, 112, 115–16, 118.

102. Hawa Coulibaly et al., interview, April 19, 2010.

103. Molodo-Bamana also had its own small market once a week. N'Faly Samaké, personal communication, Molodo-Bamana, April 9, 2010.

104. Information for food plants introduced by colonial botanists and later sold in Office markets is found in a variety of botanical reports located in the Office du Niger Archives. Report for Kokry, n.d. [ca. 1948–49], AON 367/8; Dossier Roberty botanique, AON 30; Dossier botanique (suite), AON 93; "Introduction de plantes à Soninkoura," AON 85.

105. Hawa Diarra, interview, April 30, 2010. In the 1950s, 500 FR was worth around one US dollar in 2021.

106. Oumou Dembélé, interview by author, Kouyan N'Goloba, April 13, 2010.
107. Daouda Bouaré and Bintu Dembélé, interview by author, Sokorani, June 1, 2010. Hawa Diarra similarly remembered that by the 1960s women were purchasing and cultivating onions for the new onion sauce. See Hawa Diarra, interview, April 30, 2010.
108. Wassa Dembélé and Kalifa Dembélé, interview, May 5, 2010.
109. Wassa Dembélé and Kalifa Dembélé, interview, May 5, 2010.
110. Djenebu Coulibaly, interview, April 9, 2010; Djewari Samaké, interview, April 9, 2010; Oumou Dembélé, interview, April 13, 2010; Fatoumata Coulibaly, Kadiatou Mallé, and Maïssa Sountura, interview, April 14, 2010; Fati Kindo, interview, May 11, 2010; and Mariam "Mamu" Coulibaly, interview, May 4, 2010.
111. Hawa Diarra, interview, April 30, 2010; Wassa Dembélé and Kalifa Dembélé, interview, May 5, 2010; Kono Dieunta, interview, May 3, 2010; and Assane Coulibaly, interview, April 28, 2010.
112. Fati Kindo, interview, May 11, 2010.
113. Mamadou Djiré, interview by author, Niono market, August 7, 2010.
114. Rokia Diarra, interview, April 29, 2010.
115. Assane Coulibaly, interview, April 28, 2010.
116. Aramata Diarra, interview, March 30, 2010.
117. Alimata Dembélé, Moussa Coulibaly, and Nana Dembélé, interview, May 3, 2010; and Wassa Dembélé and Kalifa Dembélé, interview, May 5, 2010.
118. Mariam "Mamu" Coulibaly, interview, May 4, 2010; and Alimata Dembélé, Moussa Coulibaly, and Nana Dembélé, interview, May 3, 2010, May 3, 2010.
119. Dossier confidentiel, Colonisation 1953/6 (affaire Sangaré).
120. Van Beusekom, Negotiating Development, 96.
121. See Schreyger, L'Office du Niger au Mali, 135.
122. For a comparison of Office prices to regional markets see Van Beusekom, Negotiating Development, 131–45.
123. Rokia Diarra, interview, April 29, 2010. During the interview Rokia lamented the fact that she could no longer find the same large fish and joked that they all went somewhere; she just didn't know where.
124. Van Beusekom, Negotiating Development, 132.
125. Informal discussion with Bintu Dieunta, Kokry, April 30, 2010.
126. Critique de R. Dumont (Intéressant le service des recherches), February 5, 1951, AON 138/2.
127. Sékou Coulibaly and Kadiatou Traoré, interview, May 31, 2010.
128. Fatoumata Coulibaly and Oumou Sow, interview, May 10, 2010.
129. Wassa Dembélé and Kalifa Dembélé, interview, May 5, 2010. In the 1950s the sum of 10,000 FR would have been worth around twenty US dollars today.
130. Wassa Dembélé and Kalifa Dembélé, interview, May 5, 2010.

131. Many women laughed when viewing pictures of women at the Office without duloki (blouses). Most often their comments were also unprompted. General questions were asked about the labor women were performing in the photos, but many women remarked that this economic difference was reflected in the women's clothing.

132. Most women wore shirts just before the time of the first Malian president, Modibo Keita. See Hawa Coulibaly, interview, March 30, 2010; and Fatoumata Coulibaly, Kadiatou Mallé, and Maïssa Sountura, interview, April 14, 2010.

133. See Maria Grosz-Ngaté, "Memory, Power, and Performance in the Construction of Muslim Identity," *PoLAR: Political and Legal Anthropology Review* 25, no. 2 (2002): 5–20; and James C. McCann, *Stirring the Pot: A History of African Cuisine* (Athens: Ohio University Press, 2009), 35.

134. Notes et réflexions au sujet du rapport de M. Dumont, ca. 1951, Réorganisation de l'Office du Niger, Service de l'exploitation, AON 138/2.

135. Fatoumata Coulibaly, Kadiatou Mallé, and Maïssa Sountura, interview, April 14, 2010; Aïssata Mallé and Assane Pléah, interview, May 28, 2010; Kadja "Ma" Coulibaly, Fatoumata Guindo, Fatoumata Zaré Coulibaly, and Lalafacouma Tangara, interview by author, Koue-Bamana, May 26, 2010. During the latter interview, the women in Koue-Bamana insisted that the harvest parties ended because of the rise of Islam in the region and that the millet beer drinking that had characterized the parties was not acceptable to Muslims.

136. Kokry-Kolongotomo, "Rapport annuel, 1948–1949," AON 367/8.

137. Mme. Dagno Adam Bah, interview, February 6, 2010; Kono Dieunta, interview, May 3, 2010; Bintu Traoré, Mariam Doumbia, and Fanta Sogoba, interview, April 15, 2010.

138. Kadja Coulibaly, interview, May 2, 2010.

139. Daouda Bouaré and Bintu Dembélé, interview, June 1, 2010.

140. Djenebu Coulibaly and Djewari Samaké, interview by author, Kolony (km 26), March 22, 2010; Hawa Coulibaly, interview, March 30, 2010; and Mamadou Djiré, interview, August 7, 2010.

141. René Dumont remarks on a visible level of deforestation around the Office. See René Dumont, incomplete report, pp. 13–14, ca. 1950, Réorganisation de l'Office du Niger, AON 138/2. Multiple interviewees commented on women seeking alternative means of earning cash when wood suplies declined: Aïssata Mallé et al., interview, April 12, 2010 (comment by Sitan Mallé); and Hawa Coulibaly, interview, March 30, 2010.

142. Mme. Dagno Adam Bah, interview, February 6, 2010; Hawa Coulibaly, interview, March 30, 2010.

143. Djewari Samaké similarly recounted that daughters in the second Office generation collected more household goods before their wedding than previous generations, much of it stored in a *kesu* (trunk). Some of the items were purchased by the bride to be. Additional purchases (by

mothers) were supported by cash gifts from the husband to be. Djewari Samaké, interview, March 29, 2010.

144. Women also used tafew for their husbands or children's clothes.

145. Adam Bah's grandfather was a jeweler who left his daughter (Bah's mother) an inheritance of gold that she used for major purchases like her daughter's wedding gifts. The gold wealth gifted to Bah was relatively unique among women at the Office. However, her story only amplifies the strategies of young women and their mothers in preparing for a wedding and the activities of young women at the Office. Mme. Dagno Adam Bah, interview, February 6, 2010.

146. Dossier confidentiel, Colonisation 1953/6 (affaire Sangaré).

147. Monica van Beusekom similarly argues that while many farmers were impoverished by poor conditions at the Office dating to its founding, many other households fared well. This gap can be explained by the quality of different land allotments, the relative wealth of some farmers prior to coming to the Office, administrative connections, and the number of workers in the household. See Van Beusekom, *Negotiating Development*, 108–10.

148. Dumont, incomplete report; and Kono Dieunta, interview, May 3, 2010.

149. Dumont, incomplete report.

150. Bintu Traoré, Mariam Doumbia, and Fanta Sogoba, interview, April 15, 2010. During this interview Sékou Sall Ouloguem intervened to reinforce the fact that women who earned cash purchased cattle. He added that a few women even built their own homes in town. See Fatouma Coulibaly and Nièni Tangara, interview by author, Kolony, May 25, 2010; Kadja "Ma" Coulibaly et al., interview, May 26, 2010.

151. Rokia Diarra, interview, April 29, 2010.

152. Kono Dieunta, interview, May 3, 2010. On this point, the historian James McCann surmises that much attachment to specific staple crops in West Africa had to do with texture and their bulk rather than flavor. See McCann, *Stirring the Pot*, 33.

153. Assane Coulibaly, interview, April 28, 2010.

154. Schreyger, *L'Office du Niger au Mali*, 143.

155. Assane Coulibaly, interview, April 28, 2010.

156. Hawa Diarra, interview, April 30, 2010.

157. Dege is an exception. Women did not cook dege. Dege powder was prepared by pounding the grain. It was later mixed with water or milk to make a cold porridge.

158. Kono Dieunta, interview, May 3, 2010. Aïssata Mallé and Assane Pléah, interview, May 28, 2010.

159. Kono Dieunta, interview, May 3, 2010; Aïssata Mallé and Assane Pléah, interview, May 28, 2010.

160. Kono Dieunta, interview, May 3, 2010. *Chouchou* is my approximation for how to spell the name of the sauce that Kono mentions in the interview.

161. Assane Coulibaly, interview, April 28, 2010.
162. Alimata Dembélé, Moussa Coulibaly, and Nana Dembélé, interview, May 3, 2010; Mariam "Mamu" Coulibaly, interview, May 4, 2010.
163. Fatoumata Coulibaly and Oumou Sow, interview, May 10, 2010.
164. Amadou Sow and Sekou Salla Ouloguem, interview by author, Molodo-Centre, April 8, 2010.
165. Fatoumata Coulibaly, Kadiatou Mallé, and Maïssa Sountura, interview, April 14, 2010.
166. Nianzon Bouaré and Harouna Bouaré, interview, April 16, 2010. In the rice sector Moctar Coulibaly claims that farmers also refused to grow any cotton in Office fields when the instructors brought the tools for cotton planting (a specialized rake, for example). He said they would only grow the rice. Moctar Coulibaly, Mariatou Traoré, and Arouna Coulibaly, interview, April 28, 2010. Similarly, Office administrators recorded the negotiations with farmers in Sokolo to only grow rice. Coyaud to Director General Wibaux, June 12, 1957, Correspondance directeur general 1957–1960, AON 106.
167. Hawa Coulibaly, interview, March 30, 2010.
168. Hawa Coulibaly, interview, March 30, 2010. By 1958, 28 percent of cultivated land in the Niono sector was planted with vegetable crops. See Schreyger, L'Office du Niger au Mali, 162.
169. Baba Djiguiba, interview, April 10, 2010.
170. "Vade Mecum de l'instructeur de colonisation en centre cotonnier," p. 32, ca. 1954, AON 11.
171. Monica van Beusekom argued previously that the switch from cotton to rice at the Office was a result of pressure from farmers in the cotton sector. In this way, farmers negotiated some of the terms of "development" at the project. See Van Beusekom, Negotiating Development.
172. Djenebu Coulibaly, interview by author, Kolony (km 26), March 27, 2010.
173. Hawa Coulibaly, interview, March 30, 2010.
174. Oumou Dembélé, interview, April 13, 2010; Hawa Coulibaly, interview, March 30, 2010; Djenebu Coulibaly, interview, March 27, 2010.
175. Aramata Diarra, interview, March 30, 2010.
176. Aïssata Mallé and Assane Pléah, interview, May 28, 2010.
177. "Rapport à la commission parlementaire chargée d'une enquête sur l'Office du Niger: observations sur les travaux scientifiques et agronomiques relatifs à la culture du cotonnier et des plantes vivrières dans la vallée du Niger par le professeur Chevalier, membre de l'Institut, 17 janvier 1939," p. 23, Inspection générale des travaux publics, Office du Niger, ANOM FM 1 TP/878.
178. Peter, "Un exemple d'assistance technique," 13.
179. Chabert the Controller of Niono to the Head of the Service d'Exploitation in Ségou, April 9, 1955, Dossier confidentiel, Colonisation 1953/6 (affaire Sangaré), AON 111.

180. Note sur la commercialisation du riz, June 15, 1954, Inspection générale des Travaux publics, Office du Niger, ANOM FM 3TP/335.
181. L'Office du Niger: Note de présentation technique Mai 15, p. 4, 1960, AON 1.
182. Monica van Beusekom asserts that the terms of development were shared between French experts and administrators and African political leaders during this period. In this case, some Office farmers also shared in that discourse. See Van Beusekom, "Individualism, Community, and Cooperatives."
183. Sékou Coulibaly and Kadiatou Traoré, interview, May 31, 2010.
184. Wassa Dembélé and Kalifa Dembélé, interview, May 5, 2010.
185. Sékou Coulibaly and Kadiatou Traoré, interview, May 31, 2010. Women in Kolony (km 26) also vividly remember food distributions. Djenebu Coulibaly and Djewari Samaké, interview, March 22, 2010.
186. Assane Coulibaly, interview, April 28, 2010.
187. Notes et réflexions au sujet du rapport de M. Dumont by the Service de l'exploitation, ca. 1950–1951, AON 138/2.
188. Kadja Coulibaly, interview, May 2, 2010. At the time Kadja and her husband were living near Molodo-Centre because he worked for the all-mechanized sector of the Office.
189. Kono Dieunta, interview, May 3, 2010.
190. Kadja Coulibaly, interview, May 2, 2010.
191. Sékou Coulibaly and Kadiatou Traoré, interview, May 31, 2010.
192. Aïssata Mallé and Assane Pléah, interview, May 28, 2010; Nièni Tangara and Mamadou "Seyba" Coulibaly, interview, August 7, 2010.
193. From a copy of the journal article saved for the Office administration. *Les cahiers coloniaux*, July 15, 1939, AON 79/1.

CHAPTER 4: REENGINEERING THE OFFICE

1. Documentation for this trend toward increased mechanization at the Office du Niger is available in several reports from the public works administration for French West Africa. See Office du Niger, 1947/1953, Inspection générale des travaux publics, ANOM FM 3TP/172; Office du Niger, 1952/1957, Inspection générale des travaux publics, ANOM FM 3TP/334; Office du Niger, 1946/1952, Inspection générale des travaux publics, ANOM FM 2TP/104. See also Emil Schreyger, *L'Office du Niger au Mali 1932 à 1982: La problématique d'une grande entreprise agricole dans la zone du Sahel* (Paris: L'Harmattan, 1984), 136–68.
2. In attending to everyday sensory experiences, I draw from the foundational work by anthropologist Paul Stoller on the senses and the quotidian. See Paul Stoller, *The Taste of Ethnographic Things: The Senses in Anthropology* (Philadelphia: University of Pennsylvania Press, 1989).
3. Dossier Roberty botanique, AON 30; and Dossier botanique (suite), AON 93.

4. Mariam "Mamu" Coulibaly, interview by author, Kankan (formerly San-garébougou), May 4, 2010.

5. Djewari Samaké, interview by author, Kolony (km 26), March 29, 2010; Hawa Coulibaly, interview by author, Fouabougou, March 30, 2010; and Aïssata Mallé et al., interview by author, Kouyan-N'Péguèna, April 12, 2010.

6. Georges Peter, "Un exemple d'assistance technique: L'Office du Niger," 1955, AON 118/2; and Mission du botanist à Baroueli, Niénébalé, Bagu-inèda du 8 au 11 Mars 1937, Dossier botanique, AON 30/5.

7. Djenebu Coulibaly, interview by author, Kolony (km 26), April 9, 2010. On a similar point, Djewari Samaké recalled Europeans poisoning trees. Djewari Samaké, interview by author, Kolony (km 26), April 9, 2010. The high level of botanical exchange introduced by colonial botanists and agronomists may also have introduced diseases that attacked the trees that Djenebu remembers dying during her childhood.

8. Nianzon Bouaré and Harouna Bouaré, interview by author, Molodo-Bamana, April 16, 2010. Laurence C. Becker noted similar tendencies to preserve specific tree varieties among inhabitants around the Faya forest in Mali (between Bamako and Ségou). His interviewees similarly reported that "in the past" shea, tamarind, and other trees were protected. See Laurence C. Becker, "Seeing Green in Mali's Woods: Colonial Leg-acy, Forest Use, and Local Control," *Annals of the Association of Ameri-can Geographers* 91, no. 3 (2001): 504–26, here 512.

9. Mission du botaniste à Barouéli, Niénébalé, Baguinèda du 8 au 11 Mars 1937.

10. Becker, "Seeing Green in Mali's Woods," 507–8. The French policy of the forest as a site for fuel production during the war was matched by similar forest policies in France when wood was rediscovered as a major fuel source given the wartime shortages of other fuel sources. However, the Vichy government was not successful at increasing the production of wood fuel in France. See Chris Pearson, "'The Age of Wood': Fuel and Fighting in French Forests, 1940–1944," *Environmental History* 11 (2006): 775–803.

11. Hawa Coulibaly, interview, March 30, 2010.

12. Djenebu Coulibaly, interview by author, Kolony (km 26), March 27, 2010; Mariam "Mamu" Coulibaly, interview, May 4, 2010.

13. Mamadou Djiré, interview by author, Niono market, August 7, 2010. In 2010 Djiré was an elderly merchant in Niono who inherited his trade from his father, who had also worked in the colonial Office market at Niono. Djiré spoke about the history of the market in Niono specifically, but the same predominance of women in the market during the rainy season, in particular, was likely to be the case in the Kolongotomo and Kokry markets.

14. Djenebu Coulibaly, interview, March 27, 2010.

15. Barbara E. Frank, "Marks of Identity: Potters of the Folona (Mali) and Their 'Mothers,'" *African Arts* 40, no. 1 (2007): 30–41.

16. Djewari Samaké, interview by author, Kolony (km 26), June 2, 2010.

17. This time frame for the introduction of metal pots reflects the chronology of market purchases by women at the Office du Niger. Other markets in the Soudan, such as Bamako, offered imported metal goods earlier than 1940, but the women interviewed for this book did not recall seeing many of these heavy iron pots or hearing about metal pot purchases until the years roughly after the Second World War. Based on demographic statistics collected by researchers and Office officials, households ranged in size from seven to forty members from the late 1940s to the 1960s. The women interviewees confirmed this range of household sizes, but their recollections of cooking more often described large households, or cooking for large parties of people attending festivals or parties.

18. Hawa Diarra, interview by author, Nara, April 30, 2010.

19. Fatoumata Coulibaly et al., interview by author, Kouyan-Kura, May 27, 2010.

20. Women who purchased metal goods were described by the merchant Mamdou Djiré as broadly "middle-aged" or "old," which likely means women old enough to have marriageable daughters. Djiré apprenticed with his father, who was in charge of their market stall during the period covered in this interview, and thus remembers age from the vantage point of a young boy and teenager. Most women interviewees noted that women in their generation tended to marry around the age of twenty. Mamadou Djiré, interview, August 7, 2010.

21. Barbara Cooper refers to the displays women made with these items in their own rooms. The women interviewed for this book remembered the parade of goods made during the bride's trip to her new home. Cooper's observations about demonstrating social value translate well into this public display of goods. See Barbara M. Cooper, *Marriage in Maradi: Gender and Culture in a Hausa Society in Niger, 1900–1989* (Portsmouth, NH: Heinemann, 1997), 103.

22. Today young women purchase enamel pots for similar purposes. Jerimy Cunningham studies this pattern in the Djenné area and found that women emphasized the importance of purchasing the newest and trendiest styles of enamelware. Women used some of these items (notably serving dishes), but most of the collection was regifted to other women. His analysis emphasizes the perceived value imbued in these items on display and in their social exchange rather than on their daily use. See Jerimy J. Cunningham, "Pots and Political Economy: Enamel-Wealth, Gender, and Patriarchy in Mali," *Journal of the Royal Anthropological Institute* 15, no. 2 (2009): 276–94.

23. Mariam "Mamu" Coulibaly, interview, May 4, 2010.

24. Kadja Coulibaly, interview by author, Kokry, May 2, 2010.

25. Hawa Diarra, interview, April 30, 2010.

26. For more on women's status and marriage, see Maria Grosz-Ngaté, "Hidden Meanings: Explorations into a Bamanan Construct of Gender," *Ethnology* 28, no. 2 (1989): 167–83.
27. On this point, Cunningham argues that the contemporary possession of enamel pots helps young women demonstrate their positive work ethic as new brides today purchase many of their own household goods. Indeed, many young women work hard to accumulate such goods before marriage. See Cunningham, "Pots and Political Economy."
28. Comments from Fatoumata during Fatoumata Coulibaly and Oumou Sow, interview by author, Sabula, May 10, 2010.
29. When women were asked how they learned to cook rice if they came from predominantly millet-producing areas, they were often surprised by the question. To them the answer was easy: they just learned. When pressed further, many explained that an older woman in the household would show them what to do (see chapter 3).
30. Assane Pléah, interview by author, Kouyan-N'Péguèna, May 28, 2010.
31. Assane Pléah, interview, May 28, 2010.
32. Tanja Winther made a similar observation about the change from clay to aluminum pots during her research in contemporary rural Zanzibar. See Tanja Winther, *The Impact of Electricity: Development, Desires and Dilemmas* (New York: Berghahn, 2008), 197–208.
33. Assane Coulibaly, interview by author, Sirakoro, April 28, 2010.
34. Jack Goody, *Cooking, Cuisine, and Class: A Study in Comparative Sociology* (New York: Cambridge University Press, 1982).
35. Tanja Winther similarly noted that the taste of food prepared over electronic cookers in late twentieth-century Zanzibar was markedly different than the taste of food prepared over wood fires. Women and men noted the change, but women appreciated the possibility of saving labor time by using the electronic cookers. Her findings differ in that her informants believed that they could not cook *all* of the same things with electricity. See Winther, *Impact of Electricity*, 198–208.
36. Women's cash-earning activities are discussed in chapter 3 of this book.
37. Judith A. McGaw, "Reconceiving Technology: Why Feminine Technologies Matter," in *Gender and Archaeology*, ed. Rita P. Wright (Philadelphia: University of Pennsylvania Press, 1996), 25.
38. Kadja "Ma" Coulibaly et al., interview by author, Koué-Bamana, May 26, 2010.
39. Based on my participant observation, cooks in rural Mali still use this method for maintaining their metal pots.
40. McGaw, "Reconceiving Technology," 20.
41. Emily Lynn Osborn, "Casting Aluminum Pots: Labour, Migration and Artisan Production in West Africa's Informal Sector, 1945–2005," *African Identities* 7, no. 3 (2009): 376.
42. Jacques Marseille, *Empire colonial et capitalisme: Histoire d'un divorce* (Paris: Albin Michel, 1984), 49–55.

43. Osborn, "Casting Aluminum Pots," 376.
44. Fatoumata Coulibaly et al., interview, May 27, 2010.
45. Female potters ultimately lost a great number of their regular consumers in these years, but they continued to produce and sell other necessary items such as water jars and steamers. See Cunningham, "Pots and Political Economy," 281–84.
46. Osborn, "Casting Aluminum Pots," 377.
47. For more on Markala as a site of manufacture and metalworking, see Birama Diakon, *Office du Niger et pratiques paysannes: Appropriation technologique et dynamique sociale* (Paris: L'Harmattan Mali, 2012), 52–61, 82–94.
48. By this point, colonial French West Africa was undergoing political restructuring as the colony gained greater autonomy from France. African political leaders also held a measure of power in the new government structure. See Tony Chafer, *The End of Empire in French West Africa: France's Successful Decolonization?* (New York: Berg, 2002).
49. Director General to the Ministre du Commerce de l'Industrie et des Transports, January 29, 1959, Correspondance directeur general 1957–1960, AON 106. Most women interviewed for this book who arrived at the Office around 1960 recall bringing with them at least one metal pot purchased by their mother to their husband's household.
50. Director General to the Ministre du Commerce de l'Industrie et des Transports, January 29, 1959.
51. Monica van Beusekom points to the general situation of indebtedness to the Office. See Monica M. van Beusekom, *Negotiating Development: African Farmers and Colonial Experts at the Office du Niger, 1920–1960* (Portsmouth, NH: Heinemann, 2002), 103–10.
52. Director General to the Ministre du Commerce de l'Industrie et des Transports, January 29, 1959.
53. Ester Boserup, *Woman's Role in Economic Development* (London: Earthscan, 1989).
54. Bintu Traoré, Mariam Doumbia, and Fanta Sogoba, interview by author, Molodo-Centre, April 15, 2010. Several men were present during this interview including Sekou Sall Ouloguem, who enthusiastically recalled that Sogoba and her sister were widely known for their plowing skills.
55. Moctar Coulibaly, Mariatou Traoré, and Arouna Coulibaly, interview by author, Sirakoro, April 28, 2010.
56. Diakon, *Office du Niger et pratiques paysannes*, 41–42.
57. Wendy Faulkner, "The Technology Question in Feminism: A View from Feminist Technology Studies," *Women's Studies International Forum* 24, no. 1 (2001): 84. See also Jessica Smith Rolston, "Talk about Technology: Negotiating Gender Difference in Wyoming Coal Mines," *Signs: Journal of Women in Culture and Society* 35, no. 4 (2010): 893–918. Rolston demonstrates how talking about the use of male-associated industrial

technology helped women to negotiate the workplace but often reinforced the association of this technology with masculinity.

58. Comment from Fanta Sogoba during Bintu Traoré, Mariam Doumbia, and Fanta Sogoba, interview, April 15, 2010. Fanta was looking at a posed picture of a family from the colonial Office and what looked to the author like a young woman (because of her dress) holding the family's plow. Fanta objected and explained that the arms of the person holding the plow looked too strong to be a woman's.

59. Nianzon Bouaré and Hawa Coulibaly, interview by author, Molodo-Bamana, May 29, 2010.

60. Rapport de visite à l'Office du Niger de Mr. le Professeur Réné Dumont, ca. 1950, AON 138/2. Dumont also visited the Office in 1961. The file contains documents pertaining to both visits.

61. Hawa Coulibaly, interview, March 30, 2010; Fatoumata Coulibaly, Kadiatou Mallé, and Maïssa Sountura, interview by author, Kouyan-Kura, April 14, 2010.

62. Hawa Coulibaly, interview, March 30, 2010; Rokia Diarra, interview by author, Nara, April 29, 2010; Fatoumata Coulibaly et al., interview, May 27, 2010. In response to the author's question about how the change came about, some of the women jokingly said that the men who started collecting wood for their wives did so because they loved them too much. The women interviewed for this book only mentioned married men owning carts. Archival evidence of the number of carts, viewed by the author, only recorded numbers per household, not the identity, age, or marital status of the men purchasing carts.

63. Based on participant observation in regions of Ségou outside of the Office, this labor shift was not replicated in other rural areas where women were collecting wood and products from the wilds during the author's fieldwork in 2009–10.

64. The Office in particular was anxious to settle herders whose animals crossed the northern portion of Office lands every year. See Office du Niger, Rapport sur l'élevage et la colonisation indigène en terres irriguées 1934, ANM FR 1R 65.

65. Mariam "Mamu" Coulibaly, interview, May 4, 2010.

66. Mamadou Djiré, interview, August 7, 2010.

67. Mamadou Djiré, interview, August 7, 2010.

68. Fatouma Coulibaly and Nièni Tangara, interview by author, Kolony (km 26), May 25, 2010; Daouda Bouaré and Bintu Dembélé, interview by author, Sokorani, June 1, 2010; Brahmin Fané et al., interview by author, Markala-Kirango, June 3, 2010; and Mamadou Djiré, interview, August 7, 2010.

69. Hawa Fomba, personal communication, Kalaké-Bamana, October 2, 2010. Kalaké-Bamana is in the Ségou region south of the Office area but shares the cultural food history of most Office families and was a recruitment zone for the Office.

70. Fatoumata Coulibaly et al., interview, May 27, 2010.
71. Both the net and the basket required the fisher to enter the water. Women dropped the basket into the water and grabbed fish through the larger hole in the top or bottom of the basket with their hands, while the fish swam through the smaller side openings. Both techniques were described to the author by Sékou Coulibaly and observed while living at the Office. See Sékou Coulibaly and Kadiatou Traoré, interview by author, Nyamina, May 31, 2010.
72. Sékou Coulibaly and Kadiatou Traoré, interview, May 31, 2010.
73. Fatouma Coulibaly, interview by author, Kolony (km 26), March 24, 2010; Kono Dieunta, interview by author, Kokry, May 3, 2010.
74. A baarakolo is a large calabash with a small hole cut into the top for carrying water.
75. Bintu Traoré, Mariam Doumbia, and Fanta Sogoba, interview, April 15, 2010; Alimata Dembélé, Moussa Coulibaly, and Nana Dembélé, interview by author, Koutiala-Kura, May 3, 2010.
76. How domestic technologies fit into a woman's daily work routine impacts which technologies are adopted. This dynamic was explored by Joy Parr for Canadian women who, unlike their American counterparts, did not adopt automatic washers until the 1960s either because their households were not equipped with electricity or running water, or out of concern for excess water use. Parr points out that successful household technologies fit into the larger technological system of the home. See Joy Parr, "What Makes Washday Less Blue? Gender, Nation, and Technology Choice in Postwar Canada," *Technology and Culture* 38, no. 1 (1997): 153–86.
77. Fatouma Coulibaly was also one of the rare owners of a small metal mortar and pestle. She said she made this relatively expensive purchase because she knew the items would last, unlike the wooden mortar and pestle that many women still preferred. Her daughter-in-law, Nièni Tangara, made the point about the small gourd spoons breaking easily and that it was necessary for cooking equipment and eating utensils to last. Fatouma Coulibaly and Nièni Tangara, interview, May 25, 2010.
78. Office du Niger documentation 1937–1946, ANM FR 1R 42; Monica van Beusekom, *Negotiating Development*, 164.
79. Peter, "Un exemple d'assistance technique." The notion that the irrigation system of the Office transformed a desert into productive farmland was a rhetorical claim dating to before the project opened. However, this was a common rhetorical description used by Office promoters into the 1960s. The same claim was the major theme of the promotional booklet produced by the Office called "La delta réussucite," ca. 1961 (from AON 6 bis). This picture of the region as a desert is highly misleading as the region now hosting the Office du Niger was, before the project's founding, one of the largest grain-producing regions in French West Africa.
80. The shift to mechanization was noted by Emile Schreyger. See Schreyger, *L'Office du Niger au Mali*, 2, 136–37.

81. Georges Spitz, *Sansanding: Les irrigations du Niger* (Paris: Société d'Editions Géographiques, Maritimes, et Coloniales, 1949), 63–70. Schreyger, *L'Office du Niger au Mali*, 154–58. See Van Beusekom, *Negotiating Development*, 70–71, 140.

82. August 1946, Correspondance directeur general 1957–1960, AON 106.

83. Folder 2, Office du Niger 1946/1952, Inspection générale des travaux publics, ANOM, 2TP/104; also see Schreyger, *L'Office du Niger au Mali*, 154–55.

84. CCTA-Ségou, Aperçu sur l'Office du Niger, p. 2, report, 1961, AON 1/1.

85. See Programme de développement economique et social de l'A.O.F., 1949–1950, ANM RFD 171. FIDES stands for Fonds d'investissments pour le developppment economiques et sociales.

86. Revillon was on the council overseeing a reorganization of the Office du Niger following the Second World War. Several letters between Revillon and other government officials attest to his support for the institution and his interest in reinforcing its technical capacities. See Dossier confidentiel, Colonisation 1953/6 (affaire Sangaré), AON 111.

87. One other reason for the losses was the repayment of loans. Note sur la situation de l'Office du Niger, unnumbered document, ca. 1957, AON 1–4.

88. Mesures à envisager pour le developpement de la production oléagineux en Afrique Occidentale, arachides, April 1942, Dossier mission oléagineux, Institut colonial de Marseille commission des matières grasses, AON 79/1.

89. "Recherches sur le Karité," report, Instutit colonial de Marseille, Dossier mission oléagineux, AON 79/2; and correspondence between A.R.D.I.C. and the Director of Agricultural Services, May 7, 1942, Dossier mission oléagineux, AON 79/2.

90. M. Peissi, report from a conference at the Institut technique de batiment et des traveaux publics, April 30, 1946, Silos en acier, AON 93/4.

91. Peter, "Un exemple d'assistance technique," 16.

92. "Sansanding: Possibilitiés d'equipment hydro-electrique," Inspection générale des travaux publics, Office du Niger, 1947/1953, ANOM FM 3TP/172.

93. CCTA-Ségou, Aperçu sur l'Office du Niger. See also Schreyger, *L'Office du Niger au Mali*.

94. Correspondance directeur général 1957 to 1960, AON 106.

95. For example, Apaye Serou was sent from Niono to meet with the administrator in Segou in 1953 about their request to lower the fees for land preparation and water services. See Folder 1953, Rapports politiques et rapports de tournées cercle de Segou I (1940–1959), ANM FR 1E 40.

96. Note sur la situation de l'Office du Niger, ca. 1956, AON 1.

97. On this point, see Laura Ann Twagira, "'Robot Farmers' and Cosmopolitan Workers: Divergent Masculinities and Agricultural Technology Exchange in the French Soudan (1945–68)," *Gender and History* 26, no. 3 (2014): 459. This colonial ambivalence about the impact of mechanization in rural West Africa was not unique to the Office du Niger. For

example, just as the Office was mechanizing, the long-serving colonial official Robert Delavignette questioned whether industrialization would lead to a breakdown in rural West Africa in the fictional account of a machine's impact on a small village. See Robert Delavignette, *Les paysans noirs*, 2nd ed. (Paris: Editions Stock, 1946).

98. James C. Scott, *Seeing Like a State: How Certain Schemes to Improve the Human Condition Have Failed* (New Haven, CT: Yale University Press, 1998).

99. From a copy of a newspaper column clipped for the Office administration. See "Echos de la Colonisation," *L'Essor*, no. 2676 (December 31, 1957), *L'Essor* articles 1957–1958, AON 156.

100. Dominique Zahan, *Antilopes du soleil: Arts et rites agraires d'Afrique noire* (Vienna: Edition A. Schendl, 1980). The author collected popular and ritual farming songs celebrating young men's physical labor in the fields and champion farmers. Brahmin Fané et. al., interview, June 3, 2010.

101. Zahan, *Antilopes du soleil*, 20–21, 31. Zahan made this observation about *Ciwara* societies in the 1980s after Pascal Imperato seemed to confirm that the same societies continued to decline in the 1970s and after. Both Zahan and Imperato also suggest that the rise of Islamic conversion played a role in the decline. However, Stephen Wooten has documented the more recent resiliency of *Ciwara* societies and performances. Similarly, Birama Diakon has argued that mutual aid to support the purchase of plows through farming cooperatives is a renewed form of communal agricultural practice. Pascal James Imperato, "The Dance of the Tyi Wara," *African Arts* 4, no. 1 (1970): 8–13, 71–80; Stephen Wooten, *The Art of Livelihood: Creating Expressive Agri-culture in Rural Mali* (Durham, NC: Carolina Academic Press, 2009); Diakon, *Office du Niger et pratiques paysannes*, 36–41. Brahmin Fané et al., interview, June 3, 2010.

102. I have elaborated on this point in relation to men at the Office du Niger elsewhere. See Twagira, "'Robot Farmers' and Cosmopolitan Workers." Birama Diakon also argues that men claimed the plow specifically as a masculine technology after initially assessing the tool as a threat to male prowess in the fields. See Birama Diakon, *Office du Niger et pratiques paysannes: Appropriation technologique et dynamique sociale* (Paris: L'Harmattan Mali, 2012), 34–41, 43–44.

103. Procès verbaux Kossouka, 1956, Dossier divers [ca. 1950–1960], AON 118/25.

104. Peter, "Un exemple d'assistance technique."

105. Like other large industrial machines, the threshers were also dangerous. The death of a farmer in 1960 in a threshing machine accident is recorded in the Office archives. See May to June 1960, Correspondance directeur général 1957 to 1960, AON 106.

106. Notes et réflexions au sujet du rapport de M. Dumont Service de l'exploitation, internal report from the Service d'exploitation, ca. 1951, AON 138/2.

107. Hawa Coulibaly, interview, March 30, 2010.

108. Djenebu Coulibaly and Djewari Samaké, interview by author, Kolony (km 26), March 22, 2010.

109. Bintu Traoré, Mariam Doumbia, and Fanta Sogoba, interview, April 15, 2010.

110. Bintu Traoré, Mariam Doumbia, and Fanta Sogoba, interview, April 15, 2010.

111. Kadja Coulibaly, interview, May 2, 2010.

112. Tchaka Diallo, interview by author, Markala-Diamarabougou, January 27, 2010; Fatoumata Coulibaly, Kadiatou Mallé, and Maïssa Sountura, interview by author, Kouyan-Kura, April 14, 2010.

113. Tchaka Diallo, interview, May 2, 2010.

114. Fatoumata Coulibaly, Kadiatou Mallé, and Maïssa Sountura, interview, April 14, 2010.

115. Nianzon Bouaré and Harouna Bouaré, interview, April 16, 2010.

116. Twagira, "'Robot Farmers' and Cosmopolitan Workers," 466, 68–73.

117. "Rapport d'inspection sur le ravitaillement de la colonie de l'A.O.F.," 1916, ANM FA 1Q 190.

118. In interviews, most elderly women at the Office du Niger characterized women's activities as "household duties." "Work" was more often characterized as wage labor. For them, even going to market did not always qualify as work.

119. Fatoumata Coulibaly, Kadiatou Mallé, and Maïssa Sountura, interview, April 14, 2010.

120. "Vade Mecum de l'instructeur de colonisation en centre cotonnier," ca. 1954, p. 45, AON 11.

CHAPTER 5: RICE BABIES AND FOOD AID

1. Fatoumata Coulibaly, Kadiatou Mallé, and Maïssa Sountura, interview by author, Kouyan-Kura, April 14, 2010.

2. Moussa Traoré, "Allocution du Colonel Moussa Traoré, président du Comité militaire de libération nationale, président du gouvernement, chef de l'etat à l'occasion de la réunion du Comité permanent inter-etats du lutte contre la sécheresse dans le Sahel," microfiche, p. 2, conference speech, Bamako, March 15–16, 1974, General Research Division, Stephen A. Schwarzman Building, New York Public Library. In 1968 Moussa Traoré led a military coup against the first president of Mali, Modibo Keita. For a brief overview of this political history, see Andrew F. Clark, "From Military Dictatorship to Democracy: The Democratization Process in Mali," in Democracy and Development in Mali, ed. R. James Bingen, David Robinson, and John M. Staatz (East Lansing: Michigan State University Press, 2000), 256.

3. Shortly after the end of the 1969–73 drought international observers assessed the crisis as among the worst natural disasters of the twentieth

century. See, for example, Anne de Lattre and Arthur M. Fell, *The Club du Sahel: An Experiment in International Co-operation* (Paris: Organisation for Economic Co-operation and Development [OECD], 1984), 22. In Mali, climate change scholars have similarly noted powerful popular memories of the Great Sahel Drought. See Téréba Togola, "Memories, Abstractions, and Conceptualization of Ecological Crisis in the Mande World," in *The Way the Wind Blows: Climate, History, and Human Action*, ed. Roderick J. McIntosh, Joseph A. Tainter, and Susan Keech McIntosh (New York: Columbia University Press, 2000), 187–89.

4. Comment from Nana Coulibaly in Alimata Dembélé, Moussa Coulibaly, and Nana Dembélé, interview by author, Koutiala-Kura, May 3, 2010.

5. Micah Trapp has similarly documented complaints about food aid goods in Liberia in the 1990s and 2000s and their reputation for being barely edible, arguing for the consideration of local preferences in the provision of food aid and not simply caloric counts or bare minimum nutrition requirements. See Micah M. Trapp, "You-Will-Kill-Me-Beans: Taste and the Politics of Necessity in Humanitarian Aid," *Cultural Anthropology* 31, no. 3 (2016): 412–37.

6. Togola, "Memories, Abstractions, and Conceptualization," 187. Almost all interviewees for this book, when speaking in Bamana, referred to the drought as *Nyoblékongkong*, or the "Famine of the Red Millet."

7. Lattre and Fell, *Club du Sahel*, 23.

8. Lattre and Fell, 21; Jonathan Derrick, "The Great West African Drought, 1972–1974," *African Affairs* 76, no. 305 (1977): 537–86; Jonathan Derrick, "West Africa's Worst Year of Famine," *African Affairs* 83, no. 332 (1984): 281–99; and Richard W. Franke and Barbara Chasin, *Seeds of Famine: Ecological Destruction and the Development Dilemma in the West African Sahel* (Montclair, NJ: Allanheld, Osmun, 1980).

9. Togola, "Memories, Abstractions, and Conceptualization," 187.

10. Togola, 187.

11. Neighboring Niger and Burkina Faso similarly approached the goal of food security as one of national development and sovereignty. See, for example, Benedetta Rossi, *From Slavery to Aid: Politics, Labour, and Ecology in the Nigerien Sahel, 1800–2000* (Cambridge: Cambridge University Press, 2015), 238–51. I thank Gregory Mann for pushing me to think more critically about the terminology of "food security."

12. OPAM stands for Office des produits d'agricole du Mali. For more on the history of socialism in Mali, see Daouda Gary-Tounkara, "Quand les migrants demandent la route, Modibo Keita rétorque: 'Retournez à la terre!' Les 'Baragnini' et la désertion du 'Chantier national' (1958–1968)," *Mande Studies* 5 (2003): 49–64; and Guy Martin, "Socialism, Economic Development and Planning in Mali, 1960–1968," *Canadian Journal of African Studies* 10, no. 1 (1976): 23–47.

13. Mme. Dagno Adam Bah, interview by author, Markala-Diamarabougou, January 25, 2010.
14. Mme. Dagno Adam Bah, interview, January 25, 2010.
15. Mme. Dagno Adam Bah, interview, January 25, 2010. During the same interview Adam stressed that she knew the policies were inspired by Mao in China (rather than the Soviet Union), suggesting how Keita's socialist policies were popularly interpreted.
16. Martin, "Socialism, Economic Development and Planning." The nationalist political party that took control of the government in the 1960s (US-RDA) promoted collective work groups and other socialist-leaning initiatives as early as the 1950s. See also Monica M. van Beusekom, "Individualism, Community, and Cooperatives in the Development Thinking of the Union Soudanaise-RDA, 1946–1960," *African Studies Review* 51, no. 2 (2008): 6–7.
17. Jean Marie Kohler, *Les Mosi de Kolongotomo et la collectivisation à l'Office du Niger: Notes sociologiques*, Travaux et documents de l'ORSTOM no. 37 (Paris: ORSTOM, 1947), 46.
18. Moussa Diawara, Aminata Tangaré, and Hawoyi Diawara, interview by author, Boky-Were, May 7, 2010. On the Economic Police in the 1970s, see also Chéibane Coulibaly, *Politiques agricoles et stratégies paysannes au Mali de 1910 à 2010: Mythes et réalités à l'Office du Niger*, 2nd ed. (Paris: L'Harmattan Mali, 2014), 111–12.
19. Kono Dieunta, interview by author, Kokry, May 3, 2010.
20. "L'Office du Niger: Instrument important de la réalisation du plan quinquennal," *L'Essor*, 1962 (specific publication date illegible). This article was accessed in the 1962 collection of issues from *L'Essor* in the National Archives of Mali. However, the 1962 volume is in poor condition, and some articles lack specific date references (The newspaper archive is only available in paper format.) An announcement in the January 13, 1962, paper issued the call for new farmers to request entry at the Office. *L'Essor* was first published in 1949 and was associated with the pro-labor and anticolonial political party US-RDA. It continued to operate in the 1960s and following the coup in 1968 became the official state-run newspaper.
21. Former colonial-era French experts such as Maurice Rossin and Claude de Caso continued to advise the Office administration through international research and aid channels.
22. Directives pour la culture intensive du coton campagne 1960–1961, unnumbered document, AON 5–13; Djibril Aw, Contribution à la préparation du plan de redressement et de l'assainissement de l'economie nationale, March 14, 1966, AON 61/6.
23. Indeed, farmers continued to refuse to grow cotton well into the 1960s. Aw, Contribution à la préparation. The heightened national emphasis for rice production was partly a consequence of the closing of diplomatic relations with Senegal in 1960. During the colonial era, Senegal had been

the biggest purchaser of Soudanese and Office rice. In fact, the Office suffered significant financial losses when Mali could no longer collect debts on rice for that had already been delivered to Senegal.

24. Director General to the Ministre des Finances, September 30, 1960, Correspondances directeur general 1957–1960, AON 106; Situation de l'Office du Niger au moment du transfer des pouvoirs de la République française à la République du Mali, December 1960, AON 61/7. Just one year prior, sales of rice to Senegal publicly figured in reporting on relations between the two colonies and their future relations under independent rule. "Perspective de l'Office du Niger," *L'Essor*, June 12, 1959. On the intensification of rice, see also Coulibaly, *Politiques agricoles et stratégies paysannes*, 101–13.

25. Note sur l'etat du colonat par Remy Madier, November 1960, AON 132/1. Initially, the colonial government had also held a monopoly over Office crops. Under pressure to reform in the late 1950s, the colonial administration had liberalized agricultural markets. At that point, farmers were free to sell their harvest on the free market after paying their water and service fees. In 1960, the Keita government reinstated the government monopoly in order to direct the rice harvest to national urban markets and selected export markets.

26. The CFA or West African franc is based on the former French colonial currency in the region, which in Mali was replaced by the Malian franc between 1962 and 1984.

27. Emil Schreyger, *L'Office du Niger au Mali 1932 à 1982: La problématique d'une grande entreprise agricole dans la zone du Sahel* (Paris: L'Harmattan, 1984), 275. Schreyger cites a report he found in the Office du Niger Archives: "Enquête budgétaire dans le delta central nigérien-zone inondée," pp. 61–63, Office du Niger, mission socio-économique, Paris, January 1961.

28. The author witnessed the same price disparity for basic goods while researching in the Office region. Basic foodstuffs as well as other goods like soap were priced significantly higher in Office markets than in Ségou and rural towns to the south of the Office.

29. The Traoré government also established several other rice production programs including Riz Ségou and Riz Mopti. See ADRAO (Association for the Development of Rice Growing in West Africa), "Etude prospective de l'intensification de la riziculture à l'Office du Niger (République du Mali)," p. 5, report, 1974, AON 23 bis. For more on urban consumption of rice and its cultural significance, see Dolores Koenig, "Food for the Malian Middle Class: An Invisible Cuisine," in *Fast Food/Slow Food: The Cultural Economy of the Global Food System*, ed. Richard Wilk (Lanham, MD: Altamira, 2006), 49–67.

30. At the same time, Mali continued to export some of its rice harvest up through 1964. See ADRAO, "Etude prospective," 5.

31. *Mali Office du Niger rapport d'identification* (Washington, DC: World Bank, 1978), AON 227/2. This report was prepared for the working group Cultures Irriguées of the CILSS of the meeting held September 19–22, 1978, in Dakar; See also Schreyger, *L'Office du Niger au Mali*, 287; and Salifou Bakary Diarra, John M. Staatz, and Niama Nango Dembélé, "The Reform of Rice Milling and Marketing in the Office Du Niger: Catalysts for an Agricultural Success Story in Mali," in *Democracy and Development in Mali*, ed. R. James Bingen, David Robinson, and John M. Staatz (East Lansing: Michigan State University Press, 2000).

32. Lucy E. Creevey, "The Role of Women in Malian Agriculture," in *Women Farmers in Africa: Rural Development in Mali and the Sahel*, ed. Lucy E. Creevey (Syracuse, NY: Syracuse University Press, 1986), 54; Schreyger, *L'Office du Niger au Mali*, 284.

33. ADRAO, "Etude prospective."

34. Mariam Koné, a former clerical worker for the Office du Niger, recalled that wage workers received rice for half price during the 1970s and early 1980s. Mme. Koné Mariam Diarra and Tchaka Diallo, interview by author, Markala-Diamarabougou, January 26, 2010.

35. Marcel Mauss, "Techniques of the Body," in *Incorporations*, ed. Jonathan Crary and Sanford Kwinter (New York: Zone, 1992).

36. I also draw from the analysis of gender and embodiment by Rosemary Joyce. See Rosemary A. Joyce, "Feminist Theories of Embodiment and Anthropological Imagination: Making Bodies Matter," in *Feminist Anthropology: Past, Present, and Future*, ed. Pamela L. Geller and Miranda K. Stockett (Philadelphia: University of Pennsylvania Press, 2006).

37. Recently, Joy Parr called on historians of the environment and technology to consider the intersection of these two fields from a gendered perspective. She revisits questions related to how people experienced their surroundings as the techno-nature setting changed. Her key concern in these histories is the body. I draw on her discussion of the senses in this regard. Joy Parr, "Our Bodies and Our Histories of Technology and the Environment," in *The Illusory Boundary: Environment and Technology in History*, ed. Martin Reuss and Stephen H. Cutcliff (Charlottesville: University of Virginia Press, 2010).

38. The subject of women hiding rice is also discussed by Office interviewees featured in a film produced by the French agency Institut de Recherches et d'Application des Méthodes de Développement (IRAM). See *L'Office du Niger: Du travailleur forcé au paysan syndiqué*, directed by Loïc Colin and Vincent Petit (Paris: IRAM, 2008), DVD.

39. Nièni Tangara, interviews by author, Kolony (km 26), March 31, 2010, and April 1, 2010; and Fati Kindo, interview by author, Kossouka, May 11, 2010.

40. Nièni Tangara, interviews, March 31, 2010, and April 1, 2010.

41. Nièni Tangara, interviews, March 31, 2010, and April 1, 2010.

42. Hawa Coulibaly, interview by author, Fouabougou, March 30, 2010.

43. Assane Coulibaly, interview by author, Sirakoro, April 28, 2010.
44. Correspondance directeur general 1957–1960, AON 106.
45. Djewari Samaké, interview by author, Kolony (km 26), March 29, 2010.
46. Oumou Dembélé, interview by author, Kouyan-N'Goloba, April 13, 2010.
47. This is a point made in the report supervised by Madame Correze for the Ministry of Agriculture and supporting agricultural development and research institutions. See Institut de Recherches et d'Application de Methodes de Developpement (IRAM), *Office du Niger: L'organisation collective des paysans, la situation des femmes* (Paris: IRAM, 1981), 32.
48. Aramata Diarra, interview by author, Fouabougou, March 30, 2010.
49. Gleaning is the practice of gathering leftover grains or other agricultural products from the fields after the harvest.
50. The anthropologist Maria Grosz-Ngaté noted that women in the region regularly took small amounts of grain for themselves when they winnowed the cut grains from the household fields. All of this harvest was formally controlled by the male head of the household. Women were able to take these grains because no men were allowed in the area where women winnowed. Senior men generally tolerated this practice as long as the amount taken was not excessive. See Maria Grosz-Ngaté, "Hidden Meanings: Explorations into a Bamanan Construct of Gender," *Ethnology* 28, no. 2 (1989): 177.
51. Grosz-Ngaté, "Hidden Meanings," 176.
52. IRAM, *Office du Niger*, 30, 35–36.
53. This point has also been made by scholars studying the Malian grain market. See Diarra, Staatz, and Dembélé, "Reform of Rice Milling and Marketing."
54. A 1981 report specified that this was the case for the years 1968–69 to 1978–79. See IRAM, *Office Du Niger*, 29.
55. IRAM, *Office du Niger*, 29. See also Coulibaly, *Politiques agricoles et stratégies paysannes*, 112–13.
56. Assane Coulibaly, interview, April 28, 2010.
57. Comment from Assane Pléah in Aïssata Mallé et al., interview by author, Kouyan-N'Péguèna, April 12, 2010.
58. Comment from Mamadou "Seyba" Coulibaly in Mamadou "Seyba" Coulibaly and Soumaïlla Diao, interview by author, Kolony (km 26), March 26, 2010.
59. Several other interviewees remember the requirement to show a permission slip to take rice; they also recall the labor of hand threshing the ration. Sékou Coulibaly and Kadiatou Traoré, interview by author, Nyamina, May 31, 2010; Alimata Dembélé, Moussa Coulibaly, and Nana Dembélé, interview, May 3, 2010; and Hawa Diarra, interview by author, Nara, April 30, 2010.
60. Plan de battage et commercialisation campagne 1982–1983, unpublished administrative review, January 1983, AON 237/8. In the Niono sector, men

similarly remembered four guards being stationed in their towns. See Mamadou "Seyba" Coulibaly and Soumaïlla Diao, interview, March 26, 2010.

61. Plan de campagne rizicole 1982–1983, unpublished planning document, AON 237/5.
62. Plan de campagne rizicole 1982–1983.
63. Oumou Dembélé, interview, April 13, 2010.
64. Fatoumata Coulibaly and Oumou Sow, interview by author, Sabula, May 10, 2010.
65. Djenebu Coulibaly, interview by author, Kolony (km 26), March 27, 2010.
66. Alimata Dembélé, Moussa Coulibaly, and Nana Dembélé, interview, May 3, 2010.
67. Sékou Coulibaly and Kadiatou Traoré, interview, May 31, 2010.
68. Comments from Mamadou "Seyba" Coulibaly in Mamadou "Seyba" Coulibaly and Soumaïlla Diao, interview, March 26, 2010; comment from Mamadou "Seyba" Coulibaly in Fatouma Coulibaly, interview by author, Kolony (km 26), March 24, 2010. Comments from Moussa Coulibaly in Alimata Dembélé, Moussa Coulibaly, and Nana Dembélé, interview, May 3, 2010; Nianzon Bouaré and Harouna Bouaré, interview by author, Molodo-Bamana, April 16, 2010.
69. Amadou Sow and Sekou Salla Ouloguem, interview by author, Molodo-Centre, April 8, 2010.
70. Kadja Coulibaly, interview by author, Kokry, May 2, 2010.
71. Mamadou "Seyba" Coulibaly and Soumaïlla Diao, interview, March 26, 2010; Nianzon Bouaré and Harouna Bouaré, interview, April 16, 2010.
72. Fatoumata Coulibaly, Kadiatou Mallé, and Maïssa Sountura, interview, April 14, 2010.
73. Mali Office du Niger rapport d'identification.
74. A. Ouattara, "Réflexions sur le réamenagment à l'Office du Niger: Historique de la mécanisation," unpublished report, July 1972, AON 52/5. See also Schreyger, L'Office du Niger au Mali, 338.
75. Djenebu Coulibaly and Djewari Samaké, interview by author, Kolony (km 26), March 22, 2010; and Hawa Coulibaly, interview, March 30, 2010.
76. Schreyger, L'Office du Niger au Mali, 262.
77. Note sur l'etat du colonat par Remy Madier; ADRAO, "Etude prospective."
78. Fatoumata Coulibaly, Kadiatou Mallé, and Maïssa Sountura, interview, April 14, 2010; Oumou Dembélé, interview, April 13, 2010; Hawa Diarra, interview, April 30, 2010; and Mamadou "Seyba" Coulibaly and Soumaïlla Diao, interview, March 26, 2010.
79. Comment from Seyba in Mamadou "Seyba" Coulibaly and Soumaïlla Diao, interview, March 26, 2010.
80. The same report suggested that black market sales of rice directed cereals grown at the Office outside of Mali to Mauritania. See Rapport d'activité de l'Office du Niger, internal report, 1964–1965, AON 67/1.

81. Harvesting and threshing machines in Niono were notably in poor condition as early as the 1968–69 campaign. See Rapport d'activité de l'Office du Niger au cours des trois derniers campagnes 68/69–70/71, p. 5, internal report, AON 67/1. See also "Conférence régionale des cadres rapport Office du Niger," p. 4, March 4, 1971, AON 67/2.

82. Djenebu Coulibaly and Djewari Samaké, interview, March 22, 2010; and Hawa Coulibaly, interview, March 30, 2010.

83. "Principaux problèmes de l'Office du Niger," October 12, 1972, AON 61/1.

84. Résultat de la production et de la collecte campagne: 1975–1976, AON 70/4.

85. Victor Douyon, "Rapport circonstancie portant sur quelques cas de Beri-Beri apparu à SABULA U.P. 1 secteur agricole Kolongotomo," August 30, 1981, AON 247/8.

86. David McCandless, "Beriberi," in *Thiamine Deficiency and Associated Clinical Disorders* (New York: Humana, 2010). See also Dana Simmons, "Starvation Science: From Colonies to Metropole," in *Food and Globalization: Consumption, Markets and Politics in the Modern World*, ed. Alexander Nützenadel and Frank Trentmann (New York: Berg, 2008).

87. Douyon, "Rapport circonstancie."

88. Douyon.

89. Kathleen Cloud, "Sex Roles in Food Production and Distribution Systems in the Sahel," in *Women Farmers in Africa: Rural Development in Mali and the Sahel*, ed. Lucy E. Creevey (Syracuse, NY: Syracuse University Press, 1986), 45; and Mariam Thiam, "The Role of Women in Rural Development in the Segou Region of Mali," in *Women Farmers in Africa: Rural Development in Mali and the Sahel*, ed. Lucy E. Creevey (Syracuse, NY: Syracuse University Press, 1986), 78.

90. For an elaboration of this point and discussions over appropriate international intervention, see Vincent Bonnecase, *La pauvreté au Sahel: Du savoir colonial à la mesure internationale* (Paris: Éditions Karthala, 2011), see especially chapter 5; and Gregory Mann, *From Empires to NGOs in the West African Sahel* (Cambridge: Cambridge University Press, 2015), chapter 5.

91. Some observers like René Dumont argued that such policies were rooted in the colonial era, which created a policy dynamic that led to severe economic inequalities in countries like Mali. Dumont was a former colonial agricultural specialist who had even advised the Office. René Dumont and Charlotte Paquet, *Pour l'Afrique, j'accuse: Le journal d'un agronome au Sahel en voie de destruction* (Paris: Plon, 1986). See also Amartya Sen, *Poverty and Famines: An Essay on Entitlement and Deprivation* (Oxford: Clarendon, 1981); Sara Berry, *No Condition Is Permanent: The Social Dynamics of Agrarian Change in Sub-Saharan Africa* (Madison: University of Wiconsin Press, 1993); Megan Vaughan, *The Story of an African Famine: Gender and Famine in Twentieth-Century Malawi* (Cambridge:

Cambridge University Press, 1987); and Michael Watts, *Silent Violence: Food, Famine and Peasantry in Northern Nigeria* (Berkeley: University of California Press, 1983).

92. Emergency operations, West Africa 1983–1985, annex I, in *Summary Evaluation Report on WFP Emergency Operations in West Africa, 1983–85 (Burkina Faso, Mali, Chad)*, vol. 3, CFA General 1986–1987, Food and Agricultural Organization Archives (hereafter FAO) RG 16 WFP 1/37. See also Derrick, "West Africa's Worst Year of Famine."

93. *Summary Evaluation Report*, 12, 14.

94. *Summary Evaluation Report*, 17.

95. Victor Douyon, "Compte rendu de mission effectuée par le chef du Bureau paysannat de l'Office du Niger," June 26, 1985, AON 410/7.

96. It is likely that the decision for the young men to construct the communities' lodgings was an attempt to allow the group to live in Tuareg-style housing. Douyon, "Compte rendu de mission effectuée."

97. Jeremy Swift, "Desertification: Narratives, Winners and Losers," in *The Lie of the Land: Challenging Received Wisdom on the African Environment*, ed. Melissa Leach and Robin Mearns (Portsmouth, NH: Heinemann, 1996). Also see Sara Randall and Alessandra Giuffrida, "Forced Migration, Sedentarisation and Social Change: Malian Kel Tamasheq," in *Pastoralists of North Africa and the Middle East: Entering the 21st Century*, ed. Dawn Chatty (Leiden, The Netherlands: Brill, 2006).

98. Douyon, "Compte rendu de mission effectuée."

99. In recent years, scholars concerned with food security and malnutrition have since broadened their scope in analyzing food crises. According to these new metrics, the persistent hunger and malnutrition at the Office during the drought would have raised alarm bells. See Rosamond L. Naylor, ed., *The Evolving Sphere of Food Security* (Oxford: Oxford University Press, 2014). On this point, Chéibane Coulibaly previously highlighted the lack of food as a key aspect of life at the Office in the 1980s. See Coulibaly, *Politiques agricoles et stratégies paysannes*, 123–24.

100. In the 1974 meeting in Bamako, Moussa Traoré refuted claims that Sahelian governments (and perhaps Mali in particular) were anything more than simply indifferent to the nomadic victims of famine, rather than guilty of inaction that could be characterized as genocidal. He also objected to the fact that observers said they only acted on international pressure. Clearly the nations saw their sovereignty wrapped up in their ability to respond to the crisis. See Traoré, "Allocution du colonel Moussa Traoré," 2–3.

101. World Food Aid donations for the Office were shipped first to the project's administrative center in Ségou before being distributed to each sector. See Mariam Diarra and Tchaka Diallo, interview, January 26, 2010.

102. Alimata Dembélé, Moussa Coulibaly, and Nana Dembélé, interview, May 3, 2010.

103. Rossi, *From Slavery to Aid*, 272–95.
104. Executive director's note in response to lessons learned from the African food crisis: evaluation of the WFP emergency response, CFA General 1986–1987, vol. 3, FAO RG 16 WFP 1/37.
105. *Summary Evaluation Report*, 19. At the time the World Food Program discouraged development projects along these lines unless they also offered technical advice for production, as well as access to tools and credit. On this point see page 12 of the report.
106. A diet of rice alone did not provide protein—it led to kwashiorkor—nor did it provide B vitamins, a lack of which led to beriberi. The team further diagnosed rampant impetigo, tetanus, bilharzia, and conjunctivitis among residents. See Rapport de mission 7 Avril 1981 Jeannette Sidibé et Madame Bolt, Bureau du paysannat, AON 247/31.
107. As Diana Wylie observes for South Africa, the line between malnutrition and starvation is often politically defined. At the same time, the effects of malnutrition can be severe but easily distanced from political causes. See Diana Wylie, *Starving on a Full Stomach: Hunger and the Triumph of Cultural Racism in Modern South Africa* (Charlottesville: University Press of Virginia, 2001).
108. Notes relatives au rapport "Organisation collective des paysans" de Madame Correze la commission 1981 (23 Jan), Bureau du paysannat, AON 247/4. As of 1984, the Malian government had plans to reform OPAM. On the slow distribution of supplies also see Derrick, "West Africa's Worst Year of Famine," 290.
109. Victor Douyon, Répartition de 30 tonnes sucre en poudre (600 sacs de 50kg) entre unités coopératives des secteurs rizicoles de l'Office du Niger, December 16, 1980, Bureau du paysannat, AON 247/28.
110. Friends of the Sahel, *Propositions pour une stratégie et un programme de lutte contre la sécheresse et de développement dans le Sahel: Rapport de synthèse* (Ouagadougou, Burkina Faso: CILSS/Club du Sahel, 1977), 95.
111. Comment from Aïssata Mallé in Aïssata Mallé et al., interview, April 12, 2010.
112. Comment from Oumou Sow in Fatoumata Coulibaly and Oumou Sow, interview, May 10, 2010.
113. Food aid to Mali as a whole had begun as early as 1972. One example of such aid was the donation of ten thousand tons of millet from the United States, which was reported in the national newspaper. See G. Dolo, "Les Etats-Unis font un don de cereales d'une valeur de 500 millions de francs au Mali," *L'Essor*, February 11, 1972.
114. Situation des DONS PAM reçus par l'Office du Niger, November 15, 1984, Bureau paysannat, AON 410/10.
115. Victor Douyon, Détermination des prix unitaires des différentes DONS du Programme alimentaire mondial, February 23, 1981, Bureau du paysannat, AON 247/30.

116. From the available documentation, new families received aid most likely for one to two years, as had been the practice in the colonial era.
117. In the last recorded year for World Food donations (1986–87) the numbers had fallen to only 252 families still receiving assistance. The records for the distribution of these donations are recorded in several folders holding material for Bureau du paysannat at the Office du Niger (also marked Bureau paysannat). See Repartition des dons PAM entre les nouveaux colons des secteurs rizicoles de l'Office du Niger, September 23, 1983, August 23, 1983, and October 12, 1983, Bureau du paysannat, AON 247/17c; Repartition des dons du Programme alimentaire mondial (PAM) entre les nouvelles familles colons des secteurs rizicoles de l'Office du Niger, November 12, 1984, Bureau paysannat, no. 49/BP, AON 410/3; Repartition des dons du Programme alimentaire mondial (PAM), June–September 1985, Bureau du paysannat, nos. 117/BP and 220/BP, AON 410/9; Repartition des dons du Programme alimentaire mondial (PAM) du mois de Juin 1985 entre les colons des secteurs rizicoles de l'O.N., June 1985, Bureau du paysannat, no. 131/BP, AON 410/7; Repartitions des dons PAM des mois d'Octobre, Novembre et Decembre 1985 aux nouvelles familles colons des secteurs rizicoles de l'O.N., October 14, 1985 (October to December 1985), Bureau paysannat, AON 410/3; and Distribution des dons du Programme alimentaire mondial (PAM) aux nouvelles familles colons de l'Office du Niger, August 25, 1987, Bureau du paysannat, no. 39/BP, AON 426/24.
118. Direction general décision no. 006/DG portant réorganisation du Bureau paysannat de l'Office du Niger, March 9, 1983, PAM (World Food Program), AON 248/19.
119. Funds from United States and the European countries sometimes facilitated the purchase of corn or other grains from other countries, including one purchase for ten thousand tons of corn from Honduras destined for Mali between 1984 and 1985. The WFP also purchased some grains in Niger to be transported to Burkina Faso that same year. However, local WFP officials were limited in how much aid they were able to purchase regionally. *Summary Evaluation Report*, 21; see also "Emergency Operations, West Africa 1983–1985."
120. Office du Niger records relating to the Bureau du paysannat (Bureau paysannat) in AON 247 and AON 410 include several records for World Food Program donations to the Office du Niger.
121. One recent example of this analysis addresses medical assistance. See Peter Redfield, "Cleaning Up the Cold War: Global Humanitarianism and the Infrastructure of Crisis Response," in *Entangled Geographies: Empire and Technopolitics in the Global Cold War*, ed. Gabrielle Hecht (Cambridge, MA: MIT Press, 2011), 271–86.
122. Mme. Thiam Mariam N'Diaye, interview by author, Bamako, February 1, 2010.

123. Mme. Thiam Mariam N'Diaye, interview, February 1, 2010. The Markala Women's Cooperative was established in 1975. See Susan Caughman and Mariam N'Diaye Thiam, *The Markala Cooperative: A New Approach to Traditional Economic Roles*, Seeds Pamphlet Series no. 5 (New York: Population Council, 1982).

124. Alex de Waal and other scholars of famine have made this broad point for food aid more generally to Africa and other regions. See, for example, Alex de Waal, *Famine Crimes: Politics and the Disaster Relief Industry in Africa*, African Issues (Bloomington: Indiana University Press, 1997).

125. Repartition des DONS PAM entre les nouveaux colons des secteurs rizicoles de l'Office du Niger, September 23, 1983, Bureau du paysannat, AON 247.

126. Repartition des DONS PAM des mois d'Octobre, Novembre, et Decembre 1985 aux nouvelles familles colons des secteurs rizicoles de l'O.N., October 14, 1987, Bureau paysannat, no. 239/BP, AON 410/3.

127. *Summary Evaluation Report*, 8.

128. Repartition des DONS PAM du mois de Juin 1986 aux nouvelles familles colons des secteurs rizicoles de l'Office du Niger, Bureau paysannat, AON 410/16; Distribution des dons du Programme alimentaire mondial (PAM) aux nouvelles familles colons de l'Office du Niger 25 August 1987, July-August-September 1987, Bureau paysannat suite, AON 426/24; and Distribution des dons du Programme alimentaire mondial (PAM) aux nouvelles familles colons de l'Office du Niger, 3e trimestre (Octobre-Novembre-Decembre), December 13, 1987, Bureau paysannat suite, AON 426/25.

129. *Summary Evaluation Report*, 12 (emphasis in the original report).

130. *Summary Evaluation Report*, 14.

131. Douyon, "Compte rendu de mission effectuée."

132. Drissa N. Keita, Compte rendu de mission du paysannat (conjointe Banque nationale de developpements agricole BNDA) 23 au 29 Septembre 1985, Bureau paysannat, AON 410/11.

133. Her husband could not be interviewed because he is deceased. See Kadja Coulibaly, interview, May 2, 2010.

134. The recollection of workers in Markala was that food aid products were offered to them at reduced prices. Mariam Diarra and Tchaka Diallo, interview, January 26, 2010.

135. Moussa Diawara, Aminata Tangaré, and Hawoyi Diawara, interview, May 7, 2010.

136. Fatoumata Coulibaly and Oumou Sow, interview, May 10, 2010; Oumou Dembélé, interview, April 13, 2010; and Assane Coulibaly, interview, April 28, 2010.

137. Comment from Sekou Salla Ouleguem in Bintu Traoré, Mariam Doumbia, and Fanta Sogoba, interview by author, Molodo-Centre, April 15, 2010. Sekou even blamed a cholera outbreak during the period on poor-quality red millet.

138. Fatoumata Coulibaly, Kadiatou Mallé, and Maïssa Sountura, interview, April 14, 2010.
139. Fatoumata Coulibaly and Oumou Sow, interview, May 10, 2010.
140. Moussa Diawara, Aminata Tangaré, and Hawoyi Diawara, interview, May 7, 2010.
141. Moussa Diawara, Aminata Tangaré, and Hawoyi Diawara, interview, May 7, 2010; and Fatoumata Coulibaly, Kadiatou Mallé, and Maïssa Sountura, interview, April 14, 2010.
142. Comment by Sekou Salla Ouloguem in Bintu Traoré, Mariam Doumbia, and Fanta Sogoba, interview, April 15, 2010.
143. Peter Redfield discusses what he calls the "kit culture" of medical humanitarian aid, which came out of the medical kits assembled by the humanitarian group Doctors without Borders (Médecins Sans Frontièrs). With the delivery of these kits, which included basic medical supplies assembled outside the area of intervention, the group drew away from the original impulse of many doctors to find local solutions to medical crises. The logic applies here in that the standard basic foodstuffs in and of themselves did little more than prevent outright starvation. See Redfield, "Cleaning Up the Cold War," 284.
144. Assane Coulibaly, interview, April 28, 2010.
145. Dossier mil, AON 87; Guy Roberty, "Introduction de plantes à Soninkoura, cultures vivrières, mils indigènes," 1939, AON 85/9.
146. Derrick, "West Africa's Worst Year of Famine," 283.
147. Comment by Bintou Dieunta in Assane Coulibaly, interview, April 28, 2010.
148. Comment by Sekou Salla Ouloguem in Bintu Traoré, Mariam Doumbia, and Fanta Sogoba, interview, April 15, 2010.
149. Comments by Oumou Sow in Fatoumata Coulibaly and Oumou Sow, interview, May 10, 2010.
150. This type of millet was classified by the colonial botanist Guy Roberty in his studies of local plants. Dossier mil, AON 87; Roberty, "Introduction de plantes à Soninkoura." He lists this millet separately from the type of red millet grown in fishing communities.
151. Comment from Aïssata Mallé in Aïssata Mallé et al., interview, April 12, 2010.
152. Moussa Travélé, *Proverbes et contes bambara: Accompagnés d'une traduction française et précédés d'un abrégé de droit coutumier* (Paris: Librarie Orientaliste Paul Geuthner, 1923), 86.
153. Fatoumata Guindo, who worked for the Office as a women's development coordinator, explained that even poor women created social ties and cultivated respect by sharing with others, including with well to do women civil servants like herself. Mme. Tamboura Fatoumata Guindo, interview by author, Niono, April 16, 2010.
154. Comment from Aïssata Mallé in Aïssata Mallé et al., interview, April 12, 2010.

155. Thiam, "The Role of Women in Rural Development," 76–77. I retain Thiam's orthography for *toh* in the quote.
156. Studies of migration during the drought of 1983–85 documented the particular increase in the number of female (and children) migrants. See Sally E. Findley, "Does Drought Increase Migration? A Study of Migration from Rural Mali during the 1983–1985 Drought," *International Migration Review* 28, no. 3 (1994): 539–53.
157. Some observers pointed to the inefficiency of the national marketing board OPAM as a reason for the ongoing troubles of the agricultural economy. See Diarra, Staatz, and Dembélé, "Reform of Rice Milling and Marketing."
158. "Conférence spéciale consacrée aux problèmes de l'Office du Niger," pp. 4, 19, 1979, Office du Niger, L'institut d'economie rural (hereafter IER), BKO CO 1694; and Recensement annuel population colons au 30 Juin 1981, Bureau du paysannat, AON 247/5.
159. Comment from Tafron Dembélé in Assane Coulibaly, interview, April 28, 2010.
160. Baba Djiguiba, interview by author, Molodo-Centre, April 10, 2010. Amadou Sow and Sekou Salla Ouloguem, interview, April 8, 2010. Also see Schreyger, *L'Office du Niger au Mali,* 322.
161. Fatoumata Coulibaly, Kadiatou Mallé, and Maïssa Sountura, interview, April 14, 2010.
162. Hawa Diarra, interview, April 30, 2010.
163. Oumou Dembélé, interview, April 13, 2010.
164. Comment by Sekou Sall Ouloguem in Amadou Sow and Sekou Sall Ouloguem, interview, April 8, 2010.
165. Mariam Diarra and Tchaka Diallo, interview, January 26, 2010. Fatoumata Coulibaly, Kadiatou Mallé, and Maïssa Sountura, interview, April 14, 2010. Hawa Diarra, interview, April 30, 2010.
166. Mme. Thiam Mariam N'Diaye, interview, February 1, 2010.
167. Moussa Diawara, Aminata Tangaré, and Hawoyi Diawara, interview, May 7, 2010.
168. It was not clear to me who sold the rice, it could have been individual staff members or grain merchants working on behalf of OPAM. Nianzon Bouaré and Harouna Bouaré, interview, April 16, 2010.
169. IRAM, *Office du Niger,* 28; Schreyger, *L'Office du Niger au Mali,* 322; Amadou Sow and Sekou Sall Ouloguem, interview, April 8, 2010.
170. Ségou entry, October 1, 1927, diaire 1927, 2 vols. (région du Soudan), Fonds des Soeurs Blancs (FSB), MAFR. This specific observation was recorded near the mission in Ségou.

CONCLUSION

1. Mme. Koumba Djeneba "Ba Djeneba" Diarra, interview by author, Molodo-Centre, April 19, 2010. Diarra was director of the Molodo zone.

2. Richard Schroeder offers an excellent overview of the emergence of women and development in the introduction to his book on women and environmental development in the region. See Richard A. Schroeder, *Shady Practices: Agroforestry and Gender Politics in The Gambia* (Berkeley: University of California Press, 1999).
3. Mme. Cissé Oumou Koné, interview by author, Alphalog offices (Niono), March 15, 2010; Mme. Dagnoko Bintu Kané, interview by author, Markala-Diamarabougou, March 13, 2010; Mme. Tamboura Fatoumata Guindo, interview by author, Niono, April 16, 2010. Both Mme. Dagnoko and Mme. Tamboura were employed as women's development workers for the Office in the late 1980s and were still working as an "animatrices" in 2010. See Mme. Koumba Djeneba "Ba Djeneba" Diarra, interview, April 19, 2010; Djenebu Coulibaly, interview by author, Kolony (km 26), August 7, 2010. Informal discussion during a visit to the Benkadi women's group with Mme. Dagnoko Bintu Kané, Heremakono, February 4, 2010.
4. Cheick Koné, interview by author, Niono Placi market (Bamako), December 31, 2009.
5. Djenebu Coulibaly, interview, August 7, 2010; informal discussion with Kané and the Benkadi women's group members.
6. Birama Diakon, *Office du Niger et pratiques paysannes: Appropriation technologique et dynamique sociale* (Paris: L'Harmattan Mali, 2012), 111–12.
7. Djenebu Coulibaly and Djewari Samaké, interview by author, Kolony (km 26), March 22, 2010; Aramata Diarra, interview by author, Fouabougou, March 30, 2010; and Nièni Tangara, interview by author, Kolony (km 26), March 31, 2010.
8. Fatouma Coulibaly and Nièni Tangara, interview by author, Kolony (km 26), May 25, 2010.

Bibliography

INTERVIEWS

All interviews were conducted by the author.

Cheick Koné, Niono Placi market (Bamako), December 31, 2009
Bakary Marka Traoré, Markala-Diamarabougou, January 23, 2010
Mme. Dagno Adam Bah, Markala-Diamarabougou, January 25, 2010, and February 6, 2010
Mme. Koné Mariam Diarra and Tchaka Diallo, Markala-Diamarabougou, January 26, 2010
Tchaka Diallo, Markala-Diamarabougou, January 27, 2010, and March 10, 2010
Mme. Thiam Mariam N'Diaye, Bamako, February 1, 2010
Koké Coulibaly and Fodé Traoré, Markala-Kirango, February 3, 2010
Mme. Dagnoko Bintu Kané, Markala-Diamarabougou, March 13, 2010
Mme. Cissé Oumou Koné, Alphalog offices (Niono), March, 15, 2010
Djenebu Coulibaly and Djewari Samaké, Kolony (km 26), March 22, 2010
Fatouma Coulibaly, Kolony (km 26), March 24, 2010, and April 9, 2010
Mamadou "Seyba" Coulibaly and Soumaïlla Diao, Kolony (km 26), March 26, 2010
Djenebu Coulibaly, Kolony (km 26), March 27, 2010; April 9, 2010; and August 7, 2010
Djewari Samaké, Kolony (km 26), March 29, 2010; April 9, 2010; and June 2, 2010
Hawa Coulibaly, Fouabougou, March 30, 2010
Aramata Diarra, Fouabougou, March 30, 2010
Nièni Tangara, Kolony (km 26), March 31, 2010, and April 1, 2010
Mamadou "Seyba" Coulibaly, Kolony (km 26), April 1, 2010
Daouda Coulibaly and Almamy Thiènta, Sokolo, April 5, 2010
Amadou Sow and Sekou Salla Ouloguem, Molodo-Centre, April 8, 2010
Baba Djiguiba, Molodo-Centre, April 10, 2010
Aïssata Mallé, Salimata Samaké, Assan Pléah, Mariam Sall, and Sitan Mallé, Kouyan-N'Péguèna, April 12, 2010

Oumou Dembélé, Kouyan-N'Goloba, April 13, 2010
Fatoumata Coulibaly, Kadiatou Mallé, and Maïssa Sountura, Kouyan-Kura,
  April 14, 2010
Bintu Traoré, Mariam Doumbia, and Fanta Sogoba, Molodo-Centre, April 15,
  2010
Moctar Coulibaly, Molodo-Centre, April 15, 2010
Nianzon Bouaré and Harouna Bouaré, Molodo-Bamana, April 16, 2010
Mme. Tamboura Fatoumata Guindo, Niono, April 16, 2010
Mme. Koumba Djeneba "Ba Djeneba" Diarra, Molodo-Centre, April 19, 2010
Hawa Coulibaly, Mariam Sango, Djéné Mariko, Adam Coulibaly, and Nyé
  Diarra, Molodo-Bamana, April 19, 2010
Assane Traoré, Markala-Kirango, April 24, 2010
Assane Coulibaly, Sirakoro, April 28, 2010
Moctar Coulibaly, Mariatou Traoré, and Arouna Coulibaly, Sirakoro, April 28,
  2010
Rokia Diarra, Nara, April 29, 2010
Hawa Diarra, Nara, April 30, 2010
Kadja Coulibaly, Kokry, May 2, 2010
Alimata Dembélé, Moussa Coulibaly, and Nana Dembélé, Koutiala-Kura,
  May 3, 2010
Kono Dieunta, Kokry, May 3, 2010
Mariam "Mamu" Coulibaly, Kankan (formerly Sangarébougou), May 4, 2010
Wassa Dembélé and Kalifa Dembélé, San-Kura, May 5, 2010
Moussa Diawara, Aminata Tangaré, and Hawoyi Diawara, Boky-Were, May 7,
  2010
Fatoumata Coulibaly and Oumou Sow, Sabula, May 10, 2010
Fati Kindo, Kossouka, May 11, 2010
Bintou Diarra and Aïssata Coulibaly, Kolongotomo, May 11, 2010
Fatouma Coulibaly and Nièni Tangara, Kolony (km 26), May 25, 2010
Kadja "Ma" Coulibaly, Fatoumata Guindo, Fatoumata Zaré Coulibaly, and
  Lalafacouma Tangara, Koue-Bamana, May 26, 2010
Fatoumata Coulibaly, Kadja Mallé, Mariam Mallé, Aminata Dembélé, and
  Kadja Koné, Kouyan-Kura, May 27, 2010
Aïssata Mallé and Assane Pléah, Kouyan-N'Péguèna, May 28, 2010
Assane Pléah, Kouyan-N'Péguèna, May 28, 2010
Nianzon Bouaré and Hawa Coulibaly, Molodo-Bamana, May 29, 2010
Sékou Coulibaly and Kadiatou Traoré, Nyamina, May 31, 2010
Daouda Bouaré and Bintu Dembélé, Sokorani, June 1, 2010
Tiesson Dembélé, Bambougouji, June 3, 2010
Brahmin Fané, Madu Saré, Kadja Coumaré, Bading Traoré, and Issa Saré,
  Markala-Kirango, June 3, 2010
Mamadou Djiré, Niono market, August 7, 2010
Nièni Tangara and Mamadou "Seyba" Coulibaly, Kolony (km 26), August 7, 2010
Hawa Fomba, Kalaké-Bamana, September 11, 2010

National Archives of Mali, Bamako (ANM)

I consulted documents from the Fonds Anciens, Fonds Récents, and Fonds Numérique. In the Fonds Récents the agricultural series (Series 1R) contains extensive documentation for the Office du Niger. I also consulted the political records (Series 1E) for the Ségou region and areas of heavy Office recruitment. In addition, I consulted materials pertaining to agricultural policy and food production in the correspondence of the general administrative records (Series 1D) and the economic series (Series 1Q).

Office du Niger Archives, Ségou (AON)

The institutional archives of the Office du Niger contain extensive records relating to the technical and administrative functioning of the scheme. In addition, the archive holds botanical research materials pertaining to central and southern Mali for the early twentieth century.

Archives Nationales d'Outre Mer, Aix-en-Provence (ANOM)

The section of the French national archives dealing with its colonial territories contains extensive documentation for the Office du Niger up to 1961.

Food and Agricultural Organization Archives, Rome (FAO)

I consulted materials for the World Food Program dating to the 1970s and 1980s, as well as materials relating to international agricultural development projects, specifically related to rice production for the 1940s and 1950s.

Archives de la Société des Missionaires d'Afrique, Rome (MAFR)

The archive holds material for the White Father missionaries posted in the Ségou region dating to the 1890s, as well as the records for the mission at Kolongotomo in the Office du Niger. I consulted diaries for both missions, as well as the diaries of White Sisters posted in the French Soudan.

Schomburg Center for Research in Black Culture, New York City

While a fellow at the Schomburg Center I consulted several folktale collections for Mali and Sahelian West Africa, as well as materials related to international aid and the Great Sahel Drought.

L'Institut d'Economie Rural, Bamako (IER)

The library holds several unpublished reports relating to agricultural development at the Office du Niger.

Adas, Michael. *Machines as the Measure of Men: Science, Technology, and Ideologies of Western Dominance*. Ithaca, NY: Cornell University Press, 1989.

Akyeampong, Emmanuel Kwaku. *Between the Sea and the Lagoon: An Ecosocial History of the Anlo of Southeastern Ghana, c. 1850–Recent Times.* Athens: Ohio University Press, 2001.

———. *Drink, Power, and Cultural Change: A Social History of Alcohol in Ghana, C. 1800–Recent Times.* Portsmouth, NH: Heinemann, 1996.

Alaimo, Stacy. *Undomesticated Ground: Recasting Nature as Feminist Space.* Ithaca, NY: Cornell University Press, 2000.

Alaimo, Stacy, and Susan Hekman. "Introduction: Emerging Models of Materiality in Feminist Theory." In *Material Feminisms*, edited by Stacy Alaimo and Susan Hekman, 1–19. Bloomington: Indiana University Press, 2008.

Appadurai, Arjun, ed. *The Social Life of Things: Commodities in Cultural Perspective.* Cambridge: Cambridge University Press, 1986.

Arnold, David. "Europe, Technology, and Colonialism in the 20th Century." *History and Technology* 21, no. 1 (2005): 84–106.

Arnoldi, Mary Jo. "Wild Animals and Heroic Men: Visual and Verbal Arts in the Sogo Bo Masquerades of Mali." *Research in African Literatures* 31, no. 4 (2000): 63–75.

Arnoldi, Mary Jo, Christraud M. Geary, and Kris L. Hardin, eds. *African Material Culture.* Bloomington: Indiana University Press, 1996.

Austen, Ralph A. "The Problem of the Mande Creation Myth." In *Mande Mansa: Essays in Honor of David C. Conrad*, edited by Stephen Belcher, Jan Jansen, and Mohamed N'Daou, 31–39. Zürich: Lit, 2008.

Austen, Ralph A., and Daniel Headrick. "The Role of Technology in the African Past." *African Studies Review* 26, no. 3/4 (1983): 163–84.

Aw, Djibril, and Geert Diemer. *Making a Large Irrigation Scheme Work: A Case Study from Mali.* Directions in Development. Washington, DC: World Bank, 2005.

Ba Konaré, Adame. *Dictionnaire des femmes célèbres du Mali (des temps mythico-légendaires au 26 Mars 1991); précédé d'une analyse sur le rôle et l'image de la femme dans l'histoire du Mali.* Bamako, Mali: Éditions Jamana, 1993.

Barrett, Christopher B., and Daniel G. Maxwell. *Food Aid after Fifty Years: Recasting Its Role.* New York: Routledge, 2005.

Bastian, Misty L. "The Naked and the Nude: Historically Multiple Meanings of Oto (Undress) in Southeastern Nigeria." In *Dirt, Undress, and Difference: Critical Perspectives on the Body's Surface*, edited by Adeline Masquelier, 34–60. Bloomington: Indiana University Press, 2005.

Bazin, Jean. "A chacun son Bambara." In *Au coeur de l' ethnie: Ethnies, tribalism et État en Afrique*, edited by Jean-Loup Amselle and Elikia M'Bokolo, 87–125. Paris: Paris Éditions de la Découverte 1985.

———. "Genèse de l'état et formation d'un champ politique: Le royaume de Segu." *Revue Française de Science Politique* 38, no. 5 (1988): 709–19.

———. "Guerre et servitude à Ségou." In *L'esclavage en Afrique Précoloniale*, edited by Claude Meillassoux, 135–81. Paris: François Maspero, 1975.

———. "Princes désarmés, corps dangereux: Les 'rois-femmes' de la région de Segu." *Cahiers d'Études Africaines* 28, nos. 111–12 (1988): 375–441.

Becker, Laurence C. "Seeing Green in Mali's Woods: Colonial Legacy, Forest Use, and Local Control." *Annals of the Association of American Geographers* 91, no. 3 (2001): 504–26.

Belcher, Stephen. *Epic Traditions of Africa*. Bloomington: Indiana University Press, 1999.

Bélime, Emile, Auguste Chevalier, Yves Henry, and Fernand Abraham Bernard. *Les irrigations du Niger: Discussions et controverses*. Paris: Comité du Niger, 1923.

Benjaminsen, Tor A. "Natural Resource Management, Paradigm Shifts, and the Decentralization Reform in Mali." *Human Ecology* 25, no. 1 (1997): 121–43.

Berger, Iris. "Fertility as Power: Spirit Mediums, Priestesses and the Precolonial State in Interlacustrine East Africa." In *Revealing Prophets: Prophecy in Eastern African History*, edited by David M. Anderson and Douglas H. Johnson, 65–82. Athens: Ohio University Press, 1995.

Berry, Sara. *Fathers Work for Their Sons: Accumulation, Mobility, and Class Formation in an Extended Yorùbá Community*. Berkeley: University of California Press, 1985.

———. *No Condition Is Permanent: The Social Dynamics of Agrarian Change in Sub-Saharan Africa*. Madison: University of Wisconsin Press, 1993.

Bijker, Wiebe E., Thomas P. Hughes, and Trevor Pinch, eds. *The Social Construction of Technological Systems: New Directions in the Sociology of History and Technology*. Cambridge, MA: MIT Press, 2012.

Binger, Louis Gustave. *Du Niger au Golfe de Guinée par le pays de Kong et le Mossi*. 2 vols. Paris: Hachette, 1892.

Bird, Charles, and Mamadou Kanté. *Bambara-English, English-Bambara: Student Lexicon*. Bloomington: Indiana University Linguistics Club, 1977.

Boddy, Janice. *Wombs and Alien Spirits: Women, Men, and the Zar Cult in Northern Sudan*. Madison: University of Wisconsin Press, 1989.

Bogosian, Catherine. "Forced Labor, Resistance and Memory: The *Deuxième Portion* in the French Soudan, 1926–1950." PhD diss., University of Pennsylvania, 2002.

Bonnecase, Vincent. *La pauvreté au Sahel: Du savoir colonial à la mesure internationale*. Paris: Éditions Karthala, 2011.

———. "Le goût du riz: Une valeur sensorielle et politique au Burkina Faso." *Genèses* 104, no. 3 (2016): 7–29.

———. "When Numbers Represented Poverty: The Changing Meaning of the Food Ration in French Colonial Africa." *Journal of African History* 59, no. 3 (2018): 463–81.

Bonneuil, Christophe. "Development as Experiment: Science and State Building in Late Colonial and Postcolonial Africa, 1930–1970." *Osiris* 15 (2000): 258–81.

Boserup, Ester. *Woman's Role in Economic Development.* London: Earthscan, 1989.

Bourdieu, Pierre. *Distinction: A Social Critique of the Judgement of Taste.* Cambridge, MA: Harvard University Press, 1984.

Brasseur-Marion, Paule. "Pères blancs et Bambara: Une rencontre manquée?" *Mélanges de l'école française de Rome* 101, no. 2 (1989): 875–96.

Bray, Francesca. "Gender and Technology." *Annual Review of Anthropology* 36 (2007): 37–53.

———. *Technology and Gender: Fabrics of Power in Late Imperial China.* Berkeley: University of California Press, 1997.

Brett-Smith, Sarah C. *The Making of Bamana Sculpture: Creativity and Gender.* Cambridge: Cambridge University Press, 1994.

———. *The Silence of the Women: Bamana Mud Clothes.* Milan: 5 Continents Editions, 2014.

Broekhuyse, J. Th., and Andrea M. Allen. "Farming Systems Research on the Northern Mossi Plateau." *Human Organization* 74, no. 4 (Winter 1988): 330–42.

Brown, Marie Grace. *Khartoum at Night: Fashion and Body Politics in Imperial Sudan.* Stanford, CA: Stanford University Press, 2017.

Bryceson, Deborah Fahy. "Alcohol in Africa: Substance, Stimulus, and Society." In *Alcohol in Africa: Mixing Business, Pleasure, and Politics,* edited by Deborah Fahy Bryceson, 3–21. Portsmouth, NH: Heinemann, 2002.

———. "Changing Modalities of Alcohol Usage." In *Alcohol in Africa: Mixing Business, Pleasure, and Politics,* edited by Deborah Fahy Bryceson, 23–52. Portsmouth, NH: Heinemann, 2002.

Buhl, Solveig, and Katherine Homewood. "Milk Selling among Fulani Women in Northern Burkina Faso." In *Rethinking Pastoralism in Africa: Gender, Culture, and the Myth of the Patriarchal Pastoralist,* edited by Dorothy L. Hodgson, 207–26. Athens: Ohio University Press, 2000.

Burrell, Jenna. *Invisible Users: You in the Internet Cafés of Urban Ghana.* Cambridge, MA: MIT Press, 2012.

Burrill, Emily. *States of Marriage: Gender, Justice, and Rights in Colonial Mali.* Athens: Ohio University Press, 2015.

Caillié, René. *Journal d'un voyage à Tombouctou et à Jenné, dans l'Afrique centrale.* 2 vols. Paris: P. Mongie, 1830.

Carney, Judith A. *Black Rice: The Origins of Rice Cultivation in the Americas.* Cambridge, MA: Harvard University Press, 2001.

Carney, Judith A., and Richard Nicholas Rosomoff. *In the Shadow of Slavery: Africa's Botanical Legacy in the Atlantic World.* Berkeley: University of California Press, 2009.

Carney, Judith, and Michael Watts. "Disciplining Women? Rice, Mechanization, and the Evolution of Mandinka Gender Relations in Senegambia." *Signs: Journal of Women in Culture and Society* 16, no. 4 (1991): 651–81.

Caughman, Susan, and Mariam N'Diaye Thiam. *The Markala Cooperative: A New Approach to Traditional Economic Roles.* Seeds Pamphlet Series no. 5. New York: Population Council, 1982.

Chafer, Tony. *The End of Empire in French West Africa: France's Successful Decolonization?* New York: Berg, 2002.

Chalfin, Brenda. *Shea Butter Republic: State Power, Global Markets, and the Making of an Indigenous Commodity.* New York: Routledge, 2004.

Chirikure, Shadreck. "The Metalworker, the Potter, and the Pre-European African Laboratory." In *What Do Science, Technology, and Innovation Mean from Africa?*, edited by Clapperton Chakanetsa Mavhunga, 63–77. Cambridge, MA: MIT Press, 2017.

Clark, Andrew F. "From Military Dictatorship to Democracy: The Democratization Process in Mali." In *Democracy and Development in Mali*, edited by R. James Bingen, David Robinson, and John M. Staatz, 251–64. East Lansing: Michigan State University Press, 2000.

Cloud, Kathleen. "Sex Roles in Food Production and Distribution Systems in the Sahel." In *Women Farmers in Africa: Rural Development in Mali and the Sahel*, edited by Lucy E. Creevey, 19–49. Syracuse, NY: Syracuse University Press, 1986.

Cohen, William B. "Imperial Mirage: The Western Sudan in French Thought and Action." *Journal of the Historical Society of Nigeria* 7, no. 3 (1974): 417–45.

Colin, Loïc, and Vincent Petit, dir. *L'Office du Niger: Du travailleur forcé au paysan syndiqué.* Paris: IRAM, 2008. DVD.

Colleyn, Jean-Paul. "Horse, Hunter, and Messenger: The Possessed Men of the Nya Cult in Mali." In *Spirit Possession, Modernity, and Power in Africa*, edited by Heike Behrend and Ute Luig, 68–78. Madison: University of Wisconsin Press, 1999.

Conklin, Alice L. *A Mission to Civilize: The Republican Idea of Empire in France and West Africa, 1895–1930.* Stanford, CA: Stanford University Press, 1997.

Connelly, Matthew. *Fatal Misconception: The Struggle to Control World Population.* Cambridge, MA: Belknap, 2008.

Conrad, David C. "'Bilali of Faransekila': A West African Hunter and World War I Hero According to a World War II Veteran and Hunters' Singer of Mali." *History in Africa* 16 (1989): 41–70.

Cooper, Barbara M. *Marriage in Maradi: Gender and Culture in a Hausa Society in Niger, 1900–1989.* Portsmouth, NH: Heinemann, 1997.

———. "Oral Sources and the Challenge of African History." In *Writing African History*, edited by John Edward Philips, 191–215. Rochester, NY: University of Rochester Press, 2005.

Cooper, Frederick. *Colonialism in Question: Theory, Knowledge, History.* Berkeley: University of California Press, 2005.

———. *Decolonization and African Society: The Labor Question in French and British Africa.* Cambridge: Cambridge University Press, 1996.

——. "Modernizing Bureaucrats, Backward Africans, and the Development Concept." In *International Development and the Social Sciences: Essays on the History and Politics of Knowledge*, 64–92. Berkeley: University of California Press, 1997.

Cooper, Frederick, and Randall Packard. "Introduction." In *International Development and the Social Sciences: Essays on the History and Politics of Knowledge*, edited by Frederick Cooper and Randall Packard, 1–41. Berkeley: University of California Press, 1997.

Cordell, Dennis D., Joel W. Gregory, and Victor Piché. *Hoe and Wage: A Social History of a Circular Migration System in West Africa*. Boulder, CO: Westview, 1996.

Cornwall, Andrea, Elizabeth Harrison, and Ann Whitehead. "Gender Myths and Feminist Fables: The Struggle for Interpretive Power in Gender and Development." *Development and Change* 38, no. 1 (2007): 1–20.

Coulibaly, Chéibane. *Politiques agricoles et stratégies paysannes au Mali de 1910 à 2010: Mythes et réalités à l'Office du Niger*. 2nd ed. Paris: L'Harmattan Mali, 2014.

Couloubaly, Pascal Baba F. *Une société rural bambara à travers des chants de femmes*. Dakar: Editions IFAN, Université de Dakar, 1990.

Cowan, Ruth Schwartz. *More Work for Mother: The Ironies of Household Technology from the Open Hearth to the Microwave*. New York: Basic Books, 1983.

Creevey, Lucy E. "The Role of Women in Malian Agriculture." In *Women Farmers in Africa: Rural Development in Mali and the Sahel*, edited by Lucy E. Creevey, 51–66. Syracuse, NY: Syracuse University Press, 1986.

Cunningham, Jerimy J. "Pots and Political Economy: Enamel-Wealth, Gender, and Patriarchy in Mali." *Journal of the Royal Anthropological Institute* 15, no. 2 (2009): 276–94.

Daget, Jacques. "La pêche dans le Delta central du Niger." *Journal de la Société des Africanistes* 19, no. 1 (1949): 1–79.

de Laet, Marianne, and Annemarie Mol. "The Zimbabwe Bush Pump: Mechanics of a Fluid Technology." *Social Studies of Science* 30, no. 2 (2000): 225–63.

de Lattre, Anne, and Arthur M. Fell. *The Club du Sahel: An Experiment in International Co-operation*. Paris: Organisation for Economic Co-operation and Development [OECD], 1984.

Delavignette, Robert. *Les paysans noirs*. 2nd ed. Paris: Editions Stock, 1946.

Derrick, Jonathan. "The Great West African Drought, 1972–1974." *African Affairs* 76, no. 305 (1977): 537–86.

——. "West Africa's Worst Year of Famine." *African Affairs* 83, no. 332 (1984): 281–99.

de Waal, Alex. *Famine Crimes: Politics and the Disaster Relief Industry in Africa*. African Issues. Bloomington: Indiana University Press, 1997.

Diabate, Naminata. *Naked Agency: Genital Cursing and Biopolitics in Africa*. Durham, NC: Duke University Press, 2020.

Diakon, Birama. *Office du Niger et pratiques paysannes: Appropriation technologique et dynamique sociale.* Paris: L'Harmattan Mali, 2012.

Diallo, Djibril, ed. *Le Mali sous Moussa Traoré.* Bamako: La Sahélienne, 2016.

Diarra, Salifou Bakary, John M. Staatz, and Niama Nango Dembélé. "The Reform of Rice Milling and Marketing in the Office Du Niger: Catalysts for an Agricultural Success Story in Mali." In *Democracy and Development in Mali,* edited by R. James Bingen, David Robinson, and John M. Staatz, 167–88. East Lansing: Michigan State University Press, 2000.

Dias, Jill R. "Famine and Disease in the History of Angola C. 1830–1930." *Journal of African History* 22, no. 3 (1981): 349–78.

Diawara, Mamdou. "Development and Administrative Norms: The Office du Niger and Decentralization in French Sudan and Mali." *Africa* 81, no. 3 (2011): 434–54.

———. "L'Office du Niger ou l'univers sur-moderne (1920–2000)." In *African historians and globalization,* edited by Issiaka Mandé and Blandine Stefanson, 31–43. Paris: Karthala, 2005.

Diawara, Mamadou, and Ute Röschenthaler. "Green Tea in the Sahel: The Social History of an Itinerant Consumer Good." *Canadian Journal of African Studies* 46, no. 1 (2012): 39–64.

Dieterlen, Germaine. *Essai sur la religion Bambara.* Paris: Presses Universitaires de France, 1951.

———. "The Mande Creation Myth." *Africa: Journal of the International African Institute* 27, no. 2 (1957): 124–38.

Dobres, Marica-Anne. "Archaeologies of Technology." *Cambridge Journal of Economics* 34 (2010): 103–14.

Dougnon, Isaïe. *Travail de Blanc, travail de Noir: La migration des paysans dogon vers l'Office du Niger et au Ghana (1910–1980).* Paris: Karthala-Sephis, 2007.

Dubourg, Jacques. "La vie des paysans mossi: Le village de Taghalla." *Cahiers d'Outre Mer* 10 no. 40 (1957): 285–324.

Dumestre, Gérard. "De l'alimentation au Mali." *Cahiers d'Études Africaines* 36, no. 144 (1996): 689–702.

———. *Dictionnaire bambara-français: Suivi d'un index abrégé.* Paris: Éditions Karthala, 2011.

Dumont, René, and Charlotte Paquet. *Pour l'Afrique, j'accuse: Le journal d'un agronome au Sahel en voie de destruction.* Paris: Plon, 1986.

Edwards, Paul N. "Infrastructure and Modernity: Force, Time, and Social Organization in the History of Sociotechnical Systems." In *Modernity and Technology,* edited by Thomas J. Misa, Philip Brey, and Andrew Feenberg, 185–225. Cambridge, MA: MIT Press, 2003.

Edwin, Shirwin. "Subverting Social Customs: The Representation of Food in Three West African Francophone Novels." *Research in African Literatures* 39, no. 3 (2008): 39–50.

Ehrenberg, Margaret. *Women in Prehistory.* London: British Museum, 1989.

Ehret, Christopher. *The Civilizations of Africa: A History to 1800.* Charlottesville: University of Virginia Press, 2002.

Escobar, Arturo. *Encountering Development: The Making and Unmaking of the Third World*. Princeton, NJ: Princeton University Press, 1995.

Fairhead, James, and Melissa Leach. "Dessication and Domination: Science and Struggles of Environment and Development in Colonial Guinea." *Journal of African History* 41, no. 1 (2000): 35–54.

———. *Misreading the African Landscape: Society and Ecology in a Forest-Savanna Mosaic*. Cambridge: Cambridge University Press, 1996.

Falola, Toyin, and Fallou Ngom. "Introduction: Orality, Literacy and Cultures." In *Oral and Written Expressions of African Cultures*, edited by Toyin Falola and Fallou Ngom, xvii–xxxviii. Durham, NC: Carolina Academic Press, 2009.

Faulkner, Wendy. "The Technology Question in Feminism: A View from Feminist Technology Studies." *Women's Studies International Forum* 24, no. 1 (2001): 79–95.

Ferguson, James. *Anti-politics Machine: Development, Depoliticization, and Bureaucratic Power in Lesotho*. Minneapolis: University of Minnesota Press, 1994.

———. *Global Shadows: Africa in the Neoliberal World Order*. Durham, NC: Duke University Press, 2006.

Ferguson, Priscilla Parkhurst. "The Senses of Taste." *American Historical Review* 116, no. 2 (April 2011): 371–84.

Filipovich, Jean. "Destined to Fail: Forced Settlement at the Office du Niger, 1926–45." *Journal of African History* 42, no. 2 (2001): 239–60.

Findley, Sally E. "Does Drought Increase Migration? A Study of Migration from Rural Mali during the 1983–1985 Drought." *International Migration Review* 28, no. 3 (1994): 539–53.

Frank, Barbara E. *Mande Potters and Leather Workers: Art and Heritage in West Africa*. Washington, DC: Smithsonian Institution Press, 1998.

———. "Marks of Identity: Potters of the Folona (Mali) and Their 'Mothers.'" *African Arts* 40, no. 1 (2007): 30–41.

———. "Nansa Doumbia: Matriarch, Artist, and Guardian or Tradition." In *Mande Mansa: Essays in Honor of David C. Conrad*, edited by Stephen Belcher, Jan Jansen, and Mohamed N'Daou, 73–86. Münster: Lit Verlag, 2008.

———. "Reconstructing the History of an African Ceramic Tradition: Technology, Slavery and Agency in the Region of Kadiolo (Mali)." *Cahiers d'Etudes africaines* 33, no. 131 (1993): 381–401.

Franke, Richard W., and Barbara Chasin. *Seeds of Famine: Ecological Destruction and the Development Dilemma in the West African Sahel*. Montclair, NJ: Allanheld, Osmun, 1980.

Freed, Libbie. "Networks of (Colonial) Power: Roads in French Central Africa after World War I." *History and Technology* 26, no. 3 (2010): 203–23.

Freidberg, Susanne. "Postcolonial Paradoxes: The Cultural Economy of African Export Horticulture." In *Food and Globalization: Consumption Markets and Politics in the Modern World*, edited by Alexander Nützenadel and Frank Trentmann, 215–33. New York: Berg, 2008.

Friends of the Sahel. *Propositions pour une stratégie et un programme de lutte contre la sécheresse et de développement dans le Sahel: Rapport de synthèse*. Ouagadougou, Burkina Faso: CILSS/Club du Sahel, 1977.

Fuentes, Marisa J. *Dispossessed Lives: Enslaved Women, Violence, and the Archive*. Philadelphia: University of Pennsylvania Press, 2016.

Gary-Tounkara, Daouda. "Quand les migrants demandent la route, Modibo Keita rétorque: 'Retournez à la terre!' Les 'Baragnini' et la désertion du 'Chantier national' (1958–1968)." *Mande Studies* 5 (2003): 49–64.

Geiger, Susan. *TANU Women: Gender and Culture in the Making of Tanganyikan Nationalism, 1955–1965*. Portsmouth, NH: Heinemann, 1997.

Gengenbach, Heidi. "Boundaries of Beauty: Tattooed Secrets of Women's History in Magude District, Southern Mozambique." *Journal of Women's History* 14, no. 4 (2003): 106–41.

Gervais, Raymond R., and Issiaka Mandé. "How to Count the Subjects of Empire? Steps toward an Imperial Demography in French West Africa before 1946." In *The Demographics of Empire: The Colonial Order and the Creation of Knowledge*, edited by Karl Ittmann, Dennis D. Cordell, and Gregory H. Maddox, 89–112. Athens: Ohio University Press, 2010.

Giles-Vernick, Tamara. "Lives, Histories, and Sites of Recollection." In *African Words, African Voices: Critical Practices in Oral History*, edited by Luise White, Stephan F. Miescher, and David William Cohen. 194–213. Bloomington: Indiana University Press, 2001.

Gomez, Michael A. *African Dominion: A New History of Empire in Early and Medieval West Africa*. Princeton, NJ: Princeton University Press, 2018.

Goody, Jack. *Cooking, Cuisine, and Class: A Study in Comparative Sociology*. New York: Cambridge University Press, 1982.

Görög, Veronika, and Abdoulaye Diarra. *Contes bambara du Mali*. Paris: Publications Orientalistes de France, 1979.

Griaule, Marcel, and Germaine Dieterlen. "L'agriculture rituelle des Bozo." *Journal de la Société des Africanistes* 19, no. 2 (1949): 209–22.

Grosz-Ngaté, Maria. "Hidden Meanings: Explorations into a Bamanan Construct of Gender." *Ethnology* 28, no. 2 (1989): 167–83.

———. "Memory, Power, and Performance in the Construction of Muslim Identity." *PoLAR: Political and Legal Anthropology Review* 25, no. 2 (2002): 5–20.

———. "Monetization of Bridewealth and the Abandonment of 'Kin Roads' to Marriage in Sana, Mali." *American Ethnologist* 15, no. 3 (1988): 501–14.

Guyer, Jane. "Female Farming in Anthropology and African History." In *Gender at the Crossroads of Knowledge*, edited by Micaela di Leonardo, 257–77. Berkeley: University of California Press, 1991.

Guyer, Jane I., and Samuel M. Eno Belinga. "Wealth in People as Wealth in Knowledge: Accumulation and Composition in Equatorial Africa." *Journal of African History* 36, no. 1 (1995): 91–120.

Hansen, Karen Tranberg, ed. *African Encounters with Domesticity*. New Brunswick, NJ: Rutgers University Press, 1992.

Haraway, Donna J. "Otherworldly Conversations, Terran Topics, Local Terms." In *Material Feminisms*, edited by Stacy Alaimo and Susan Hekman, 157–87. Bloomington: Indiana University Press, 2008.

Hart, Jennifer. *Ghana on the Go: African Mobility in the Age of Motor Transportation*. Bloomington: Indiana University Press, 2016.

Hecht, Gabrielle. *Being Nuclear: Africans and the Global Uranium Trade*. Cambridge, MA: MIT Press, 2014.

——. "Introduction." In *Entangled Geographies: Empire and Technopolitics in the Global Cold War*, edited by Gabrielle Hecht, 1–12. Cambridge, MA: MIT Press, 2011.

Henige, David. "Oral Tradition as a Means of Reconstructing the Past." In *Writing African History*, edited by John Edward Philips, 169–90. Rochester, NY: University of Rochester Press, 2005.

Herbart, Pierre. *Le chancre du Niger*. Paris: Gallimard, 1939.

Herbert, Eugenia W. *Iron, Gender, and Power: Rituals of Transformation in African Societies*. Bloomington: Indiana University Press, 1993.

Hershatter, Gail. *The Gender of Memory: Rural Women and China's Collective Past*. Berkeley: University of California Press, 2011.

Hoffman, Barbara G. "Gender Ideology and Practice in Mande Societies and in Mande Studies." *Mande Studies* 4 (2002): 1–20.

Holmes, Christina. *Ecological Borderlands: Body, Nature, and Spirit in Chicana Feminism*. Urbana: University of Illinois Press, 2016.

Holtzman, Jon. *Uncertain Tastes: Memory, Ambivalence, and the Politics of Eating in Samburu, Northern Kenya*. Berkeley: University of California Press, 2009.

Hunter, Tera W. *To 'Joy My Freedom: Southern Black Women's Lives and Labors after the Civil War*. Cambridge, MA: Harvard University Press, 1997.

Imperato, Pascal James. "The Dance of the Tyi Wara." *African Arts* 4, no. 1 (1970): 8–13, 71–80.

Institut de Recherches et d'Application de Methodes de Developpement (IRAM). *Office du Niger: L'organisation collective des paysans, la situation des femmes*. Paris: IRAM, 1981.

Isaacman, Allen F., and Barbara S. Isaacman. *Dams, Displacement, and the Delusion of Development: Cahora Bassa and Its Legacies in Mozambique, 1965–2007*. Athens: Ohio University Press, 2013.

Jaime, Jean Gilbert Nicomède. *De Koulikoro à Tomboctou sur la cannonière "Le Mage."* Paris: Les Libraires Associés, 1894.

Jean-Baptiste, Rachel. *Conjugal Rights: Marriage, Sexuality and Urban Life in Colonial Libreville, Gabon*. Athens: Ohio University Press, 2014.

Joyce, Rosemary A. "Feminist Theories of Embodiment and Anthropological Imagination: Making Bodies Matter." In *Feminist Anthropology: Past, Present, and Future*, edited by Pamela L. Geller and Miranda K. Stockett, 43–54. Philadelphia: University of Pennsylvania Press, 2006.

Kaplan, Temma. "Female Consciousness and Collective Action: The Case of Barcelona, 1910–1918." *Signs: Journal of Women in Culture and Society* 7, no. 3 (1982): 545–66.

Kéita, Aoua. *Femme d'Afrique: La vie d'Aoua Kéita racontée par elle-même.* Paris: Présence Africaine, 1975.

Klein, Martin. *Slavery and Colonial Rule in French West Africa.* New York: Cambridge University Press, 1998.

——. "Women in Slavery in the Western Sudan." In *Women and Slavery in Africa*, edited by Claire C. Robertson and Martin A. Klein, 67–88. Portsmouth, NH: Heinemann, 1997.

Klein, Martin A., and Richard Roberts. "The Resurgence of Pawning in French West Africa during the Depression of the 1930s." *African Economic History* 16 (1987): 23–37.

Koenig, Dolores. "Food for the Malian Middle Class: An Invisible Cuisine." In *Fast Food/Slow Food: The Cultural Economy of the Global Food System*, edited by Richard Wilk, 49–67. Lanham, MD: Altamira, 2006.

Kohler, Jean Marie. *Les Mosi de Kolongotomo et la collectivisation à l'Office du Niger: Notes sociologiques.* Travaux et documents de l'ORSTOM no. 37. Paris: ORSTOM, 1947.

Kone, Kassim. "When Male Becomes Female and Female Becomes Male in Mande." *Wagadu* 1 (Spring 2002): 21–29.

Kreike, Emmanuel. *Environmental Infrastructure in African History: Examining the Myth of Natural Resource Management in Namibia.* Cambridge: Cambridge University Press, 2013.

——. "Hidden Fruits: A Social Ecology of Fruit Trees in Namibia and Angola." In *Social History and African Environments*, edited by William Beinart and JoAnn McGregor, 27–42. Athens: Ohio University Press, 2003.

Kuper, Marcel, Jean-Philippe Tonneau, and Pierre Bonneval, eds. *L'Office du Niger, grenier à riz du Mali: Succès économiques, transitions culturelles et politiques de développement.* Paris: Karthala, 2003.

Kusiak, Pauline. "'Tubab' Technologies and 'African' Ways of Knowing: Nationalist Techno-Politics in Senegal." *History and Technology* 26, no. 3 (2010): 225–49.

Labouret, Henri. *Les Manding et leur langue.* Paris: Librairie Larose, 1934.

——. *Paysans d'Afrique Occidental.* Paris: Gallimard, 1941.

Landau, Paul S. "When Rain Falls: Rainmaking and Community in a Tswana Village, C. 1870 to Recent Times." *International Journal of African Historical Studies* 26, no. 1 (1993): 1–30.

Larder, Nicolette. "Possibilities for Alternative Peasant Trajectories through Gendered Food Practices in the Office du Niger." In *Postcolonialism, Indigeneity and Struggles for Food Sovereignty: Alternative Food Networks in Subaltern Spaces*, edited by Marisa Wilson, 106–26. New York: Routledge, 2017.

Larkin, Brian. *Signal and Noise: Media, Infrastructure, and Urban Culture in Nigeria*. Durham, NC: Duke University Press, 2008.

Leach, Melissa, and Cathy Green. "Gender and Environmental History: From Representation of Women and Nature to Gender Analysis of Ecology and Politics." *Environment and History* 3, no. 3 (1997): 343–70.

Leach, Melissa, and Robin Mearns. "Challenging Received Wisdom in Africa." In *The Lie of the Land: Challenging Received Wisdom on the African Environment*, edited by Melissa Leach and Robin Mearns, 1–33. Portsmouth, NH: Heinemann, 1996.

———. "Environmental Change and Policy: Challenging Received Wisdom in Africa." In *Lie of the Land: Challenging Received Wisdom on the African Environment*, edited by Robin Mearns and Melissa Leach, 1–33. Portsmouth, NH: Heinemann, 1996.

Lecocq, Baz. "From Colonialism to Keita Comparing Pre- and Post-independence Regimes (1946–1968)." *Mande Studies* 5 (2003): 29–47.

Lentz, Carola. *Land, Mobility, and Belonging in West Africa*. Bloomington: Indiana University Press, 2013.

Lerman, Nina E., Ruth Oldenziel, and Arwen P. Mohun, eds. *Gender and Technology: A Reader*. Baltimore, MD: Johns Hopkins University Press, 2003.

Light, Jennifer, "When Computers Were Women." *Technology and Culture* 40, no. 3 (1999): 455–83.

Livingston, Julie. *Debility and the Moral Imagination in Botswana*. Bloomington: Indiana University Press, 2005.

———. *Self-Devouring Growth: A Planetary Parable as Told from Southern Africa*. Durham, NC: Duke University Press, 2019.

Luning, Sabine. "To Drink or Not to Drink: Beer Brewing, Rituals, and Religious Conversion in Maane, Burkina Faso." In *Alcohol in Africa: Mixing Business, Pleasure, and Politics*, edited by Deborah Fahy Bryceson, 231–48. Portsmouth, NH: Heinemann, 2002.

Lydon, Ghislaine. "Women in Francophone West Africa in the 1930s: Unraveling a Neglected Report." In *Democracy and Development in Mali*, edited by R. James Bingen, David Robinson, and John M. Staatz. East Lansing: Michigan State University Press, 2000.

Magasa, Amidu. *Papa-commandant a jeté un grand filet devant nous: Les exploités des rives du Niger, 1902–1962*. Paris: François Maspero, 1978.

Mage, Eugène. *Voyage dans le Soudan occidental (Sénégambie-Niger)*. Paris: Hachette, 1868.

Mamadi, Kaba, trans. and ed. *Anthologie des chants mandingues: Cote d'Ivoire, Guinée, Mali*. Paris: Harmattan, 1995.

Mandala, Elias. *The End of Chidyerano: A History of Food and Everyday Life in Malawi, 1860–2004*. Portsmouth, NH: Heinemann, 2005.

Mann, Gregory. *From Empires to NGOs in the West African Sahel*. Cambridge: Cambridge University Press, 2015.

———. "What's in an Alias? Family Names, Individual Histories, and Historical Method in the Western Sudan." *History in Africa* 29 (2002): 309–20.

Marchand, Trevor H. J. *The Masons of Djenné.* Bloomington: Indiana University Press, 2009.

Marseille, Jacques. *Empire colonial et capitalisme: Histoire d'un divorce.* Paris: Albin Michel, 1984.

Marsters, Kate Ferguson. "Introduction." In *Travels in the Interior Districts of Africa*, edited by Kate Ferguson Marsters, 1–28. Durham, NC: Duke University Press, 2000.

Martin, Guy. "Socialism, Economic Development and Planning in Mali, 1960–1968." *Canadian Journal of African Studies* 10, no. 1 (1976): 23–47.

Martin, M. S., J. H. Westwood, M. N'Diaye, A. R. Goble, D. Mullins, R. Fell, B. Dembélé, and K. Gamby. "Characterization of Foliar-Applied Potash Solution as a Non-selective Herbicide in Malian Agriculture." *Journal of Agricultural, Food, and Environmental Sciences* 2, no. 1 (2008): 1–8.

Mauss, Marcel. *The Gift: The Form and Reason for Exchange in Archaic Societies.* Translated by W. D. Halls with foreword by Mary Douglas. New York: W. W. Norton, 1990.

———. "Techniques of the Body." In *Incorporations*, edited by Jonathan Crary and Sanford Kwinter, 454–77. New York: Zone, 1992.

Mavhunga, Clapperton Chakanetsa. "Introduction: What Do Science, Technology, and Innovation Mean from Africa?" In *What Do Science, Technology, and Innovation Mean from Africa?*, edited by Clapperton Chakanetsa Mavhunga, 1–27. Cambridge, MA: MIT Press, 2017.

———. *Transient Workspaces: Technologies of Everyday Innovation in Zimbabwe.* Cambridge, MA: MIT Press, 2014.

McCandless, David. "Beriberi." In *Thiamine Deficiency and Associated Clinical Disorders*, 31–46. New York: Humana, 2010.

McCann, James C. *Maize and Grace: Africa's Encounter with a New World Crop, 1500–2000.* Cambridge, MA: Harvard University Press, 2005.

———. *Stirring the Pot: A History of African Cuisine.* Athens: Ohio University Press, 2009.

McGaw, Judith A. "Reconceiving Technology: Why Feminine Technologies Matter." In *Gender and Archaeology*, edited by Rita P. Wright, 52–75. Philadelphia: University of Pennsylvania Press, 1996.

McGregor, JoAnn. "Living with the River: Landscape and Memory in the Zambezi Valley, Northwest Zimbabwe." In *Social History and African Environments*, edited by William Beinart and JoAnn McGregor, 87–105. Athens: Ohio University Press, 2003.

McIntosh, Roderick J. "Social Memory in Mande." In *The Way the Wind Blows: Climate, History, and Human Action*, edited by Roderick J. McIntosh, Joseph A. Tainter, and Susan Keech McIntosh, 141–80. New York: Columbia University Press, 2000.

McIntosh, Roderick J., Joseph A. Tainter, and Susan Keech McIntosh. "Climate, History, and Human Action." In *The Way the Wind Blows: Climate, History, and Human Action*, edited by Roderick J. McIntosh, Joseph A. Tainter, and Susan Keech McIntosh, 1–42. New York: Columbia University Press, 2000.

McKittrick, Meredith. "Forsaking Their Fathers? Colonialism, Christianity, and Coming of Age in Ovamboland, Northern Namibia." In *Men and Masculinities in Modern Africa*, edited by Lisa A. Lindsay and Stephan F. Miescher, 33–51. Portsmouth, NH: Heinemann, 2003.

McNaughton, Patrick R. *The Mande Blacksmiths: Knowledge, Power, and Art in West Africa*. Bloomington: Indiana University Press, 1993.

Meillassoux, Claude. *Maidens, Meal and Money: Capitalism and the Domestic Community*. Cambridge: Cambridge University Press, 1981.

Merchant, Carolyn. *The Death of Nature: Women, Ecology, and the Scientific Revolution*. Reprint ed. New York: Harper One, 1989.

Miller, Joseph C. "The Significance of Drought, Disease, and Famine in the Agriculturally Marginal Zones of West-Central Africa." *Journal of African History* 23, no. 1 (1982): 17–61.

Mitchell, Timothy. *Rule of Experts: Egypt, Techno-Politics, Modernity*. Berkeley: University of California Press, 2002.

Mohanty, Chandra Talpade. "Under Western Eyes: Feminist Scholarship and Colonial Discources." *Boundary* 2, no. 1 (1984): 333–58.

Monson, Jamie. *Africa's Freedom Railway: How a Chinese Development Project Changed Lives and Livelihoods in Tanzania*. Bloomington: Indiana University Press, 2009.

Monteil, Charles. *Les Bambara du Ségou et du Kaarta: Étude historique, ethnographique et littéraire d'une peuplade du Soudan français [1923]*. Paris: G. P. Maisonneuve et LaRose, 1976.

———. *Contes soudanais*. Paris: Ernest Leroux, 1905.

———. *Le coton chez les noirs*. Paris: Émile Larose, 1927.

Moon, Suzanne. "Place, Voice, Interdisciplinarity: Understanding Technology in the Colony and the Postcolony." *History and Technology* 26, no. 3 (2010): 189–201.

Morgan, Jennifer L. *Laboring Women: Reproduction and Gender in New World Slavery*. Philadelphia: University of Pennsylvania Press, 2004.

Mutongi, Kenda. *Worries of the Heart: Widows, Family, and Community in Kenya*. Chicago: University of Chicago Press, 2007.

Nast, Heidi J. *Concubines and Power: Five Hundred Years in a Northern Nigerian Palace*. Minneapolis: University of Minnesota Press, 2004.

Naylor, Rosamond L., ed. *The Evolving Sphere of Food Security*. Oxford: Oxford University Press, 2014.

Nnaemeka, Obioma. "Mapping African Feminisms." In *Readings in Gender in Africa*, edited by Andrea Cornwall, 31–41. Bloomington: Indiana University Press, 2005.

Nordeide, M. B., A. Hatløy, M. Følling, E. Lied, and A. Oshaug. "Nutrient Composition and Nutritional Importance of Green Leaves and Wild Food Resources in an Agricultural District, Koutiala, in Southern Mali." *International Journal of Food Sciences and Nutrition* 47 (1996): 455–68.

Odumosu, Toluwalogo. "Making Mobiles African." In *What Do Science, Technology, and Innovation Mean from Africa?*, edited by Clapperton Chakanetsa Mavhunga, 137–50. Cambridge, MA: MIT Press, 2017.

Oldenziel, Ruth. *Making Technology Masculine: Men, Women and Modern Machines in America 1870–1945.* Amsterdam: Amsterdam University Press, 1999.

Oldenziel, Ruth, and Karin Zachman. "Kitchens as Technology and Politics: An Introduction." In *Cold War Kitchen: Americanization, Technology, and European Users*, edited by Ruth Oldenziel and Karin Zachman, 1–29. Cambridge, MA: MIT Press, 2009.

Ortoli, Henri. "Le gage des personnes au Soudan français." *Bulletin de l'Institut Français d'Afrique Noire* 1 (1939): 313–24.

Osborn, Emily Lynn. "Casting Aluminum Pots: Labour, Migration and Artisan Production in West Africa's Informal Sector, 1945–2005." *African Identities* 7, no. 3 (2009): 373–86.

———. "Containers and Mobility in West Africa." *History and Anthropology* 29, no. 1 (2018): 21–31.

———. *Our New Husbands Are Here: Households, Gender, and Politics in a West African State from the Slave Trade to Colonial Rule.* Athens: Ohio University Press, 2011.

Osseo-Asare, Abena Dove. *Atomic Junction: Nuclear Power in Africa after Independence.* Cambridge: Cambridge University Press, 2019.

Osseo-Asare, Fran. *Food Culture in Sub-Saharan Africa.* Westport, CT: Greenwood, 2005.

Oudshoorn, Nelly, and Trevor Pinch. "Introduction: How Users and Nonusers Matter." In *How Users Matter: The Co-construction of Users and Technology*, edited by Nelly Oudshoorn and Trevor Pinch, 1–25. Cambridge, MA: MIT Press, 2003.

Paques, Viviana. *Les Bambara.* Paris: Presses Universitaires de France, 1954.

Park, Mungo. *Travels in the Interior Districts of Africa [1799].* Edited by Kate Ferguson Marsters. Durham, NC: Duke University Press, 2000.

Parr, Joy. "Our Bodies and Our Histories of Technology and the Environment." In *The Illusory Boundary: Environment and Technology in History*, edited by Martin Reuss and Stephen H. Cutcliff, 26–42. Charlottesville: University of Virginia Press, 2010.

———. "What Makes Washday Less Blue? Gender, Nation, and Technology Choice in Postwar Canada." *Technology and Culture* 38, no. 1 (1997): 153–86.

Pearson, Chris. "'The Age of Wood': Fuel and Fighting in French Forests, 1940–1944." *Environmental History* 11 (2006): 775–803.

Perinbam, B. Marie. *Family Identity and the State in the Bamako Kafu, c.1800–c.1900*. Boulder, CO: Westview, 1997.

Peterson, Brian J. *Islamization from Below: The Making of Muslim Communities in Rural French Soudan, 1880–1960*. New Haven, CT: Yale University Press, 2011.

Randall, Sara, and Alessandra Giuffrida. "Forced Migration, Sedentarisation and Social Change: Malian Kel Tamasheq." In *Pastoralists of North Africa and the Middle East: Entering the 21st Century*, edited by Dawn Chatty, 421–63. Leiden, The Netherlands: Brill, 2006.

Redfield, Peter. "Cleaning Up the Cold War: Global Humanitarianism and the Infrastructure of Crisis Response." In *Entangled Geographies: Empire and Technopolitics in the Global Cold War*, edited by Gabrielle Hecht, 267–91. Cambridge, MA: MIT Press, 2011.

Richards, Paul. *Indigenous Agricultural Revolution: Ecology and Food Production in West Africa*. Boulder, CO: Westview, 1985.

Roberts, Richard. "The Case of Faama Mademba Sy and the Ambiguities of Legal Jurisdiction in Early Colonial French Soudan." In *Law in Colonial Africa*, edited by Kristin Mann and Richard Roberts, 185–201. Portsmouth, NH: Heinemann, 1991.

——. "The Coercion of Free Markets: Cotton, Peasants, and the Colonial State in the French Soudan, 1924–1932." In *Cotton, Colonialism, and Social History in Sub-Saharan Africa*, edited by Allen Isaacman and Richard Roberts, 221–43. Portsmouth, NH: Heinemann, 1995.

——. "Fishing for the State: The Political Economy of the Middle Niger Valley." In *Modes of Production in Africa: The Precolonial Era*, edited by Donald Crummey and C. C. Stewart, 175–203. Beverly Hills, CA: SAGE, 1981.

——. *Litigants and Households: African Disputes and Colonial Courts in the French Soudan, 1895–1912*. Portsmouth, NH: Heinemann, 2005.

——. *Two Worlds of Cotton: Colonialism and the Regional Economy in the French Soudan, 1800–1946*. Stanford, CA: Stanford University Press, 1996.

——. *Warriors, Merchants, and Slaves: The State and the Economy in the Middle Niger Valley, 1700–1914*. Stanford, CA: Stanford University Press, 1987.

Roberty, Guy. *Les associations végétales de la vallee moyenne du Niger*. Bern, Switzerland: Verlag Hans Huber, 1946.

Robinson, David. *The Holy War of Umar Tal*. Oxford: Oxford University Press, 1988.

Rolston, Jessica Smith. "Talk about Technology: Negotiating Gender Difference in Wyoming Coal Mines." *Signs: Journal of Women in Culture and Society* 35, no. 4 (2010): 893–918.

Rossi, Benedetta. *From Slavery to Aid: Politics, Labour, and Ecology in the Nigerien Sahel, 1800–2000*. Cambridge: Cambridge University Press, 2015.

Rottenburg, Richard. *Far-Fetched Facts: A Parable of Development Aid*. Cambridge, MA: MIT Press, 2009.

Sacré-Coeur, Marie-André du, Sr. "La femme Mossi, sa situation juridique." *Ethnographie* 33–34 (1937): 15–33.

Saidi, Christine. "Pots, Hoes, and Food: Women in Technology and Production." In *Women's Authority and Society in Early East-Central Africa*, 128–46. Rochester, NY: University of Rochester Press, 2010.

Saikia, Yasmin. *Women, War, and the Making of Bangladesh: Remembering 1971*. Durham, NC: Duke University Press, 2011.

Salo, Samuel. "Les Voltaïques sur les chantiers coloniaux (1895–1960)." In *African Historians and Globalization*, edited by Issiaka Mandé and Blandine Stefanson, 45–61. Paris: Karthala, 2005.

Şaul, Mahir. "Money in Colonial Transition: Cowries and Francs in West Africa." *American Anthropologist* 106, no. 1 (2004): 71–84.

Savineau, Denise. *La famille en A.O.F.: Condition de la femme*. Dakar: Gouvernement de l'A.O.F., 1938.

Scheper-Hughes, Nancy, and Margaret M. Lock. "The Mindful Body: A Prolegomenon to Future Work in Medical Anthropology." *Medical Anthropology Quarterly (New Series)* 1, no. 1 (1987): 6–41.

Scheub, Harold. "A Review of African Oral Traditions and Literature." *African Studies Review* 28, no. 2/3 (1985): 1–72.

Schiebinger, Londa. *Nature's Body: Gender in the Making of the Modern Science*. New Brunswick, NJ: Rutgers University Press, 2004.

Schmidt, Elizabeth. *Peasants, Traders, and Wives: Shona Women in the History of Zimbabwe, 1870–1939*. Portsmouth, NH: Heinemann, 1992.

Schreyger, Emil. *L'Office du Niger au Mali 1932 à 1982: La problématique d'une grande entreprise agricole dans la zone du Sahel*. Paris: L'Harmattan, 1984.

Schroeder, Richard A. *Shady Practices: Agroforestry and Gender Politics in The Gambia*. Berkeley: University of California Press, 1999.

Scott, James C. *Seeing Like a State: How Certain Schemes to Improve the Human Condition Have Failed*. New Haven, CT: Yale University Press, 1998.

Sen, Amartya. *Poverty and Famines: An Essay on Entitlement and Deprivation*. Oxford: Clarendon, 1981.

Serlin, David. "Confronting African Histories of Technology: A Conversation with Keith Breckenridge and Gabrielle Hecht." *Radical History Review* 127 (2017): 87–102.

Shipton, Parker. "African Famines and Food Security: Anthropological Perspectives." *Annual Review of Anthropology* 19 (1990): 353–94.

Silla, Eric. *People Are Not the Same: Leprosy and Identity in Twentieth-Century Mali*. Portsmouth, NH: Heinemann, 1998.

Simmons, Dana. "Starvation Science: From Colonies to Metropole." In *Food and Globalization: Consumption, Markets and Politics in the Modern*

World, edited by Alexander Nützenadel and Frank Trentmann, 173–91. New York: Berg, 2008.

Slobodkin, Yan. "Famine and the Science of Food in the French Empire, 1900–1939." *French Politics, Culture, and Society* 36, no. 1 (2018): 52–75.

Smith, Bonnie G. *The Gender of History: Men, Women, and Historical Practice*. Cambridge, MA: Harvard University Press, 1998.

Sorel, François. "L'alimentation des indigènes en Afrique occidentale française." In *L'alimentation indigène dans les colonies françaises, protectorats et territoires sous mandat*, edited by Georges Hardy and Charles Richet, 155–76. Paris: Vigot Frères, 1933.

Spitz, Georges. *Sansanding: Les irrigations du Niger*. Paris: Société d'Editions Géographiques, Maritimes, et Coloniales, 1949.

Stoller, Paul. *Embodying Colonial Memories: Spirit Possession, Power, and the Hauka in West Africa*. New York: Routledge, 1995.

———. *The Taste of Ethnographic Things: The Senses in Anthropology*. Philadelphia: University of Pennsylvania Press, 1989.

Swift, Jeremy. "Desertification: Narratives, Winners and Losers." In *The Lie of the Land: Challenging Received Wisdom on the African Environment*, edited by Melissa Leach and Robin Mearns, 73–90. Portsmouth, NH: Heinemann, 1996.

Thiam, Mariam. "The Role of Women in Rural Development in the Segou Region of Mali." In *Women Farmers in Africa: Rural Development in Mali and the Sahel*, edited by Lucy E. Creevey, 67–79. Syracuse, NY: Syracuse University Press, 1986.

Thiriet, E. *Au Soudan français: Souvenirs 1892–1894, Macina—Tombouctou*. Paris: André Lesot, 1932.

Thomas, Lynn M. "Historicising Agency." *Gender and History* 28, no. 2 (August 2016): 324–39.

———. *Politics of the Womb: Women, Reproduction, and the State in Kenya*. Berkeley: University of California Press, 2003.

Tilley, Helen. *Africa as a Living Laboratory: Empire, Development, and the Problem of Scientific Knowledge, 1870–1950*. Chicago: University of Chicago Press, 2011.

Togola, Téréba. "Memories, Abstractions, and Conceptualization of Ecological Crisis in the Mande World." In *The Way the Wind Blows: Climate, History, and Human Action*, edited by Roderick J. McIntosh, Joseph A. Tainter, and Susan Keech McIntosh, 181–92. New York: Columbia University Press, 2000.

Toulmin, Camilla. *Cattle, Women and Wells: Managing Household Survival in the Sahel*. Oxford: Oxford University Press, 1992.

Trapp, Micah M. "You-Will-Kill-Me-Beans: Taste and the Politics of Necessity in Humanitarian Aid." *Cultural Anthropology* 31, no. 3 (2016): 412–37.

Travélé, Moussa. *Petit dictionnaire français-bambara et bambara-français*. Paris: Librairie Paul Geuthner, 1913.

———. *Petit manuel français-bambara*. Paris: Librairie Orientaliste Paul Geuthner, 1955.

———. *Proverbes et contes bambara: Accompagnés d'une traduction française et précédés d'un abrégé de droit coutumier*. Paris: Librarie Orientaliste Paul Geuthner, 1923.

Twagira, Laura Ann. "Introduction: Africanizing the History of Technology." *Technology and Culture* 61, no. S2 (April 2020): S1–19.

———. "Machines That Cook or Women Who Cook? Lessons from Mali on Technology, Labor, and Women's Things." *Technology and Culture* 61, no. S2 (April 2020): S77–103.

———. "Peopling the Landscape: Colonial Irrigation, Technology, and Demographic Crisis in the French Soudan, ca. 1926–1944." *PSAE Research Series* 10 (2012): 1–29.

———. "'Robot Farmers' and Cosmopolitan Workers: Divergent Masculinities and Agricultural Technology Exchange in the French Soudan (1945–68)." *Gender and History* 26, no. 3 (2014): 459–77.

van Beusekom, Monica M. "From Underpopulation to Overpopulation: French Perceptions of Population, Environment, and Agricultural Development in French Soudan (Mali), 1990–1960." *Environmental History* 4, no. 2 (1999): 198–219.

———. "Individualism, Community, and Cooperatives in the Development Thinking of the Union Soudanaise-RDA, 1946–1960." *African Studies Review* 51, no. 2 (2008): 1–25.

———. *Negotiating Development: African Farmers and Colonial Experts at the Office du Niger, 1920–1960*. Portsmouth, NH: Heinemann, 2002.

Vaughan, Megan. *The Story of an African Famine: Gender and Famine in Twentieth-Century Malawi*. Cambridge: Cambridge University Press, 1987.

Viguier, Pierre. *Sur les traces de René Caillié: Le Mali de 1828 revisité*. Versailles: Éditions Quae, 2008.

Wajcman, Judy. *Technofeminism*. Cambridge, UK: Polity, 2004.

Ware, Rudolph T. *The Walking Qur'an: Islamic Education, Embodied Knowledge, and History in West Africa*. Chapel Hill: University of North Carolina Press, 2014.

Watts, Michael. *Silent Violence: Food, Famine and Peasantry in Northern Nigeria*. Berkeley: University of California Press, 1983.

Webb, James L. A., Jr. *Desert Frontier: Ecological and Economic Change along the Western Sahel, 1600–1850*. Madison: University of Wisconsin Press, 1995.

———. "Ecology and Culture in West Africa." In *Themes in West Africa's History*, edited by Emmanuel Kwaku Akyeampong, 33–51. Athens: Ohio University Press, 2006.

White, Luise. *Speaking with Vampires: Rumor and History in Colonial Africa*. Berkeley: University of California Press, 2000.

White, Luise, Stephan F. Miescher, and David William Cohen, eds. *African Words, African Voices: Critical Practices in Oral History.* Bloomington: Indiana University Press, 2001.

Wickens, G. E. *Role of Acacia Species in the Rural Economy of Dry Africa and the Near East.* Rome: Food and Agricultural Organization of the United Nations, 1995.

Wilder, Gary. "Framing Greater France between the Wars." *Journal of Historical Sociology* 14, no. 2 (2001): 198–225.

Winther, Tanja. *The Impact of Electricity: Development, Desires and Dilemmas.* New York: Berghahn, 2008.

Wooten, Stephen. *The Art of Livelihood: Creating Expressive Agri-culture in Rural Mali.* Durham, NC: Carolina Academic Press, 2009.

——. "Colonial Administration and the Ethnography of the Family in the French Soudan." *Cahiers d'Études Africaines* 33, no. 131 (1993): 419–46.

——, ed. *Wari Matters: Ethnographic Explorations of Money in the Mande World.* Münster, Germany: Lit Verlag, 2005.

——. "Where Is My Mate? The Importance of Complementarity: A Bamana Headdress (Ciwara)." In *See the Music, Hear the Dance: Rethinking Art at the Baltimore Museum of Art,* ed. Frederick John Lamp, 168–71. Munich: Prestel, 2004.

Worboys, Michael. "The Discovery of Colonial Malnutrition between the Wars." In *Imperial Medicine and Indigenous Societies,* edited by David Arnold, 208–25. Manchester, UK: Manchester University Press, 1988.

Wylie, Diana. "The Changing Face of Hunger in Southern African History 1880–1980." *Past and Present* 122 (1989): 159–99.

——. *Starving on a Full Stomach: Hunger and the Triumph of Cultural Racism in Modern South Africa.* Charlottesville: University Press of Virginia, 2001.

Zahan, Dominique. *Antilopes du soleil: Arts et rites agraires d'Afrique noire.* Vienna: Edition A. Schendl, 1980.

——. *The Bambara.* Leiden, The Netherlands: E. J. Brill, 1974.

——. "Problèmes sociaux posés par la transplantation des Mossis sur les terres irriguées de l'Office du Niger." In *African Agrarian Systems,* edited by Daniel Biebuyck, 392–403. Oxford: Oxford University Press, 1963.

Zimmerman, Andrew. "A German Alabama in Africa: The Tuskegee Expedition to German Togo and the Transnational Origins of West African Cotton Growers." *American Historical Review* 110, no. 5 (2005): 1362–98.

Index

The letters *f* or *m* following a page number denote a figure or a map, respectively.

abundance: associated with beer production, 23, 26, 54–60, 134–35, 243n127; association with women, 26; celebration of agricultural, 54–60, 132, 243n127; of cloth, 104; colonial association with Western technology, 17; European observations of agricultural, 33–34, 60; of fish, 114–15, 161; in Malian state propaganda, 182; of milk, 159; in Office propaganda, 102; regional reputation for agricultural, 89; of water, 111–12, 114–15. *See also* breadbasket

Adas, Michael, 76

aesthetics: colonial, 94, 166; of food, 52, 184, 203, 212; of pots, 147, 152; rural, 69; of women's technological work, 21, 142. *See also* taste

agency (capacity for action), 12, 18

agricultural festivals, 23, 26, 54–56, 111, 132, 138–41, 227n81, 243n127. *See also* social life

Agricultural Service, 40, 41, 61, 76

Alphalog, 212. *See also* development

Antonetti, Raphaël, 59

Association Cotonnière Coloniale, 65

Austen, Ralph, 12–13

B1 (Office town also called Niobougou), 127

Ba, Fanta (mythical hunter), 33

baarakolo, 49, 161, 240n97, 276n74

baby carrying, 10, 13, 30, 34–36, 46, 185f, 209, 223n42

Baguinèda (Office town), 39, 39f, 75, 87, 99, 140

Bah, Adam (Mme. Dagno), 119, 133, 180–81

Ba Konaré, Adame, 218n13

balanzan tree, 32, 232n25

Bamako, 33m, 34, 37, 60–61, 75, 77, 88, 153, 181, 200, 212

Bamako-Koura (Office town), 122m

Bamana: agricultural practices of, 62, 136, 204, 238n68; canal building, 111–12; cuisine, 47, 130, 135; cultural identity, 234n38, 234n39; cultural region, 34, 233n28; gender dynamics among, 63, 86; oral traditions of, 25, 53, 62, 111, 202, 205, 229n2, 230n10, 231n12, 245n161. *See also* Bambara

Bamana canal, 111–12

Bambara, 29, 41, 58, 64, 101, 234n38, 238n78, 244n142. *See also* Bamana

Banamba, 35. *See also* slavery

Banankourou, 37–38, 237n66

Banbugu, 111–12

Banque Nationale de Developpement Agricole (BNDA), 201–2

baobab, 25, 34, 45, 47, 52, 121, 127–28, 145, 173. *See also namugu*

Barry, Fatoumata, 108

Bary, N'Fa, 102

basi, 47, 52, 198, 239n87

Bauzil, Vincent, 89

beer: associated with wealth, 26, 55–60, 131–32, 134, 243n127; association with cultural or religious identity, 58, 132, 234n38, 267n135; brewers, 106, 131, 159;

beer (cont.)
   brewing, 48, 54–57, 59–69, 106,
      131–32, 135; colonial ban on
      production of, 59–60; drinking as a
      leisure activity, 23, 26, 54–60, 131–32;
      reduced consumption of, 132; sales
      of, 41, 56, 58–59, 131; in Segu state
      politics, 57; surveyed in regional
      consumption patterns, 59–60, 88,
      134, 243n127; symbolic importance
      of, 57–58, 60; taste of, 4, 56, 131. See
      also djisongo; dolo
Bélime, Emil: agricultural observations
      by, 59, 66, 92, 242n118; claims about
      the successes and promises of the
      Office du Niger, 80, 110; criticisms
      of, 219n20, 248n11; director of the
      colonial Textile and Hydrology
      Service for the AOF, 246n181; as
      founder of the Office du Niger,
      8, 17, 73–79, 100, 159, 176, 247n7,
      251n6; management of the Office
      du Niger by, 82, 167; role in the
      local politics of the Office du Niger,
      259n6. See also Office du Niger
Belinga, Samuel M. Eno, 103
bénéfing (wild sesame), 61
béré (famine food), 145, 242n124
beriberi, 90, 193, 196–98, 253n96, 288n106
Bida (mythical snake), 33
Binger, Louis, 50, 59
biopolitics (female body power), 4, 12.
      See also body
blacksmiths, 10–12, 33, 153–56, 214, 222n39
Blanc, Mr. (Office agent), 85, 102
Bo, 119
body: colonial body politics, 23, 78–83;
      fatigue, 44; food security and, 178,
      211–12; gendering the body, 29,
      223n43, 231n13; and material expe-
      rience, 4–5, 30–31, 218n9, 218n10,
      283n37; nakedness and clothing,
      131; and ritual, 12, 208–9; as a site
      for knowledge production, 21; state
      intrusion into, 190–91, 208, 211; state
      surveillance of, 184–86; symbolic
      value of, 31, 208–9, 211; "technique
      of the body," 3, 148–51, 184; and
      women's agency, 4–5, 30. See also
      biopolitics
bogoda, 146–47. See also cooking pots
Boky-Were (Office town), 54, 56, 71, 120,
      120m, 122m, 128, 172, 207

Bonnecase, Vincent, 20, 88
boombo (sauce ingredient), 128
Boserup, Ester, 6, 156
Bouaré, Harouna, 145
Bouaré, Moussa, 70
Bozo, 33, 62, 126, 128, 136, 203, 234n39
breadbasket, 5, 138
Brett-Smith, Sarah, 12
buckets, 11, 24, 45f, 142–43, 146–48,
      153–54, 160–62, 171, 176. See also
      modest technologies
budgeting, 130–32, 175
bulldozers, 124, 144, 144f, 164, 169
Burkina Faso, 8, 195, 238n71, 242n117,
      244n150, 262n35, 280n11, 289n119
bush: dangers in, 44; loss of, 107, 115,
      130, 145, 205; replacing the, 138–39;
      space for hiding grains in, 39; as
      space for ritual dance, 56; as space
      outside of society, 51, 217n5; wild
      foods found in, 3, 30, 39, 42–44, 46,
      48–49, 54, 122
Caillié, René, 34, 50
calabashes: definition of, 239n84; dura-
      bility of, 142, 161–62; eaten during
      famine, 198; for food gathering, 42,
      44, 46, 238n81; and marriage, 146–48;
      as a measure, 90, 132, 173, 187; sale
      and purchase of, 128, 146; as a serving
      vessel, 58, 198; for storage, 49; for use
      in food preparation, 30, 46; for water
      management, 123, 161–62; women's
      cultivation of, 64, 123, 125
canned food, 139, 198–200, 202, 205
Carbou, Inspector, 90–91
Carde, Jules, 75
carts, 85, 116, 133, 156, 158–60, 174, 190,
      260n11, 275n62
cash-crop farming: as agricultural policy,
      17, 182; in contrast to food produc-
      tion, 18; expansion of, 107; field
      preparation for, 84; and the plow,
      156, 170; as a source of income, 24;
      as a source of rice for consumption,
      136, 138
cash earnings (women's), 110, 125, 130–34,
      137, 149, 151, 158, 168, 173
Catholic missionaries, 22, 38, 53, 59, 193,
      237n66, 243n127, 257n147
cattle, 34, 71, 115, 129, 134, 159, 164–65,
      179, 268n150
Centre du Riz Mécanisé (CRM), 163–64,
      166, 171, 174

Chevalier, Auguste, 90, 137, 219n20, 248n11
childbirth, 2, 12, 30, 80, 148, 178, 184, 208–9, 231n15. *See also* fertility
China, 16, 181, 281n15
Chirikure, Shadreck, 12
circumcision, 12, 231n13, 231n15, 278n101
Ciwara, 29, 29f, 30. *See also* agricultural festivals
climate change, 32, 280n3
cloth: as currency, 64; dying of, 240n100; as a gift, 64, 104–5, 133, 140, 146; imported cloth, 153; man's head covering, 190; and marriage, 147; mud cloth, 12; Office market for, 154; as a part of women's wealth, 49; production of, 64, 66; as a protest demand, 104; purchases of, 128, 131, 134, 146; and rice babies, 184–91; as a technology for wrapping, 24, 42, 44, 177–78, 186, 188, 208; women's cash earnings from role in producing, 246n172. *See also* modest technologies
Clozel (lieutenant-governor), 39–40, 65
Cold War, 16
collective action, 2, 4, 22, 42–43, 48, 55, 73, 180–81, 281n16
collectivization, 15, 180–81
colonial export market: in contrast to local concerns for consumption, 9, 23, 40, 94; cotton as part of, 65–66, 171; French Soudan as part of, 37, 40–41, 46, 61; Office du Niger as part of, 8, 14, 40, 170; policy for, 74; rice as part of, 17, 60, 75, 87, 91–92
conscripted labor, 8, 14, 38, 77, 80, 82, 89, 116. *See also* forced labor
cooking oil, 41, 104–5, 127, 139, 153, 173, 196, 198–201, 240n100, 241n110
cooking pots: and buckets, 162, 176; clay pots (*bogoda*), 46, 49, 142, 146, 149–51, 153, 158, 162, 273n35; cost of, at the Office du Niger, 154–55; enamel pots, 272n22, 273n27; in folktales, 1–4, 25–26; gendered as women's technology, 176, 179; and generosity, 25–27; Generous Cooking Pot, 25, 28, 45–46, 57; iron pots, 146–53; labor of cleaning, 113, 151–52, 161, 273n39; metal (*negeda*), 6, 13, 24–25, 142–43, 146–55, 152f, 157, 160, 171, 176, 272n17, 273n35,

274n49; as an object of study, 11–13, 22; as part of women's material technological world, 3, 30–31; and ritual, 38; symbolic value of, 1, 3; used to transform famine food, 203. *See also* modest technologies
cooking technique, 93–94, 135–36, 148–51, 202–3, 273n29
Cooper, Barbara, 22, 147
Correze, Mme., 197
Costes (researcher), 59, 77
Côte d'Ivoire, 82, 89, 182
cotton: cloth production tasks, 41, 64, 128, 132, 137, 158, 256n132, 256n171; confiscation of, 181; cultivation, 8, 16–17, 40, 44, 64–67, 71, 73, 75, 77–79, 84, 88, 105–6, 123, 125–26, 130, 133, 135, 169, 171, 182; exchange and sale of, 14, 74, 91, 104–7, 118, 128–29, 130, 167, 183, 187, 246n172; mechanized processing, 165, 172f; mosquito net made from, 146; oil made from, 139; refusal to cultivate, 99, 136, 269n166; sector at the Office du Niger, 76, 80–81, 85, 98–99, 104, 110; women's labor and, 49, 63–64, 66–67, 84, 99, 132, 137, 164, 171, 175, 182, 192, 212, 240n100, 240n184
Coulibaly, Assane, 128, 135, 139, 188, 203
Coulibaly, Bogoba, 71
Coulibaly, Bokari, 102
Coulibaly, Chéibane, 8, 287n99
Coulibaly, Djenebu, 144–45, 238n68
Coulibaly, Fatouma, 214–15, 214f
Coulibaly, Fatoumata (Kouyan-Kura resident), 177–78
Coulibaly, Fatoumata (Sabula resident), 123, 131, 148
Coulibaly, Hawa, 137
Coulibaly, Kadja (Kokry), 139, 173, 190
Coulibaly, Mamdou "Seyba," 192
Coulibaly, Mariam "Mamu," 106–8, 122, 129, 131, 134, 159, 259n6
Coulibaly, Moussa, 188, 196–97
Coulibaly, Nana, 179
Coulibaly, Nene, 104
Coulibaly, Nianzon, 100
Coulibaly, N'Tio, 104
Coulibaly, Sékou, 160–61
Coulibaly, Semougou, 102
Couloubaly, Pascal, 53
Dakar, 39, 74–75, 92, 109, 153, 174, 200
Danel (rice merchant), 92

Niger (neighboring country), 21, 197, 227n84, 280n11
Niger River: French exploration of, 35; political import of, 233n35, 234n36; in regional ecology, 33, 37, 60; in regional mythology, 32–33, 232n24; as source of irrigation, 8, 18, 40, 67, 73–74, 219n20
Niono: center for cotton production, 80–81, 110, 119, 130, 132–33, 139, 163, 175; destination for hunger migrants, 206; flood of Gruber Canal in sector, 85, 87; gardening at, 125, 212, 269n168; guards at, 284n60; market, 54, 116, 124, 126–28, 132–33, 146, 188, 205, 242n120, 271n13; Mossi settlement at, 85–86; Office du Niger parties at, 140; perception of sector as without people, 98; poor condition of agricultural machines in sector, 286n81; settler unrest at, 100, 104–5, 258n168, 277n95; shift to rice cultivation at, 136–38; site of cotton processing factory, 139, 171, 172f; town, 114f, 116, 140; women's market at, 91
Nnaemeka, Obioma, 31
nutrition: colonial perceptions of nutrition in regional diet, 48; failure to address, 88, 90, 197, 199–200; and food security, 20, 212, 280n5; nutritional value of fish, 126; nutritional value of millet, 90, 137, 253n96; nutritional value of wild foods, 47, 54, 143, 239n89; rice, 90, 94–95, 253n96; women's awareness of, 3, 9, 46, 94, 211–12, 240n100
Nyamina (Office town), 118–19, 138, 160
Office des produits agricoles au Mali (OPAM), 177, 180, 197, 280n12, 288n108, 292n157, 292n168
Office du Niger: establishment of, 8–9, 13–15, 40, 73–78; expansion of, 107, 119–20; as a "national worksite," 182; as a postcolonial state project, 15–16, 181–83, 191–92; promotion as a development scheme, 8, 18–19, 138, 156, 163–65; recruitment for, 9, 14, 23, 77–83, 85, 86, 88, 96, 101, 182, 198, 275n69; settlement at, 9, 23, 70, 75, 78–79, 81, 83–85, 109, 195. See also irrigation
okra, 1, 46, 123, 127–28, 136, 264n80

oral history, 21–22
oral tradition, 22, 25, 28–33, 56, 111, 112, 208, 229n92, 230n10, 231n12, 231n13, 232n23
Ortoli, Henri, 101, 257n144
Osborn, Emily, 153
Ouagadougou, 59, 244n145, 257n147
Ouahigouya (region), 33m, 60, 82–83, 85, 88, 101, 123, 244n150
Oudshoorn, Nelly, 11
Ouloguem, Sekou Salla, 202–3, 206, 274n54
parboiled rice, 93–94
Park, Mungo, 34, 49–51, 57–59, 234n40
pawnship, 36, 46, 78, 101, 236n59, 257n144, 257n147, 257n148
peanuts: and cotton harvest, 66; cultivation at the Office du Niger, 71, 79, 84, 126, 251n67, 264n80; for export, 41, 64, 75; grilled, 35; in markets, 35, 126, 173, 183; possibilities for mechanized cultivation, 164; in production of cooking oil, 139, 173; sauce, 127–28, 158; as women's crop, 38, 41, 49, 239n86. See also tegedege
Pemba (deity), 28, 32
pestle, 3, 9, 46, 49, 53, 68–69, 146, 214, 276n77. See also mortar
Peter, Georges, 137, 163, 165
Pinch, Trevor, 11
Pléah, Assane, 98, 149–50, 188
plow: adoption of, 15, 39f, 81 ; costs of, 155; and fertilizer, 249n28; hunger, association with, 77–78; and land claims, 157; mechanization, 163, 167–68; men, association with, 10, 39f, 86, 156–57, 167–68, 275n58, 278n102; in oral interviews, 22; promotion of, 76–77, 81–82, 88, 155–56; sale of, 134; in technology studies, 10–13, 222n32; women's use of, 156–57
potters (female), 12, 31, 132, 146, 153, 274n45
pounding labor, 1, 48–49, 52, 68–69, 93, 97f, 145, 161, 206, 242n114, 268n157
pregnancy, 24, 184, 208
Pruvost (inspector), 104
red millet, 62, 179, 181, 195, 197–98, 202–5, 208, 290n137, 291n150
red rice, 92–93, 95
"red thing," 194, 202–3
Revillon, Tony, 164

rice as payment (to women), 132, 173, 187
rice babies, 24, 178, 184–91, 208–9, 213
rice toh, 135, 150
Rimaibe, 88
ritual, 12, 29–30, 37–38, 55–57, 60, 96,
    132, 140, 167, 208–9, 222n39
Riziam (Office town), 139
roads, 8, 14–15, 83, 110, 115–16, 165–66,
    169, 188–90, 192, 206, 223n50,
    223n51. See also transportation
    infrastructure
Roberty, Guy, 49, 61–62, 144–45, 203,
    245n169, 291n150
Rocca-Serra, Joseph, 82, 85, 99, 145
Rodger, George, 144f, 172f
Rossi, Benedetta, 21, 227n84, 280n11
saba fruits, 44, 45f
Sabula (Office town), 123, 131, 148, 193–98
Sahel, 6, 17, 20, 32–34, 177–79, 194–95,
    197, 200–201, 232n26
Sahel Canal (Canal du Sahel), 74m,
    75–77, 80, 110, 163
salt, 34, 49, 53–54, 126, 128–29, 139,
    154–55, 178, 187, 197, 205
Samaké, Djewari, 99, 100–101, 105, 124, 186
Samaké, Koké, 102
Samake, Seriba, 102
San, 33m, 82, 88, 99, 116, 128
Sangarébougou (Office town also known
    as Kankan), 76, 80, 95, 98, 102, 122,
    122m, 159, 259n6
San-Kura (Office town), 113, 116, 122m,
    123–24, 128, 131, 166
Sansanding, 33m, 36, 58–59, 65, 73, 74m,
    75, 101, 117, 121, 128, 236n58, 236n60
Sarraut, Albert, 74–75
Savineau, Denise, 41, 83–84, 98
Scott, James, 110, 166
Second World War, 9, 14, 107, 109, 123,
    146, 154, 223n46, 253n96, 263n53,
    272n17, 277n86
"sedentarization" policy, 196
Ségou: region, 82–83, 100–102, 128, 144;
    town, 33m, 36, 37, 59, 81, 109, 135,
    138, 195
Segou-Koura (Office town), 166
Segu (capital), 33m, 34, 50, 111
Segu Empire, 34, 42, 44, 55–57, 111,
    233n35, 234n39, 235n43
Segu-Sekora, 56
Sen, Amartya, 19
Senegal, 17, 75, 87, 153–54, 233n28,
    234n232, 281n23, 282n24

seri, 47
Service temporaire des irrigations du
    Niger (STIN). See also Office du
    Niger
serving bowls, 22, 49, 113, 133, 160, 214,
    272n22
sexual abuse, 100, 184, 186, 208, 211
sexual allure, 1–2, 43, 52, 147, 230n10,
    241n107
shea: butter, 35, 41, 47, 49, 56, 63, 91,
    127–28, 129, 134, 139, 164, 183, 197,
    212, 229n2; colonial interest in indus-
    trial processing, 164–65; market, 41;
    nuts, 32, 41, 44–45, 49, 139, 232n23,
    239n94; shea oil, 47, 128–29, 139,
    229n2, 240n100; trees, 34, 44, 121,
    122m, 127, 128, 271n8, 271n8
Shipton, Parker, 19, 226n72
Sibila, 117, 120m, 128
Siengo (market town), 74m129, 129
Sikasso, 88, 92, 235n48, 257n150
silo, 165–66
Simon, M. (rice merchant), 60–61, 92
Sirakora (Office town), 122m, 128, 135,
    139, 242n124, 251n67
slavery: conflation with pawnship,
    144; end of, 46, 234n40; enslaved
    women, 10, 50, 59; expectation
    for formerly enslaved peoples to
    migrate to the Office du Niger, 78;
    flight from, 36; French military
    capture of ex-slaves, 235n52; French
    tolerance of, 236n54; negotiating
    status, 36; in Sansanding, 236n60; in
    the Segu Empire, 111, 234n39; slave
    labor in textile industry, 256n132;
    "slaves of the whites," 78
social life, 10–11, 55, 96, 105, 116–19, 140,
    263n62. See also agricultural festivals
Sogoba, Fanta, 156–57, 274n54, 275n58
Sokolo (town), 36, 54, 115, 236n58,
    237n62, 269n166
Sokorani (Office town), 119, 127, 156
solar dryers, 212–13, 215
Somono, 62, 77, 91, 128, 234n39
Sorel, François, 60
sorghum, 57, 61, 66, 90, 137
Sotuba, 33, 73, 75, 79–80
soumbala, 47, 47f, 49, 91, 127, 128, 129,
    139, 212, 239n89. See also nere
sounds of cooking, 142, 148–49, 158
Sountura, Maïssa, 190
Soviet Union, 16, 281n15

Sow, Oumou, 198
state monopoly, 182, 187, 192, 282n25
Stoller, Paul, 52, 241n109
surveillance (of Office residents), 87, 100,
    107, 122–23, 184–91
Sy, Mademba, 36, 59, 65, 236n58,
    236n60
tamarind, 35, 145, 271n8
Tangara, Ba, 104
Tangara, Bandiené, 124–25
Tangara, Fassoum, 103–4
Tangara, Kariba, 95–96
Tangara, N'Golo, 86
Tangara, Nièni, 186, 214–15, 214f,
    276n77
Tangara, Sine, 71
Tangara, Youssef, 71
taste: colonial effort to standardize,
    92–95; cultural values, 9, 20–21, 23,
    46, 54, 72, 184, 227n81; flavor, 9, 20,
    23, 46, 52, 56, 61–62, 72, 94, 107, 131,
    134, 135, 139, 140, 150–51, 160, 161,
    184, 205, 211, 241n110, 273n35; food
    preferences, 5, 20, 23, 45–46, 48, 54,
    88–89, 91, 94–95, 126, 137–38, 150,
    203; memory of, 160; reshaping,
    127, 135–38, 150–51, 158; sensorium,
    20–21, 52, 94, 131, 135, 184, 211,
    241n110; unappetizing, 179, 181,
    202–4, 241n109, 242n124, 280n5. See
    also aesthetics
tegedege, 127, 128, 158, 212
Terasson (lieutenant-governor), 67
termite mounds (tungo), 42–43, 122m,
    123–24
Thiam, Mariam N'Diaye, 199–200, 207
Thiriet, E., 68–69
Thomas, Lynn, 12
threshers: and consumption, 169–71,
    175, 189, 194; costs of, 166, 169; dan-
    gers of, 278n105; in disrepair, 183;
    food security, 168–69; masculine
    prestige and, 168; mechanization,
    143; noise of, 22; produced in Mali,
    10; women and, 9–10, 24, 169–71,
    175–76
threshing (by hand), 133, 169, 175,
    188–89, 191–93, 284n59
Tièmadeli, 118
Tienfala, 77, 91
Timbo, Bakary, 88–89
tobacco, 124–25, 163
Togoba, Mamady, 103

Togola, Téréba, 32, 180
Togolo-Kura (Office town), 135
toh: changing definition of, 150–51;
    made from red millet, 203, 205;
    men's production of, 62; rice
    toh, 135; staple dish, 47, 51, 52,
    105, 130, 134, 146, 160–61; texture
    of, 30, 47, 52, 55–56, 136, 150;
    women's cooking technique for,
    142, 148–51
Tombouctou (region), 33m, 37, 237n63
Tomi, 120, 120m, 121m, 128
Tongoloba (Office town), 123, 131
Tongolo-Koura (Office town), 182
Tougan, 33m, 82, 101, 251n53
Toure, Mama, 97–98
tractors: and debt, 162–64; in disrepair,
    183; drivers, 100, 118–19, 147; and
    hunger, 168–69; masculinity and,
    167–68; mechanization, 16, 143,
    162–69; rural change and, 162, 176;
    and wood fuel, 143–44
transportation infrastructure, 40, 143. See
    also roads
Traoré, Bintu, 171–73
Traoré, Fodé, 98
Traoré, Lanciné, 72–73, 230n6
Traoré, Mamadou, 100
Traoré, Moussa (general), 15–16, 179–81,
    183, 287n100
Traoré, Moussa (Office farmer), 97
Travélé, Moussa, 42, 204
tree planting, 40, 121, 122m
Tuareg settlers, 195–96, 287n96
Umar Tal, 34–35
United States, 15–16, 199–200, 288n113,
    289n118
van Beusekom, Monica, 16–17, 137,
    259n9, 260n15, 264n80, 268n147,
    269n171, 270n182
Viguier, Pierre, 163
Volta territory, 33m
Vuillet, Jean-François, 91, 93–94
Wagadu (myth), 33
wage workers, 67, 78, 109, 118–19, 139,
    143, 163, 168, 168f, 171, 172f, 173–74,
    260n11, 283n34
Wajcman, Judy, 12
water collection, 112–13, 160–62, 165
watermelons, 84
"wealth-in-people," 103
western Soudan, 35, 49–50, 217n1, 232n25
White, Luise, 22

white rice, 40, 90, 92–95, 135, 193
Wibaux, François, 154–55
wild foods, 3, 19, 29, 56, 120m, 126, 143, 169, 198, 239n89. *See also* food gathering
winnowing, 10, 30, 134, 158, 171–75, 187, 206, 213, 284n50
women's labor value, 1, 3, 21, 25, 31, 73, 171, 176

wood fuel, 3, 6–7, 13, 120m, 121, 133, 142–47, 155, 158–60, 227n80, 232n25, 240n100, 271n10
World Bank, 16, 191
World Food Program (WFP), 7f, 23, 185f, 195–201, 205, 288n105, 289n119
World War I. *See* First World War
World War II. *See* Second World War
Wylie, Diana, 19, 48, 288n107

Printed by Printforce, United Kingdom